With love from

1979

D0927832

THE

NATURAL HISTORY

OF

SELBORNE.

BY THE LATE REV. GILBERT WHITE, M.A.

WITH ADDITIONAL NOTES,

BY THE REV. J. G. WOOD, M.A.

AUTHOR OF THE ILLUSTRATED NATURAL HISTORY, ETC.

ILLUSTRATED WITH ENGRAVINGS ON WOOD.

GRESHAM BOOKS

First published 1860
This facsimile edition published 1979

ISBN 0 905418 22 0

Gresham Books,
Unwin Brothers Limited,
The Gresham Press,
Old Woking,
Surrey, England

Reproduced from the original edition and printed and bound
in Great Britain by A. Wheaton & Co. Ltd., Exeter.

INTRODUCTION TO THIS EDITION

by Hockley Clarke
Editor, "Birds and Country" magazine

SOME young bird watchers tend to ignore the older naturalists, such as Gilbert White, feeling that their observations and records are out-of-date with the vast amount of new knowledge which has been accumulated since their days. This is not wholly true, and however we regard the matter, there is much interest in the writings of the old masters, going back to the time of that eccentric but competent naturalist, Charles Waterton, and coming forward to the days of Richard Jefferies and W. H. Hudson.

Gilbert White holds a special place in the archives of bird literature. His carefully recorded observations and investigations, built up over the years, provide highlights on bird life even today, and they retain the stamp of his individuality.

My bird watching extends over many years and I still turn, with pleasure, as I have done so many times, to look up what G.W. says about some bird; rarely does he fail. His book is not one to sit down and try to read from cover to cover; its value is to be turned to in the way that I have mentioned, to see what he writes about some point that interests us.

He was an assiduous note-maker, and beginning as a gardener his notes extended to natural history generally, and in particular to birds. He was humble enough not to rely on his own conclusions. He sought not only the help of local people who bought specimens to him, and reported items of interest seen in the course of their daily activities about the woods and fields of

Selborne, also, G.W. had the advantage of correspondence with such eminent men as Thomas Pennant and the Hon. Daines Barrington.

The cumulative result of all this formed the basis for his book, written with no thought at the time of a book. Herein lies much of the character and the charm of *The Natural History of Selborne*. Had he set out to write a book it is doubtful if he would have produced anything as worthwhile. The spontaneity of his writing is apparent — he has someone in mind and not a nebulous audience.

It is his unassuming, and often slightly amusing phrases that bring the book to life, and make it acceptable as a great work over the world.

James Russel Lowell, American statesman and eminent literary man, visited Selborne in 1880. He wrote, after describing White's book as "One of the most delightful books in my father's library" . . . "These creatures whom we affect to look down upon as the drudges of instinct, are members of a commonwealth whose constitution rests on immovable bases".

The great Darwin himself, when about 16 years old, writes, "from reading White's Selborne, I took much pleasure in watching the habits of birds, and even made notes on the subject."

Thomas Carlyle wrote, "Of all the books of Natural History I ever read, White, in his healthy, general limitation, pleases me most".

We have thought it timely to write this in the hope of encouraging young, would-be bird watchers not to ignore G.W. and his book simply because it is from a past age. It will give them pleasure, and form a basis for their own observations. To open this window, as it were, into the past is complementary to all that has been written since, and to the aids which are now available to those who wish to explore this satisfying and rewarding field of study and recreation.

There have been many different editions of this celebrated book, and the present volume is a facsimile of an 1860 edition, which was annotated by the Rev. J. G. Wood of Merton College, Oxford, himself a prolific nature writer of his day. Engravings of nearly every creature mentioned were included in this particular edition, which greatly enhanced its appeal to the general reader and younger naturalist and, indeed, still does so today. A few minor passages which might have offended young minds in the context of the 1860 period are omitted from this book, but Wood – "J.G.W." – provided useful additional observations and included the scientific names of every natural object mentioned. In the present facsimile the original frontispiece has been moved to face page 237. A small map showing the location of places of special interest in Selborne is added, and the compact size of this book makes it specially convenient for the modern visitor to Selborne. The map is repeated on the back of the dustjacket as a further aid to the visitor. Much of the charm and fascination of Gilbert White for the modern reader of his work is that Selborne is so little changed today from its 18th century appearance. So the visitor can see and enjoy in a unique personal way the surroundings and the tranquil atmosphere that so pleased and inspired the great G.W.

BIOGRAPHY.

WHEN a beam of light shines forth in darkness, it throws its brightest rays on the objects nearest to it, while objects at a distance are scarcely illuminated at all. But the light of genius is of a different character, for it often happens that he whose brilliant intellect throws light on the darkened minds of men over the whole surface of the earth, is unknown to those immediately surrounding him, and is even rendered the subject of contemptuous pity by those whose mental vision is no more capable of receiving the light of his intellect, than their corporeal vision of enduring the glory of the meridian sun.

So was it with White. He was widely known as a philosopher, in the highest sense of the word, but he was so known only to the world without. His own village could not understand him, and little did its inhabitants suppose that their insignificant little Selborne should become a world-known name by means of him, whose peaceful life was spent in retirement, and whose only eulogy, from a surviving fellow-parishioner, was, "That he was a still, quiet body, and there wasn't a bit of harm in him, there wasn't indeed."

The following short account of this celebrated man was drawn up by a relation, and published in the first edition enriched with Markwick's notes.

" GILBERT WHITE was the eldest son of John White of Selborne, Esq., and of Ann, the daughter of Thomas Holt, rector of Streatham, in Surrey.   He was born at Selborne on July 18, 1720, and received his school education at Basingstoke, under the Rev. Thomas Warton, vicar of that place, and father of those two distinguished literary characters, Dr. Joseph Warton, master of Winchester School, and Mr. Thomas Warton, poetry professor at Oxford.   He was admitted at Oriel College, Oxford, in December 1739, and took his degree of Bachelor of Arts in June 1743.   In March 1744 he was elected fellow of his college.   He became Master of Arts in October 1746, and was admitted one of the senior [1] proctors of the University in April 1753.   Being of an unambitious temper, and strongly attached to the charms of rural scenery, he early fixed his residence in his native village, where he spent the greater part of his life in literary occupations, and especially in the study of nature. This he followed with patient assiduity, and a mind ever open to the lessons of piety and benevolence which such a study is so well calculated to afford.   Though several occasions offered of settling upon a college living, he could never persuade himself to quit the beloved spot, which was, indeed, a peculiarly happy situation for an observer.   He was much esteemed by a select society of intelligent and worthy friends, to whom he paid occasional visits.   Thus his days passed, tranquil and serene, with scarcely any other vicissitudes than those of the seasons, till they closed at a mature age, on June 26, 1793."

(1) By "one of the senior proctors," is probably meant one of the " proctors " themselves ; that is, he was not a " pro-proctor." There are only two Proctors in the University, a Senior, and a Junior.—J. G. W.

POEMS

OF

THE REV. GILBERT WHITE.

INVITATION TO SELBORNE.

SEE, Selborne spreads her boldest beauties round
The varied valley, and the mountain ground,
Wildly majestic! What is all the pride
Of flats, with loads of ornaments supplied?—
Unpleasing, tasteless, impotent expense,
Compared with Nature's rude magnificence.

Arise, my stranger, to these wild scenes haste;
The unfinish'd farm awaits your forming taste:
Plan the pavilion, airy, light, and true;
Through the high arch call in the length'ning view;
Expand the forest sloping up the hill;
Swell to a lake the scant, penurious rill;
Extend the vista; raise the castle mound
In antique taste, with turrets ivy-crown'd:
O'er the gay lawn the flow'ry shrub dispread,
Or with the blending garden mix the mead;
Bid China's pale, fantastic fence delight;
Or with the mimic statue trap the sight.

Oft on some evening, sunny, soft, and still,
The Muse shall lead thee to the beech-grown hill,
To spend in tea the cool, refreshing hour,
Where nods in air the pensile, nest-like bower;[1]
Or where the hermit hangs the straw-clad cell,[2]
Emerging gently from the leafy dell,
By fancy plann'd; as once th' inventive maid
Met the hoar sage amid the secret shade:
Romantic spot! from whence in prospect lies
Whate'er of landscape charms our feasting eyes,—
The pointed spire, the hall, the pasture plain,
The russet fallow, or the golden grain,

(1) A kind of arbour on the side of a hill.
(2) A grotesque building, contrived by a young gentleman, who used on occasion
to appear in the character of a hermit.

The breezy lake that sheds a gleaming light,
Till all the fading picture fail the sight.
  Each to his task; all different ways retire :
Cull the dry stick ; call forth the seeds of fire ;
Deep fix the kettle's props, a forky row,
Or give with fanning hat the breeze to blow.
  Whence is this taste, the furnish'd hall forgot,
To feast in gardens, or th' unhandy grot ?
Or novelty with some new charm surprises,
Or from our very shifts some joy arises.
Hark, while below the village bells ring round,
Echo, sweet nymph, returns the soften'd sound ;
But if gusts rise, the rushing forests roar,
Like the tide tumbling on the pebbly shore.
  Adown the vale, in lone sequester'd nook,
Where skirting woods imbrown the dimpling brook,
The ruin'd convent lies : here wont to dwell
The lazy canon midst his cloister'd cell,[1]
While Papal darkness brooded o'er the land,
Ere Reformation made her glorious stand :
Still oft at eve belated shepherd swains
See the cowl'd spectre skim the folded plains.
  To the high Temple would my stranger go ?[2]
The mountain brow commands the woods below :
In Jewry first this order found a name,
When madding Croisades set the world in flame ;
When western climes, urged on by pope and priest,
Pour'd forth their millions o'er the deluged East :
Luxurious knights, ill suited to defy
To mortal fight Turcéstan chivalry.
  Nor be the parsonage by the Muse forgot—
The partial bard admires his native spot ;
Smit with its beauties, loved, as yet a child,
Unconscious why, its capes, grotesque and wild.
High on a mound th' exalted gardens stand,
Beneath, deep valleys, scoop'd by Nature's hand.
A Cobham here, exulting in his art,
Might blend the general's with the gardener's part ;
Might fortify with all the martial trade
Of rampart, bastion, fosse, and palisade,
Might plant the mortar with wide threat'ning bore,
Or bid the mimic cannon seem to roar.

(1) The ruins of a Priory, founded by Peter de Rupibus, Bishop of Winchester.
(2) The remains of a Preceptory of the Knights Templars ; at least it was a farm
dependent upon some preceptory of that order. I find it was a preceptory, called the
Preceptory of Suddington; now called Southington.

Now climb the steep, drop now your eye below
Where round the blooming village orchards grow;
There, like a picture, lies my lowly seat,
A rural, shelter'd, unobserved retreat.
　　Me far above the rest Selbornian scenes,
The pendent forests, and the mountain greens,
Strike with delight; there spreads the distant view,
That gradual fades till sunk in misty blue:
Here Nature hangs her slopy woods to sight,
Rills purl between and dart a quivering light.

SELBORNE HANGER.

A WINTER PIECE.—TO THE MISS B*****S.

THE bard, who sang so late in blithest strain
Selbornian prospects, and the rural reign,
Now suits his plaintive pipe to sadden'd tone,
While the blank swains the changeful year bemoan.
　　How fallen the glories of these fading scenes!
The dusky beech resigns his vernal greens;
The yellow maple mourns in sickly hue,
And russet woodlands crowd the dark'ning view.
　　Dim, clust'ring fogs involve the country round,
The valley and the blended mountain ground
Sink in confusion; but with tempest-wing
Should Boreas from his northern barrier spring,
The rushing woods with deaf'ning clamour roar,
Like the sea tumbling on the pebbly shore.
When spouting rains descend in torrent tides,
See the torn zigzag weep its channel'd sides:
Winter exerts its rage; heavy and slow,
From the keen east rolls on the treasured snow;
Sunk with its weight the bending boughs are seen,
And one bright deluge whelms the works of men.
Amidst this savage landscape, bleak and bare,
Hangs the chill hermitage in middle air;
Its haunts forsaken, and its feasts forgot,
A leaf-strown, lonely, desolated cot!
Is this the scene that late with rapture rang,
Where Delphy danced, and gentle Anna sang?
With fairy step where Harriet tripp'd so late,
And, on her stump reclined, the musing Kitty sate?

Return, dear nymphs ; prevent the purple spring,
Ere the soft nightingale essays to sing ,
Ere the first swallow sweeps the fresh'ning plain,
Ere love-sick turtles breathe their amorous pain ;
Let festive glee th' enliven'd village raise,
Pan's blameless reign, and patriarchal days ;
With pastoral dance the smitten swain surprise,
And bring all Arcady before our eyes.

Return, blithe maidens ; with you bring along
Free, native humour ; all the charms of song ;
The feeling heart, and unaffected ease ;
Each nameless grace and ev'ry power to please

Nov. 1, 1763.

~~~~~~~~~

ON THE RAINBOW.

" Look upon the Rainbow, and praise him that made it: very beautiful is it
in the brightness thereof."—*Eccles.* xliii. 11.

On morning or on evening cloud impress'd,
Bent in vast curve, the watery meteor shines
Delightfully, to th' levell'd sun opposed :
Lovely refraction ! while the vivid brede
In listed colours glows, th' unconscious swain,
With vacant eye, gazes on the divine
Phenomenon, gleaming o'er the illumined fields,
Or runs to catch the treasures which it sheds.
Not so the sage : inspired with pious awe,
He hails the federal arch ; [1] and looking up,
Adores that God, whose fingers form'd this bow
Magnificent, compassing heaven about
With a resplendent verge, " Thou mad'st the cloud,
" Maker omnipotent, and thou the bow ;
" And by that covenant graciously hast sworn
" Never to drown the world again : [2] henceforth,
" Till time shall be no more, in ceaseless round,
" Season shall follow season : day to night,
" Summer to winter, harvest to seed time,
" Heat shall to cold in regular array
" Succeed."—Heav'n taught, so sang the Hebrew bard.[3]

(1) Gen. ix. 12—17.   (2) Gen. viii. 22.   (3) Moses.

A HARVEST SCENE.

Waked by the gentle gleamings of the morn,
Soon clad, the reaper, provident of want,
Hies cheerful-hearted to the ripen'd field :
Nor hastes alone : attendant by his side
His faithful wife, sole partner of his cares,
Bears on her breast the sleeping babe ; behind,
With steps unequal, trips her infant train ;
Thrice happy pair, in love and labour join'd !
   All day they ply their task ; with mutual chat,
Beguiling each the sultry, tedious hours.
Around them falls in rows the sever'd corn,
Or the shocks rise in regular array.
   But when high noon invites to short repast,
Beneath the shade of sheltering thorn they sit,
Divide the simple meal, and drain the cask :
The swinging cradle lulls the whimpering babe
Meantime ; while growling round, if at the tread
Of hasty passenger alarm'd, as of their store
Protective, stalks the cur with bristling back,
To guard the scanty scrip and russet frock.

~~~~~ ~~~

ON THE DARK, STILL, DRY, WARM WEATHER,

OCCASIONALLY HAPPENING IN THE WINTER MONTHS.

Th' imprison'd winds slumber within their caves,
Fast bound : the fickle vane, emblem of change,
Wavers no more, long settling to a point.
   All Nature nodding seems composed : thick streams,
From land, from flood up-drawn, dimming the day,
" Like a dark ceiling stand : " slow through the air
Gossamer floats, or, stretch'd from blade to blade,
The wavy net-work whitens all the field.
   Push'd by the weightier atmosphere, up springs
The ponderous mercury, from scale to scale
Mounting, amidst the Torricellian tube.[1]
   While high in air, and poised upon his wings,
Unseen, the soft, enamour'd woodlark runs

(1) The barometer.

Through all his maze of melody; the brake,
Loud with the blackbird's bolder note, resounds.
  Soothed by the genial warmth, the cawing rook
Anticipates the spring, selects her mate,
Haunts her tall nest-trees, and with sedulous care
Repairs her wicker eyrie, tempest-torn.
  The ploughman inly smiles to see upturn
His mellow glebe, best pledge of future crop:
With glee the gardener eyes his smoking beds;
E'en pining sickness feels a short relief.
  The happy schoolboy brings transported forth
His long-forgotten scourge, and giddy gig:
O'er the white paths he whirls the rolling hoop,
Or triumphs in the dusty fields of taw.
  Not so the museful sage:—abroad he walks
Contemplative, if haply he may find
What cause controls the tempest's rage, or whence,
Amidst the savage season, Winter smiles.
  For days, for weeks, prevails the placid calm.
At length some drops prelude a change: the sun
With ray refracted, bursts the parting gloom,
When all the chequer'd sky is one bright glare.
  Mutters the wind at eve; th' horizon round
With angry aspect scowls: down rush the showers,
And float the deluged paths, and miry fields.

THE

Natural History of Selborne,

IN

A SERIES OF LETTERS

ADDRESSED TO THOMAS PENNANT, ESQ.

AND THE

HON DAINES BARRINGTON

PART I.—TO THOMAS PENNANT, ESQ.

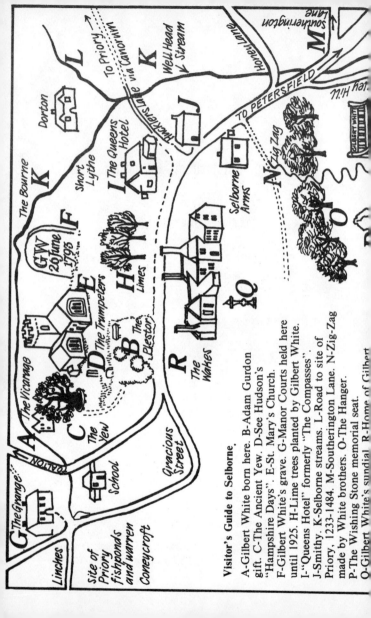

Visitor's Guide to Selborne.

A-Gilbert White born here. B-Adam Gurdon
gift. C-The Ancient Yew. D-See Hudson's
"Hampshire Days". E-St. Mary's Church.
F-Gilbert White's grave. G-Manor Courts held here
until 1925. H-Lime trees planted by Gilbert White.
I-"Queens Hotel" formerly "The Compasses".
J-Smithy. K-Selborne streams. L-Road to site of
Priory, 1233-1484. M-Southerington Lane. N-Zig-Zag
made by White brothers. O-The Hanger.
P-The Wishing Stone memorial seat.
Q-Gilbert White's sundial. R-Home of Gilbert

THE BEECH.

LETTER I.

GEOGRAPHICAL AND GEOLOGICAL SKETCHES OF SELBORNE.

THE parish of Selborne lies in the extreme eastern corner of
the county of Hampshire, bordering on the county of Sussex,
and not far from the county of Surrey; is about fifty miles
south-west of London, in latitude 51, and near midway
between the towns of Alton and Petersfield. Being very
large and extensive, it abuts on twelve parishes, two of which
are in Sussex, viz. Trotton and Rogate. If you begin from
the south, and proceed westward, the adjacent parishes are
Emshot, Newton, Valence, Faringdon, Harteley, Mauduit,
Great Ward-le-ham, Kingsley, Hedleigh, Bramshot, Trotton,
Rogate, Lysse, and Greatham. The soils of this district are

almost as various and diversified as the views and aspects. The high part to the south-west consists of a vast hill of chalk, rising three hundred feet above the village; and is divided into a sheep-down, the high wood, and a long hanging wood, called the Hanger. The covert of this eminence is altogether beech,[1] the most lovely of all forest trees, whether we consider its smooth rind or bark, its glossy foliage, or graceful pendulous boughs. The down, or sheep-walk, is a pleasing park-like spot, of about one mile by half that space, jutting out on the verge of the hill-country, where it begins to break down into the plains, and commanding a very engaging view, being an assemblage of hill, dale, woodlands, heath, and water. The prospect is bounded to the south-east and east by the vast range of mountains, called the Sussex Downs; by Guild-down, near Guildford, and by the Downs round Dorking, and Ryegate in Surrey, to the north-east; which altogether, with the country beyond Alton and Farnham, form a noble and extensive outline.

At the foot of this hill, one stage, or step from the uplands, lies the village, which consists of one single straggling street, three quarters of a mile in length, in a sheltered vale, and running parallel with the Hanger. The houses are divided from the hill by a vein of stiff clay, (good wheat land,) yet stand on a rock of white stone, little in appearance removed from chalk; but seems so far from being calcareous, that it endures extreme heat. Yet, that the freestone still preserves somewhat that is analogous to chalk, is plain, from the beeches, which descend as low as those rocks extend, and no farther, and thrive as well on them, where the ground is steep, as on the chalks.

The cart-way of the village divides, in a remarkable manner, two very incongruous soils. To the south-west is a rank clay, that requires the labour of years to render it mellow; while the gardens to the north-east, and small enclosures behind, consist of a warm, forward, crumbling mould, called black malm, which seems highly saturated with vegetable and animal manure; and these may perhaps have been the original site of the town; while the woods and coverts might extend down to the opposite bank.

(1) *Fagus sylvatica.*—J. G. W.

At each end of the village, which runs from south-east to north-west, arises a small rivulet: that at the north-west end frequently fails; but the other is a fine perennial spring, little influenced by drought or wet seasons, called Well-head.[1] This breaks out of some high grounds adjoining to Nore Hill, a noble chalk promontory, remarkable for sending forth two streams into two different seas. The one to the south becomes a branch of the Arun, running to Arundel, and so falling into the British Channel; the other to the north. The Selborne stream makes one branch of the Wey; and, meeting the Blackdown stream at Hedleigh, and the Alton and Farnham stream at Tilford-bridge, swells into a considerable river, navigable at Godalming; from whence it passes to Guildford, and so into the Thames at Weybridge; and thus at the Nore into the German Ocean.

Our wells, at an average, run to about sixty-three feet, and, when sunk to that depth, seldom fail; but produce a fine limpid water, soft to the taste, and much commended by those who drink the pure element, but which does not lather well with soap.

To the north-west, north, and east, of the village, is a range of fair enclosures, consisting of what is called a white malm, a sort of rotten or rubble stone, which, when turned up to the frost and rain, moulders to pieces, and becomes manure to itself.[3]

(1) This spring produced, September 14, 1781, after a severe hot summer, and a preceding dry spring and winter, nine gallons of water in a minute, which is five hundred and forty in an hour, and twelve thousand nine hundred and sixty, or two hundred and sixteen hogsheads in twenty-four hours, or one natural day. At this time many of the wells failed, and all the ponds in the vales were dry.

(2) Water obtained from wells is termed "hard," in contradistinction to rain-water, which is termed "soft." Hard water is the best for drinking, as rain-water is always vapid, and, moreover, is forced to pass over the root of the house before entering the tub, but it does not answer for washing purposes or tea-making until a little soda has been mixed with it, the carbonic acid of which unites with the lime, and forms chalk, a substance insoluble in water, and which falls to the bottom as a white sediment. The earthy salts which cause the hardness are very troublesome in steam-engines, as they form an incrustation upon the interior, which requires to be constantly cleared away. Charcoal, however, has been lately discovered to act as a preservative. It should be made from some hard wood and broken into small pieces not larger than half an inch cube. These will absorb the salts of lime, alkaline earth, salts of iron, &c., and prevent them from accumulating on the boiler. The charcoal must be renewed monthly.—J. G. W.

(3) This soil produces good wheat and clover.

Still on to the north-east, and a step lower, is a kind of white land, neither chalk nor clay, neither fit for pasture nor for the plough, yet kindly for hops, which root deep into the freestone, and have their poles and wood for charcoal growing just at hand. This white soil produces the brightest hops.

As the parish still inclines down towards Wolmer Forest, at the juncture of the clays and sand, the soil becomes a wet, sandy loam, remarkable for timber, and infamous for roads. The oaks of Temple and Blackmoor stand high in the estimation of purveyors, and have furnished much naval timber; while the trees on the freestone[1] grow large, but are what workmen call *shakey*, and so brittle as often to fall to pieces in sawing. Beyond the sandy loam the soil becomes a hungry lean sand, till it mingles with the forest; and will produce little without the assistance of lime and turnips.

LETTER II.

ON THE ELM AND OAK TREES.

In the court of Norton farm-house, a manor farm to the north-west of the village, on the white malms, stood within these twenty years a broad-leaved elm, or wych hazel,[2] *ulmus folio latissimo scabro* of Ray, which, though it had lost a considerable leading bough in the great storm in the year 1703, equal to a moderate tree, yet, when felled, contained eight loads of timber; and being too bulky for a carriage, was sawn off at seven feet above the butt, where it measured near eight feet in diameter. This elm I mention, to show to what a bulk planted elms may attain; as this tree must certainly have been such, from its situation. In the centre of the village, and near the church, is a square piece of ground, surrounded by houses, and vulgarly called the Plestor.[3] In the midst of this spot stood, in olden times, a vast oak,[4] with a short squat body, and huge horizontal arms,

(1) The freestone here mentioned is part of the green sand formation.—J. G. W.

(2) *Ulmus montana.*—J. G. W.

(3) A corruption of Playstow, or playing place.—J. G. W.

(4) *Quercus robur.* The chief value of the oak lies in the timber and bark. The bark is exceedingly valuable for tanning, the bark of a single tree having once sold for six hundred pounds; and the timber is extensively used for ship-building, not only on account of its soundness and strength, but also on account of the shape of

extending almost to the extremity of the area. This vene-
rable tree, surrounded with stone steps, and seats above them,
was the delight of old and young, and a place of much resort
in summer evenings ; where the former sat in grave debate,

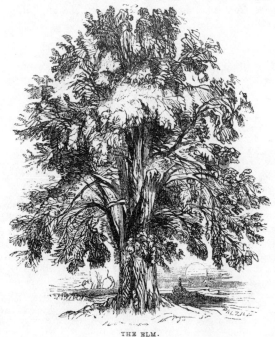

THE ELM.

while the latter froliced and danced before them. Long
might it have stood, had not the amazing tempest in 1703
overturned it at once, to the infinite regret of the inhabitants,

its branches. It is always a great object in ship building to get the grain of the
wood to follow the curve of the beams, especially in such which, like the "knees,"
are bent at nearly right angles. The gnarled branches of the oak-tree afford more
beams having the requisite curvature than do those of any other tree. The process
of sawing the tree for such purposes is termed "converting."—J. G. W.

and the vicar, who bestowed several pounds in setting it in
its place again: but all his care could not avail; the tree
sprouted for a time, then withered and died. This oak I
mention, to show to what a bulk planted oaks also may
arrive; and planted this tree must certainly have been, as
appears from what is known concerning the antiquities of
the village.

THE OAK

On the Blackmore estate there is a small wood called
Losel's, of a few acres, that was lately furnished with a set
of oaks of a peculiar growth and great value: they were tall
and taper like firs, but, standing near together, had very
small heads,—only a little brush, without any large limbs.
About twenty years ago, the bridge at the Toy, near Hampton Court, being much decayed, some trees were wanted, for
the repairs, that were fifty feet long without bough, and

would measure twelve inches diameter at the little end. Twenty such trees did a purveyor find in this little wood, with this advantage, that many of them answered the description at sixty feet. These trees were sold for £20 a-piece.

In the centre of this grove there stood an oak, which though shapely and tall on the whole, bulged out into a large excrescence about the middle of the stem. On this a pair of ravens had fixed their residence for such a series of years, that the oak was distinguished by the title of the Raven

THE RAVEN.[1]

Tree. Many were the attempts of the neighbouring youths to get at this eyrie: the difficulty whetted their inclinations, and each was ambitious of surmounting the arduous task.

But when they arrived at the swelling, it jutted out so in their way, and was so far beyond their grasp, that the most daring lads were awed, and acknowledged the undertaking to be too hazardous. So the ravens built on, nest upon

(1) *Corvus corax.*—J. G. W.

nest, in perfect security, till the fatal day arrived in which the wood was to be levelled. It was in the month of February, when those birds usually sit. The saw was applied to the butt, the wedges were inserted into the opening, the woods echoed to the heavy blows of the beetle, or mallet, the tree nodded to its fall; but still the dam sat on. At last, when it gave way, the bird was flung from her nest; and, though her parental affection deserved a better fate, was whipped down by the twigs, which brought her dead to the ground.[1]

LETTER III.

THE FOSSIL SHELLS AND STONE OF SELBORNE.

THE fossil shells of this district, and sorts of stone, such as have fallen within my observation, must not be passed over in silence. And, first, I must mention, as a great curiosity, a specimen that was ploughed up in the chalky fields, near

---

(1) An example lately came under my notice, where the parent bird showed equal affection for her young, although fortunately she did not fall a victim to her maternal attachment as did the hapless raven. A Blue Tomtit (*Parus cœruleus*) had built its nest between the stones forming the platform of the Shrivenham Station. Those who have stood on a platform while an express train passes can form some idea of the courage of this little bird. Although the fiery engine with its accompaniments of noise, steam, and rattling of wheels passed by hourly, making the entire platform tremble as it rushed by, there sat the little tom-tit calm and fearless, within two feet of the blazing furnace, and actually brushed by the waste steam. To the honour of the railway officials, they would neither touch the nest themselves nor suffer others to do so, and the courageous mother hatched and brought up her young ones in safety, and finally led them forth into the wide world, when they were doubtlessly bewildered at the absence of the periodical visits of the fiery monster to which they had been accustomed from the egg.

The maternal instinct strong in death was shown in a most affecting manner by a large Orang-outan, which had been pursued by sailors. After searching for an orang for some time, they discovered one as she was sitting on a branch with her young one. On firing a volley at her they wounded her mortally. When she felt the fatal shot she immediately appeared to lose all thoughts of herself, but placing her young one as high as her arms would reach among the branches, directed it in its escape, and kept her eyes fixed upon it, regardless of the nearer approach of the hunters, who with their next volley brought her dead to the ground.

The story of the Arctic Bear who was shot while mourning over the bodies of her murdered cubs is too well known to be related here.

The Raven is now almost extinct in England, and will probably follow the fate of the bear, the wolf, and the beaver.—J. G. W.

the side of the Down, and given to me for the singularity of its appearance, which, to an incurious eye, seems like a petrified fish of about four inches long, the cardo passing for a head and mouth. It is in reality a bivalve of the Linnæan genus of *mytilis*, and the species of *crista galli;* called by Lister, *rastellum;* by Rumphius, *ostreum plicatum minus;* by D'Argenville, *auris porci,* s. *crista galli;* and by those who make collections, cock's comb.[1] Though I applied to several such in London, I never could meet with an entire specimen; nor could I ever find in books any engraving from a perfect one. In the superb museum at Leicester

OSTREA CARINATA.

House, permission was given me to examine for this article; and though I was disappointed as to the fossil, I was highly gratified with the sight of several of the shells themselves in high preservation. This bivalve is only known to inhabit the Indian Ocean, where it fixes itself to a zoophyte, known by the name *gargonia*.

Cornua ammonis[2] are very common about this village. As we were cutting an inclining path up the Hanger, the

(1) *Ostrea carinata.* This shell is found in the chalk, marl, and firestone in many parts of Hampshire, and also rather plentifully at Dover. It is not identical with the recent cockscomb oyster as White appears to think, but closely resembles it.— J. G. W.

(2) The *Cornua ammonis*, or Ammonites as they are generally called, are very abundant in many parts of England. They are found throughout the whole secondary formations, but have not been discovered in the tertiary deposits. They vary exceedingly in size, some being not more than half an inch in diameter, while others measure no less than four feet. In general they are only casts of a shell which has decayed, but sometimes, especially in the galt of the chalk and the clays of the lias

labourers found them frequently on that steep, just under the soil, in the chalk, and of a considerable size. In the lane above Wellhead, in the way to Emshot, they abound in the bank, in a darkish sort of marl; and are usually very small and soft; but in Clay's Pond, a little farther on, at the end of the pit, where the soil is dug out for manure, I have occasionally observed them of large dimensions, perhaps fourteen or sixteen inches in diameter. But as these did not consist of firm stone, but were formed of a kind of *terra lapidosa,* or hardened clay, as soon as they were exposed to the rains and frost, they mouldered away. These seemed as if they were a very recent production. In the chalk-pit, at the north-west end of the Hanger, large *nautili* are sometimes observed.

In the very thickest strata of our freestone, and at considerable depths, well-diggers often find large scallops, or pectines, having both shells deeply striated, and ridged and furrowed alternately. They are highly impregnated with, if not wholly composed of, the stone of the quarry.

LETTER IV.

THE FREESTONE OF SELBORNE.

As, in my last letter, the freestone of this place has been only mentioned incidentally, I shall here become more particular.

This stone is in great request for hearth-stones, and the

the shell is found still in tolerable preservation. When the fossil is taken from the bed where it has lain for so many years, the shell splits, leaving the outside layer adhering to its bed while the inner remains on the fossil itself. In some soils, such as the shales of the lias, the shell becomes coated with a rich metallic brown lustre composed of sulphuret of iron. The Ammonites form part of the Cephalopoda, and are allied to the Nautilus, which they greatly resemble. In many parts of England they are called "snake-stones," from their resemblance to a snake curled up, on which lineage the country people strenuously insist. St. Hilda is reputed to be the destroyer of these snakes, each of whom lost its head as it fell over the cliffs into the sea.

There are about two hundred species of ammonites known, which are distinguished from each other principally by the situation of the siphon or tube, and the appearance of the septa or foliated marks. In some parts of England the ammonites are so common that they may be seen built into the walls of houses.—J. G. W.

beds of ovens; and in lining of lime-kilns it turns to good account; for the workmen use sandy loam instead of mortar; the sand of which fluxes,[1] and runs, by the intense heat, and so cases over the whole face of the kiln with a strong vitrified coat like glass, that it is well preserved from injuries of weather, and endures thirty or forty years. When chiselled smooth, it makes elegant fronts for houses, equal in colour and grain to the Bath stone; and superior in one respect, that, when seasoned, it does not scale. Decent chimney-pieces are worked from it, of much closer and finer grain than Portland; and rooms are floored with it; but it proves rather too soft for this purpose. It is a freestone, cutting in all directions; yet has something of a grain parallel with the horizon, and therefore should not be surbedded, but laid in the same position that it grows in the quarry.[2] On the ground abroad this fire-stone will not succeed for pavements, because, probably, some degree of saltness prevailing within it, the rain tears the slabs to pieces.[3] Though this stone is too hard to be acted on by vinegar, yet both the white part, and even the blue rag, ferment strongly in mineral acids. Though the white stone will not bear wet, yet in every quarry, at intervals, there are thin strata of blue rag, which resist rain and frost, and are excellent for pitching of stables, paths, and courts, and for building of dry walls against banks, a valuable species of fencing, much in use in this village; and for mending of roads. This rag is ragged and stubborn, and will not hew to a smooth face: but is very durable: yet, as these strata are shallow, and lie deep, large quantities cannot be procured but at considerable expense. Among the blue rags turn up some blocks tinged with a stain of yellow, or rust colour, which seem to be nearly as lasting as the blue; and every now and then balls of a friable substance, like rust of iron, called rust balls.

(1) There may, probably, be also in the chalk itself, that is burnt for lime, a proportion of sand; for few chalks are so pure as to have none.

(2) To surbed stone is to set it edgewise, contrary to the posture it had in the quarry, says Dr. Plot.—Oxfordsh. p. 77. But surbedding does not succeed in our dry walls; neither do we use it so in ovens, though he says it is best for Teynton stone.

(3) "Firestone is full of salts, and has no sulphur; must be close grained, and ave no interstices. Nothing supports fire like salts; saltstone perishes exposed to wet and frost."—Plot's Staff. p. 152.

In Wolmer Forest, I see but one sort of stone, called by the workmen sand, or forest stone. This is generally of the colour of rusty iron, and might probably be worked as iron ore; is very hard and heavy, and of a firm, compact texture, and composed of a small roundish crystalline grit, cemented together by a brown terrene, ferruginous matter; will not cut without difficulty, nor easily strike fire with steel. Being often found in broad flat pieces, it makes good pavement for paths about houses, never becoming slippery in frost or rain; is excellent for dry walls, and is sometimes used in buildings. In many parts of that waste, it lies scattered on the surface of the ground; but is dug on Weaver's Down, a vast hill on the eastern verge of that forest, where the pits are shallow, and the stratum thin.

This stone is imperishable.

From a notion of rendering their work the more elegant, and giving it a finish, masons chip this stone into small fragments about the size of the head of a large nail, and then stick the pieces into the wet mortar along the joints of their freestone walls. This embellishment carries an odd appearance, and has occasioned strangers sometimes to ask us pleasantly, "Whether we fastened our walls together with tenpenny nails?"

LETTER V.

THE ROADS AND LANES OF THE MANOR OF SELBORNE.—DESCRIPTION
OF THE INHABITANTS.

AMONG the singularities of this place, the two rocky hollow lanes, the one to Alton, and the other to the forest, deserve our attention. These roads, running through the malm lands, are, by the traffic of ages, and the fretting of water, worn down through the first stratum of our freestone, and partly through the second; so that they look more like water-courses than roads; and are bedded with naked rag for furlongs together. In many places they are reduced sixteen or eighteen feet beneath the level of the fields; and after floods, and in frosts, exhibit very grotesque and wild appearances, from the tangled roots that are twisted among the strata, and from the torrents rushing down their broken sides ; and especially when those cascades are frozen into icicles, hanging in all the fanciful shapes of frost-work. These rugged gloomy scenes affright the ladies when they peep down into them, from the paths above, and make timid horsemen shudder while they ride along them ; but delight the naturalist with their various botany, and particularly with their curious filices, with which they abound.[1]

The manor of Selborne, were it strictly looked after, with all its kindly aspects, and all its sloping coverts, would swarm with game: even now, hares, partridges, and pheasants, abound ; and in old days, woodcocks were as plentiful. There are few quails,[2] because they more affect open fields than enclosures ; after harvest some few land-rails are seen.

The parish of Selborne, by taking in so much of the forest, is a vast district. Those who tread the bounds are employed part of three days in the business, and are of opinion that the outline, in all its curves and indentings, does not comprise less than thirty miles.

The village stands in a sheltered spot, secured by the Hanger from the strong westerly winds. The air is soft, but rather moist from the effluvia of so many trees ; yet perfectly healthy, and free from agues.

The quantity of rain that falls on it is very considerable, as may be supposed in so woody and mountainous a district.

(1) Many of the lanes in Derbyshire are distinguished by the same characteristics. There is one road, a dismal one, leading from Ashbourn to Derby, which certainly, as White says, looks much more like a watercourse than a road. Houses are cut in the rock, the chimney rising about two feet from the fields above. There are legends of cattle being killed by wandering over the edge of the field, and finding themselves unexpectedly hurled down a precipice upon large broken pieces of rock below. Where human beings had not hewn out houses for themselves, the sand-martin had taken possession and completely honeycombed the entire surface of the rock.—J. G. W.

(2) *Coturnix communis.* The Quail is seldom seen single, but is accustomed to congregate in large flocks which spread over the whole of Southern Europe, from whence our markets are chiefly supplied. It is tolerably common in England. The account of the flight of quails visiting the Israelites is known to all, but it is not generally known how exact the scripture narrative is. " And it came to pass that *at even* the quails came up and covered the camp." Now the flocks of quails always fly by night, an instinct which is probably implanted in them to enable them, defenceless as they are, to avoid the various birds of prey who would inevitably attack them were their flight to take place by day. The nest of the quail is a rude and inartificial structure, placed on the ground, and containing from seven to twelve eggs.—J. G. W.

As my experience in measuring the water is but of short date, I am not qualified to give the mean quantity.[1] I only know that

	Inch.	Hund.
From May 1, 1779, to the end of the year, there fell . .	28	37 !
From Jan. 1, 1780, to Jan. 1, 1781	27	32
From Jan. 1, 1781, to Jan. 1, 1782	30	71
From Jan. 1, 1782, to Jan. 1, 1783	50	26 !
From Jan. 1, 1783, to Jan. 1, 1784	33	71
From Jan. 1, 1784, to Jan. 1, 1785	33	80
From Jan. 1, 1785, to Jan. 1, 1786	31	55
From Jan. 1, 1786, to Jan. 1, 1787	39	57

The village of Selborne, and large hamlet of Oakhanger, with the single farms, and many scattered houses along the verge of the forest, contain upwards of six hundred and seventy inhabitants.

We abound with poor; many of whom are sober and industrious, and live comfortably, in good stone or brick cottages, which are glazed, and have chambers above stairs; mud buildings we have none. Besides the employment from husbandry, the men work in hop gardens, of which we have many; and fell and bark timber. In the spring and summer the women weed the corn; and enjoy a second harvest in September by hop-picking. Formerly, in the dead months, they availed themselves greatly by spinning wool, for making of barragons, a genteel corded stuff, much in vogue at that time for summer wear; and chiefly manufactured at Alton, a neighbouring town, by some of the people called Quakers. The inhabitants enjoy a good share of health and longevity; and the parish swarms with children.

(1) A very intelligent gentleman assures me, (and he speaks from upwards of forty years' experience), that the mean rain of any place cannot be ascertained till a person has measured it for a very long period. "If I had only measured the rain," says he, "for the first four years from 1740 to 1743, I should have said the mean rain at Lyndon was 16½ inches for the year; if from 1740 to 1750, 18½ inches. The mean rain before 1763 was 20¼; from 1763 and since, 25½; from 1770 to 1780, 26. If only 1773, 1774, and 1775 had been measured, Lyndon mean rain would have been called 32 inches,—increasing from 16·6 to 32."

LETTER VI.

WOLMER FOREST—BLACK GAME—THE RED DEER—QUEEN ANNE'S VISIT.

SHOULD I omit to describe with some exactness the Forest of Wolmer, of which three-fifths perhaps lie in this parish, my account of Selborne would be very imperfect, as it is a district abounding with many curious productions, both animal and vegetable; and has often afforded me much entertainment, both as a sportsman and as a naturalist.

The royal Forest of Wolmer is a tract of land of about seven miles in length, by two and a half in breadth, running nearly from north to south, and is abutted on—to begin to the south, and so to proceed eastward—by the parishes of Greatham, Lysse, Rogate, and Trotton, in the county of Sussex; by Bramshot, Hedleigh, and Kingsley. This royalty consists entirely of sand, covered with heath and fern; but is somewhat diversified with hills and dales, without having one standing tree in the whole extent. In the bottoms, where the waters stagnate, are many bogs, which formerly abounded with subterraneous trees; though Dr. Plot says positively,[1] "that there never were any fallen trees hidden in the mosses of the southern counties." But he was mistaken; for I myself have seen cottages on the verge of this wild district, whose timbers consisted of a black hard wood, looking like oak, which the owners assured me they procured from the bogs by probing the soil with spits, or some such instruments; but the peat is so much cut out, and the moors have been so well examined, that none has been found of late.[2] Besides the oak, I have also been shown pieces of

(1) See his *Hist. of Staffordshire.*

(2) Old people have assured me, that, on a winter's morning, they have discovered these trees, in the bog, by the hoar frost, which lay longer over the space where they were concealed than on the surrounding morass. Nor does this seem to be a fanciful notion, but consistent with true philosophy. Dr. Hales saith, "That the warmth of the earth, at some depth under ground, has an influence in promoting a thaw, as well as the change of the weather from a freezing to a thawing state, is manifest, from this observation, viz. Nov. 29, 1731, a little snow having fallen in the night, it

fossil-wood, of a paler colour, and softer nature, which the
inhabitants called fir; but, upon a nice examination, and
trial by fire, I could discover nothing resinous in them:
and, therefore, rather suppose that they were parts of a
willow or alder, or some such aquatic tree.

This lonely domain is a very ageeable haunt for many
sorts of wild fowls, which not only frequent it in the winter,
but breed there in the summer; such as lapwings, snipes,
wild-ducks, and, as I have discovered within these few years,
teals. Partridges in vast plenty are bred in good seasons
on the verge of this forest, into which they love to make
excursions; and in particular, in the dry summer of 1740
and 1741, and some years after, they swarmed to such a
degree, that parties of unreasonable sportsmen killed twenty
and sometimes thirty brace in a day.

But there was a nobler species of game in this forest, now
extinct, which I have heard old people say abounded much
before shooting flying became so common, and that was the
heath-cock, or black game.[1] When I was a little boy, I
recollect one coming now and then to my father's table.
The last pack remembered was killed about thirty-five years
ago; and within these ten years one solitary grey hen was
sprung by some beagles, in beating for a hare. The sportsman cried out, " A hen pheasant !" but a gentleman present,

was, by eleven the next morning, mostly melted away on the surface of the earth,
except in several places in Bushy Park, where there were drains dug and covered
with earth, on which the snow continued to lie, whether those drains were full of
water or dry; as also where elm-pipes lay under ground: a plain proof this, that
those drains intercepted the warmth of the earth from ascending from greater depths
below them; for the snow lay where the drain had more than four feet of earth over
it. It continued also to lie on thatch, tiles. and the tops of walls."--See Hales's *Hæma-*
statics, p. 360. Quere, Might not such observations be reduced to domestic use, by
promoting the discovery of old obliterated drains and wells about houses; and, in
Roman stations and camps, lead to the finding of pavements, baths, and graves, and
other hidden relics of curious antiquity?

In Dr. Buckland's museum, at Oxford, there is part of a trunk of a tree, about
six feet long and a foot in diameter, which has been converted into a substance
closely resembling agate.—J. G. W.

(1) *Tetrao tetrix.* The male bird only has the peculiarly forked tail, the hen bird's
tail being of the ordinary fan-shape. Hybrids between the black grouse and other
birds have several times been found. The Pheasant appears to be the usual inter-
mixture, but mixed breeds have been known between this bird and the subalpine
ptarmigan, with the capercaillie, and once with the common domestic fowl.

In the Edinburgh Museum is a bird stated to be a hybrid between the black and
the r. d grouse.—J. G. W.

who had often seen black game in the north of England, assured me that it was a grey hen.

BLACK GAME.

Nor does the loss of our black game prove the only gap in the *Fauna Selborniensis*, or "Natural History of Selborne;" for another beautiful link in the chain of beings is wanting,—I mean the red-deer,[1] which, toward the beginning

(1) *Cervus elaphus.* The horns of the deer tribe differ essentially from those of any other horned creatures. The so called horns of the giraffe are processes from the skull, and covered with skin. The horns of the oxen and antelopes are permanent, hollow, and affixed to the head by a bony nucleus upon which they are placed, and which penetrates for some distance into the hollow of the horn. The horns, however, of the deer tribe are deciduous, falling off and being renewed annually. Moreover, they only belong to the males, and are solid masses of bone without any horny covering at all.

The growth of the horns is as follows:—About the beginning of March, the external carotid artery sends out branches which afford a constant supply of blood to the parts where the horn is to be. The peculiar substance which forms and at the same time defends the tender horn itself, is called the "velvet," from its appearance, and is extremely vascular. The substance of the horn is secreted with astonishing rapidity, ten weeks being sufficient for its entire development. All this time the horn is covered with the delicate and sensitive velvet, which must be got rid of to enable the deer to use its weapons without injury to itself. A very curious circumstance then takes place. The base of the horn becomes wider, and gradually encloses the arteries which supply the blood. By degrees the horn becomes thicker, and the arteries are in consequence gradually contracted, until at last they are entirely obliterated. The "velvet" then being deprived of its nourishment soon dries up, and is rubbed off by the deer against the trees.

The object in the gradual closing of the arteries is evident, as if so large a supply

of this century, amounted to about five hundred head, and made a stately appearance. There is an old keeper, now alive, named Adams, whose great-grandfather (mentioned in a perambulation taken in 1635), grandfather, father, and self, enjoyed the head keepership of Wolmer Forest in succession, for more than a hundred years. This person assures me, that his father has often told him that Queen Anne, as she was journeying on the Portsmouth road, did not think the Forest of Wolmer beneath her royal regard. For she came out of the great road at Liphock, which is just by, and reposing herself on a bank, smoothed for that purpose, lying about half a mile to the east of Wolmer Pond, and still called Queen's bank, saw with great complacency and satisfaction the whole herd of red-deer brought by the keepers along the vale before her, consisting then of about five hundred head. A sight this, worthy the attention of the greatest sovereign! But he further adds, that, by means of the Waltham blacks, or, to use his own expression, as soon as they began blacking, they were reduced to about fifty head, and so continued decreasing till the time of the late Duke of Cumberland. It is now more than thirty years ago that his highness sent down a huntsman, and six yeomen prickers, in scarlet jackets laced with gold, attended by the stag-hounds, ordering them to take every deer in this forest alive, and to convey them in carts to Windsor. In the course of the summer they caught every stag, some of which showed extraordinary diversion; but, in the following winter, when the hinds were also carried off, such fine chases were

of blood were to be immediately cut off from the circulation, the animal would be much injured. I lately saw a rein-deer in the Zoological Gardens busily engaged in ridding his horns of the velvet, and so eager was he to complete his task, that he had commenced rubbing off the velvet before the circulation had quite ceased, and his horns presented in consequence a very unpleasant sight. They were covered with blood, and strips of the still bleeding velvet were hanging from them on every side.

The names of the red deer vary according to the age, and according to the shape of the horns. Thus, in the first year, the deer has no horns at all, and is called a calf. In the second year the horns appear in the shape of short rounded knobs, and the deer is consequently termed a knobber. The "knobs" fall off, and are next year replaced by longer horns called "dags," and the deer takes the name of brocket, and so on until the full age is reached, when the top of the horn expands into a crown, and the stag is then called a hart, or a stag of ten, in allusion to the number of branches.—J. G. W.

exhibited as served the country people for matter of talk and wonder for years afterwards. I saw myself one of the yeoman prickers single out a stag from the herd, and must confess that it was the most curious feat of activity I ever beheld,—superior to anything in Mr. Astley's riding-school. The

THE RED DEER

exertions made by the horse and deer much exceeded all my expectations, though the former greatly excelled the latter in speed. When the devoted deer was separated from his companions, they gave him, by their watches, law, as they called it, for twenty minutes; when, sounding their horns, the stop-dogs were permitted to pursue, and a most gallant scene ensued.

LETTER VII.

WOLMER FOREST—THE WALTHAM BLACKS.

THOUGH large herds of deer do much harm to the neighbourhood, yet the injury to the morals of the people is of
more moment than the loss of their crops. The temptation
is irresistible; for most men are sportsmen by constitution:
and there is such an inherent spirit for hunting in human
nature, as scarce any inhibitions can restrain. Hence, towards
the beginning of this century, all this country was wild about
deer-stealing. Unless he were a hunter, as they affected to
call themselves, no young person was allowed to be possessed
of manhood or gallantry. The Waltham blacks at length
committed such enormities, that Government was forced to
interfere with that severe and sanguinary act called the Black
Act, which now comprehends more felonies than any law that
ever was framed before; and, therefore a late bishop of
Winchester, when urged to re-stock Waltham chase,[1] refused,
from a motive worthy of a prelate, replying, that "It had
done mischief enough already."

Our old race of deer-stealers are hardly extinct yet. It
was but a little while ago that, over their ale, they used to
recount the exploits of their youth; such as watching the
pregnant hind to her lair, and when the calf was dropped,
paring its feet with a penknife to the quick, to prevent its
escape, till it was large and fat enough to be killed; the
shooting at one of their neighbours with a bullet, in a turnip-
field, by moonshine, mistaking him for a deer; and the losing
a dog in the following extraordinary manner:—Some fellows,
suspecting that a calf new-fallen was deposited in a certain
spot of thick fern, went with a lurcher to surprise it; when
the parent hind rushed out of the brake, and taking a vast
spring, with all her feet close together, pitched upon the neck
of the dog, and broke it short in two.

(1) This chase remains unstocked to this day; the bishop was Dr. Hoadley.

Another temptation to idleness and sporting, was a number of rabbits,[1] which possessed all the hillocks and dry places; but these being inconvenient to the huntsmen, on account of their burrows, when they came to take away the deer, they permitted the country people to destroy them all.

RABBIT'S TEETH.

Such forests and wastes, when their allurements to irregularities are removed, are of considerable service to neighbourhoods that verge upon them, by furnishing them with peat and turf for their firing; with fuel for the burning their lime; and with ashes for their grasses; and by maintaining their geese and their stock of young cattle at little or no expense.

(1) *Lepus cuniculus.* The Rabbit and Hare are the largest of our rodentia, or gnawing animals. The constant wear of the teeth of the rodentia require that some provision should be made for the renewal of the worn portions. To attain this object, there is a constant increase of substance at the root of the tooth which forces it forward as fast as it is worn out. Now and then from some accident, the teeth do not meet, and then the growth, being uninterrupted, proceeds so rapidly, that the poor animal cannot use its jaws, and it consequently dies of hunger. In the above cut, an example is shown of this nature. The sketch was taken from a preparation in the Anatomical Museum, Oxford.

The sense of pain appears to be rather deficient in these animals. Mr. Rowell of Oxford, exhibited at a meeting of the Ashmolean Society, a number of ordinary steel traps, in which were legs of rabbits. To the disjointed members, several sinews of considerable length were left attached, which must have been drawn out by the force employed by the rabbit in making its efforts to escape. If the sense of pain had been great, the animal could not have voluntarily undergone such torture as the forcible tearing away of sinews must have occasioned. If a wild cat is caught in a steel trap, it appears to suffer such pain that it dare not struggle with sufficient violence to break off and tear away the captured limb, but the rabbits had done so not only once, but on several occasions.

When domesticated the rabbit loses many of its wild habits, and learns those of civilized life. Among the civilized habits attained by some rabbits of my own, was the eating of various sweet substances in which schoolboys delight, under the names of hardbake, toffy, and all the royal family of Rocks. The rabbits got on very well with their toffy as long as it remained hard, but as soon as it softened they got their teeth tightly fastened into it, and used to go through the most complicated evolutions to free their imprisoned jaws. One rabbit was extremely fond of tallow candles.—J. G. W.

The manor farm of the parish of Greatham has an admitted claim, I see, by an old record taken from the Tower of London, of turning all live stock on the forest, at proper seasons, *bidentibus exceptis,* "sheep excepted."[1] The reason, I presume, why sheep are excluded is, because, being such close grazers, they would pick out all the finest grasses, and hinder the deer from thriving.

Though (by statute 4 and 5 William and Mary, c. 23) "to burn on any waste, between Candlemas and Midsummer, any grig, ling, heath and furze, gorse, or fern, is punishable with whipping, and confinement in the house of correction ;" yet, in this forest, about March or April, according to the dryness of the season, such vast heath fires are lighted up, that they often get to a masterless head, and, catching the hedges, have sometimes been communicated to the underwoods, woods, and coppices, where great damage has ensued. The plea for these burnings is, that, when the old coat of heath, &c. is consumed, young will sprout up, and afford much tender browse for cattle ; but where there is large old furze, the fire, following the roots, consumes the very ground ; so that for hundreds of acres nothing is to be seen but smother and desolation, the whole circuit round looking like the cinders of a volcano ; and, the soil being quite exhausted, no traces of vegetation are to be found for years. These conflagrations, as they take place usually with a north-east or east wind, much annoy this village with their smoke, and often alarm the country ; and once, in particular, I remember that a gentleman, who lives beyond Andover, coming to my house, when he got on the downs between that town and Winchester, at twenty-five miles distance, was surprised much with smoke and a hot smell of fire ; and concluded that Alresford was in flames ; but when he came to that town, he then had apprehensions for the next village, and so on to the end of his journey.

On two of the most conspicuous eminences of this forest, stand two arbours, or bowers, made of the boughs of oaks ; the one called Waldon Lodge, the other Brimstone Lodge ; these the keepers renew annually on the feast of St. Barnabas,

(1) For this privilege the owner of that estate used to pay to the king annually seven bushels of oats.

taking the old materials for a perquisite. The farm called
Blackmoor, in this parish, is obliged to find the posts and
brushwood for the former ; while the farms at Greatham, in
rotation, furnish for the latter ; and are all enjoined to cut
and deliver the materials at the spot. This custom I mention,
because I look upon it to be of very remote antiquity.

LETTER VIII.

THE LAKES AND PONDS OF WOLMER FOREST—THE HOLT—CURIOUS
DISCOVERY OF COINS.

ON the verge of the forest, as it is now circumscribed, are
three considerable lakes ; two in Oakhanger, of which I have
nothing particular to say ; and one called Bin's, or Bean's
Pond, which is worthy the attention of a naturalist or a
sportsman ; for, being crowded at the upper end with willows,
and with the *carex cespitosa*, " clumpy sedge,"[1] it affords such
a safe and pleasant shelter to wild duck, teals, snipes, &c. that
they breed there. In the winter this covert is also frequented
by foxes, and sometimes by pheasants ; and the bogs produce
many curious plants.

By a perambulation of Wolmer Forest and the Holt, made
in 1635, and the eleventh year of Charles the First, (which
now lies before me,) it appears that the limits of the former
are much circumscribed.   For, to say nothing of the further
side, with which I am not so well acquainted, the bounds on
this side, in old times, came into Binswood, and extended to
the ditch of Ward-le-ham Park, in which stands the curious
mount, called King John's Hill, and Lodge Hill, and to the
verge of Hartley Mauduit, called Mauduit-hatch ; compre-
hending also Shortheath, Oakhanger, and Oak-woods; a large
district, now private property, though once belonging to the
royal domain.

It is remarkable, that the term *purlieu* is never once mentioned in this long roll of parchment.   It contains, besides
the perambulation, a rough estimate of the value of the

(1) I mean that sort which, rising into tall hassocks, is called by the foresters
torrets; a corruption I suppose of turrets.

timbers, which were considerable, growing at that time in the
district of the Holt; and enumerates the officers, superior and
inferior, of those joint forests, for the time being, and their
ostensible fees and perquisites.   In those days, as at present,
there were hardly any trees in Wolmer Forest.

Within the present limits of the forest are three considerable lakes, Hogmer, Cranmer, and Wolmer; all of which are

THE TENCH.[1]

stocked with carp, tench, eels, and perch : but the fish do not
thrive well, because the water is hungry, and the bottoms are
a naked sand.

A circumstance respecting these ponds, though by no
means peculiar to them, I cannot pass over in silence : and
that is, that instinct by which in summer all the kine, whether
oxen, cows, calves, or heifers, retire constantly to the water
during the hotter hours; where, being more exempt from
flies, and inhaling the coolness of that element, some belly
deep, and some only to mid-leg, they ruminate and solace
themselves from about ten in the morning till four in the
afternoon, and then return to their feeding.   During this
great proportion of the day, they drop much dung, in which
insects nestle, and so supply food for the fish, which would be
poorly subsisted but for this contingency.   Thus Nature,
who is a great economist, converts the recreation of one
animal to the support of another!   Thomson, who was a nice

(1) *Tinca vulgaris.*—J. G. W.

observer of natural occurrences, did not let this pleasing circumstance escape him. He says in his " Summer,"—

" A various group the herds and flocks compose :
——————————————— on the grassy bank,
Some ruminating lie ; while others stand
Half in the flood, and, often bending, sip
The circling surface."

Wolmer Pond, so called, I suppose, for eminence sake, is a vast lake for this part of the world, containing in its whole circumference 2,649 yards, or very near a mile and a half. The length of the north-west and opposite side is about 704 yards, and the breadth of the south-west end about 456 yards. This measurement, which I caused to be made with good exactness, gives an area of about sixty-six acres, exclusive of a large irregular arm at the north-east corner, which we did not take into the reckoning.

THE WILD DUCK.[1]

On the face of this expanse of waters, and perfectly secure from fowlers, lie all day long, in the winter season, vast flocks of ducks, teals, and widgeons, of various denominations; where they preen, and solace, and rest themselves, till towards sunset, when they issue forth in little parties—for in their natural state they are all birds of the night—to feed in the brooks and meadows : returning again with the dawn of the morning. Had this lake an arm or two more, and were it planted round with thick covert (for now it is perfectly naked), it might make a valuable decoy.

(1) *Anas boschas.*—J. G. W

Yet neither its extent, nor the clearness of its water, nor the resort of various and curious fowls, nor its picturesque groups of cattle, can render this *mere* so remarkable, as the great quantity of coins that were found in its bed about forty years ago.[1]

LETTER IX.

By way of supplement, I shall trouble you once more on this subject, to inform you that Wolmer, with her sister forest Ayles Holt, *alias* Alice Holt,[2] as it is called in old records, is held by grant from the crown for a term of years.

The grantees that the author remembers, are,—Brigadier-General Emanuel Scroope Howe, and his lady, Ruperta, who was a natural daughter of Prince Rupert, by Margaret Hughs; a Mr. Mordaunt, of the Peterborough family, who married a dowager Lady Pembroke; Henry Bilson Legge and lady; and now Lord Stawel, their son.

The lady of General Howe lived to an advanced age, long surviving her husband; and, at her death, left behind her many curious pieces of mechanism of her father's constructing, who was a distinguished mechanic and artist,[3] as well as warrior: and, among the rest, a very complicated clock, lately in possession of Mr. Elmer, the celebrated game painter at Farnham, in the county of Surrey.

Though these two forests are only parted by a narrow range of enclosures, yet no two soils can be more different;

---

(1) The "decoy" for wild ducks may be seen in its perfection in the fens of Lincolnshire. It is a kind of enormous tube or rather tunnel, very wide at one end and very narrow at the other. By means of decoy ducks, who are trained to the sport, and afterwards by sprinkled hempseed, the wild ducks are induced to enter the wide end of the tunnel, and when fairly in the trap are driven onwards into the small end, where they are captured and destroyed. Certain advantages of position are required to make a good decoy.—J. G. W.

(2) "In Rot. Inquisit. de statu forest. in Scaccar. 36 Ed. III.," it is called Aisholt. In the same, "Tit. *Woolmer* and *Aisholt* Hantisc. Dominus Rex habet unam capellam in *haia* suâ de Kingesle." " *Haia, sepes, sepimentum, parcus:* a Gall. *haie* and *haye.*"—SPELMAN's *Glossary.*

(3) This prince was the inventor of mezzotinto.—WHITE. Not the inventor, but the introducer into England.—J. G. W.

for the Holt consists of a strong loam, of a miry nature, carrying a good turf, and abounding with oaks that grow to be large timber; while Wolmer is nothing but a hungry, sandy, barren waste.

The former, being all in the parish of Binsted, is about two miles in extent from north to south, and near as much from east to west, and contains within it many woodlands and lawns, and the Great Lodge where the grantees reside, and a smaller lodge called Goose Green; and is abutted on by the parishes of Kingsley, Frinsham, Farnham, and Bentley, all of which have right of common.

One thing is remarkable, that, though the Holt has been of old well stocked with fallow-deer, unrestrained by any pales or fences more than a common hedge, yet they were never seen within the limits of Wolmer; nor were the red deer of Wolmer ever known to haunt the thickets or glades of the Holt.

At present the deer of the Holt are much thinned and reduced by the night-hunters, who perpetually harass them, in spite of the efforts of numerous keepers, and the severe penalties that have been put in force against them, as often as they have been detected, and rendered liable to the lash of the law. Neither fines nor imprisonments can deter them; so impossible is it to extinguish the spirit of sporting, which seems to be inherent in human nature.

THE BOAR.

General Howe turned out some German wild boars and sows in his forests, to the great terror of the neighbourhood;

and, at one time, a wild bull, or buffalo: but the country rose upon them, and destroyed them.

A very large fall of timber, consisting of about one thousand oaks, has been cut this spring (viz. 1784), in the Holt forest; one-fifth of which, it is said, belongs to the grantee, Lord Stawel. He lays claim also to the lop and top; but the poor of the parishes of Binsted and Frinsham, Bentley and Kingsley, assert that it belongs to them; and, assembling in a riotous manner, have actually taken it all away. One man, who keeps a team, has carried home for his share forty stacks of wood. Forty-five of these people his lordship has served with actions. These trees, which were very sound, and in high perfection, were winter cut, viz. in February and March, before the bark would run. In old times the Holt was estimated to be eighteen miles, computed measure, from water carriage, viz. from the town of Chertsey, on the Thames; but now it is not half that distance, since the Wey is made navigable up to the town of Godalming, in the county of Surrey.

LETTER X.

THE SWALLOW, SWIFT, MARTIN, FLY-CATCHER, BLACK-CAP, AND WATER-RAT.

August 4, 1767.

IT has been my misfortune never to have had any neighbours whose studies have led them towards the pursuit of natural knowledge; so that, for want of a companion to quicken my industry and sharpen my attention, I have made but slender progress in a kind of information to which I have been attached from my childhood.

As to swallows (*hirundines rusticæ*) being found in a torpid state during the winter, in the Isle of Wight, or any part of this country, I never heard any such account worth attending to.[1] But a clergyman, of an inquisitive turn, assures

(1) This question is by no means settled. It certainly has occurred that swallows have been found in this country during the winter months, but as one swallow cannot be asserted to prove a summer, neither can one or two particular instances prove a general fact.—J. G. W.

me, that when he was a great boy, some workmen, in pulling down the battlements of a church tower early in the spring, found two or three swifts (*hirundines apodes*) among the rubbish, which were at first appearance dead; but, on being carried toward the fire, revived. He told me that, out of his great care to preserve them, he put them in a paper bag, and hung them by the kitchen fire, where they were suffocated.

Another intelligent person has informed me that, while he was a schoolboy at Brighthelmstone, in Sussex, a great fragment of the chalk cliff fell down, one stormy winter, on the beach, and that many people found swallows among the rubbish; but, on my questioning him whether he saw any of those birds himself, to my no small disappointment he answered me in the negative; but that others assured him they did.

Young broods of swallows began to appear this year on July the 11th, and young martins (*hirundines urbicæ*) were then fledged in their nests. Both species will breed again once; for I see by my Fauna of last year, that young broods came forth so late as September the 18th. Are not these late hatchings more in favour of hiding than migration?

THE MARTIN.[1]

Nay, some young martins remained in their nests last year so late as September the 29th; and yet they totally disappeared with us by the 5th of October.

(1) *Chelidon urbica.*—J. G. W.

How strange it is, that the swift, which seems to live exactly the same life with the swallow and house-martin, should leave us before the middle of August invariably! while the latter stay often till the middle of October; and once I saw numbers of house-martins on the 7th of November. The martins and red-wing fieldfares were flying in sight together; an uncommon assemblage of summer and winter birds!

A little yellow bird[1] (it is either a species of the *alauda trivialis,* or rather, perhaps, of the *motacilla trochilus*) still continues to make a sibilous shivering noise in the tops of tall woods. The *stoparola* of Ray (for which we have as yet

THE FLY-CATCHER[2]

no name in these parts) is called, in your Zoology, the fly-catcher. There is one circumstance characteristic of this bird, which seems to have escaped observation; and that is, it takes its stand on the top of some stake, or post, from whence it springs forth on its prey, catching a fly in the air, and hardly ever touching the ground, but returning still to the same stand for many times together.

(1) *Calamodyta locustella,* according to Jesse.—J. G. W.
(2) The Fly-catcher, *Muscicapa grisola,* is a very pleasing bird in its habits. The custom here mentioned by White I have oftened witnessed, and once under rather peculiar circumstances. While waiting in a gig for a friend who had appointed to meet me in Nuneham Park, and passing away the time in the intellectual amusement of

I perceive there are more than one species of the *motacilla trochilus:* Mr. Derham supposes, in Ray's Philosophical Letters, that he has discovered three. In these, there is again an instance of some very common birds that have as yet no English name.

THE BLACK-CAP.[1]

Mr. Stillingfleet makes a question whether the black-cap (*motacilla atricapilla*) be a bird of passage or not. I think there is no doubt of it; for in April, in the first fine weather, they come trooping all at once in these parts, but are never seen in the winter. They are delicate songsters.

Numbers of snipes breed every summer in some moory ground on the verge of this parish. It is very amusing to see the cock bird on wing at that time, and to hear his piping and humming notes.

I have had no opportunity yet of procuring any of those mice which I mentioned to you in town. The person that

flicking off the white clover flowers, a fly-catcher came and seated itself on a branch of an oak close by, apparently for the purpose of taking the flies which were put up by the strokes of the whip. From its custom of placing its nest on the beams of outhouses, it is frequently called the beam-bird.—J. G. W.

(1) *Sylvia atricapilla.* The Blackcap is easily distinguishable from the other warblers by the peculiarity from which it derives its name. The male divides with his mate the duty of sitting upon the eggs, and has been known frequently to sing during this task. The blackcap has occasionally been seen darting into the air after insects, like the fly-catcher.—J. G. W.

brought me the last says they are plenty in harvest, at which
time I will take care to get more; and will endeavour to put
the matter out of doubt, whether it be a nondescript species
or not.

I suspect much there may be two species of water-rats.
Ray says, and Linnæus after him, that the water-rat is web-
footed behind.   Now, I have discovered a rat on the banks of

WATER RAT

our little stream that is not web-footed, and yet is an excellent swimmer and diver: it answers exactly to the *mus amphibius* of Linnæus (see *Syst. Nat.*), which, he says, "*natat in
fossis et urinatur,*" "swims and dives in the water."   I should
be glad to procure one "*plantis palmatis,*" "with webbed
feet."   Linnæus seems to be in a puzzle about his *mus amphibius,* "amphibious mouse," and to doubt whether it differs
from his *mus terrestris,* "land mouse," which, if it be, as he
allows, the "*mus agrestis capite grandi brachyuros,*" "short-
tailed, large-headed field-mouse" of Ray, is widely different
from the water-rat, both in size, make, and manner of life.

As to the *falco,* which I mentioned in town, I shall take
the liberty to send it down to you into Wales; presuming
on your candour, that you will excuse me if it should appear
as familiar to you as it is strange to me.   Though mutilated,
"*qualem dices . . . antehac fuisse, tales cum sint reliquiæ?*"

(1) *Arvicola amphibius.* The Water-rat or Water-vole is said to eat frogs, fish, and
water-plants.   I have often seen it eating the water-plants, particularly the reeds,
which they strip of their green covering, but I never saw them eat any living thing,
nor have I ever seen fish-bones in their holes except when a kingfisher had taken
possession.   The abandoned hole of a water-rat is also often occupied by crawfish.—
J. G. W.

"what would you say it was before, when such are the remains?"

It haunted a marshy piece of ground in quest of wild ducks and snipes; but, when it was shot, had just knocked down a rook, which it was tearing in pieces. I cannot make it answer to any of our English hawks; neither could I find any like it at the curious exhibition of stuffed birds in Spring Gardens. I found it nailed up at the end of a barn, which is the countryman's museum.

The parish I live in is a very abrupt uneven country, full of hills and woods, and therefore full of birds.

LETTER XI.

THE HOOPOE AND GROSBEAK—BROWN OWL AND BATS.

THE HOOPOE.

SELBORNE, *September* 9, 1767.

It will not be without impatience that I shall wait for your thoughts with regard to the *falco*; as to its weight. breadth,

&c., I wish I had set them down at the time; but, to the best of my remembrance, it weighed two pounds and eight ounces, and measured, from wing to wing, thirty-eight inches. Its *cere* and feet were yellow, and the circle of its eyelids a bright yellow. As it had been killed some days, and the eyes were sunk, I could make no good observation on the colour of the pupils and the *irides*.

The most unusual birds I ever observed in these parts were a pair of hoopoes, (*upupa*,) [1] which came several years ago in the summer, and frequented an ornamental piece of ground, which joins to my garden, for some weeks. They used to march about in a stately manner, feeding in the walks, many times in the day; and seemed disposed to breed in my outlet; but were frightened and persecuted by idle boys, who would never let them be at rest.

THE GROSBEAK. [2]

Three grosbeaks (*loxia coccothraustes*) appeared some years ago in my fields, in the winter; one of which I shot. Since

(1) *Upupa epops.* A Hoopoe in a state of captivity lost its life in a most singular manner. In the room where it was kept was a stove, to which this bird used to creep so closely, that the two mandibles of his bill were curved by the heat so that he could not peck up his food, and consequently died of hunger.—J. G. W.

(2) *Coccothraustes vulgaris.* The Hawfinch is a very shy bird, and can with difficulty be approached by the fowler. The nest of this bird is remarkable for the presence of

that, now and then, one is occasionally seen in the same dead season.

A crossbill (*loxia curvirostra*) was killed last year in this neighbourhood.

Our streams, which are small, and rise only at the end of the village, yield nothing but the bull's head, or

THE MILLER'S THUMB.[1]

miller's thumb, (*gobius fluviatilis capitatus*), the trout (*trutta fluviatilis*), the eel (*anguilla*), the lampern[2] (*lam-*

lichens in its structure; the nest being sometimes almost entirely composed of them. The Hawfinch does not appear to be a rare bird, but is seldom seen on account of its retired habits.—J. G. W.

(1) *Cottus Gobio.* The Miller's Thumb is so called from the flatness and roundness of its head, which bears a fanciful resemblance to the thumb of the miller, rendered broad by constant trying of the meal between the thumb and forefingers. The fish may be found under stones, at least with its head under stones, and its tail wagging outside, a habit which frequently betrays it to the juvenile fisherman, who, if he be at all expert, can catch it with his hands alone.—J. G. W.

(2) The Lampern is found in many brooks and rivers. In appearance it resembles a very small lamprey, its length being from twelve to fifteen inches, and its thickness about the size of a swan-quill, or perhaps a little larger. It can swim with some rapidity, and works its way against a swift stream with apparent ease. In a brook running by Ashbourn in Derbyshire, the lamperns were very common. They used to scoop hollows in the bed of the river by carrying away the stones, which their leech-like mouth enabled them to do easily, and then they lay in the hollow thus made to the number of eighteen or twenty, their heads all fixed to the stones,

pœtra parva et fluviatilis), and the stickleback (*pisciculus aculeatus*).[1]

We are twenty miles from the sea, and almost as many from a great river, and therefore see but little of sea birds. As to wild fowls, we have a few teams of ducks bred in the moors where the snipes breed; and multitudes of widgeons and teals, in hard weather, frequent our lakes in the forest.

Having some acquaintance with a tame brown owl, I find that it casts up the fur of mice and the feathers of birds in pellets, after the manner of hawks : when full, like a dog, it hides what it cannot eat.

The young of the barn-owl are not easily raised, as they want a constant supply of fresh mice ; whereas the young of the brown owl will eat indiscriminately all that is brought; snails, rats, kittens, puppies, magpies, and any kind of carrion or offal.

The house-martins have eggs still, and squab young. The last swift I observed was about the 21st of August : it was a straggler.

Redstarts, fly-catchers, white-throats, and *reguli non cristati*, still appear ; but I have seen no black-caps lately.

and their tails waving in the stream. The inhabitants of Ashbourn were considerably afraid of them, stating that "the lampreens would vemon them," by vemon meaning venom, according to an ancient custom of the Ashbournians to transpose letters ; for instance, they always spoke of birds' nests as "brids' neesens." There was only one man in Ashbourn who had sufficient moral courage to eat them.— J. G. W.

(1) *Gasterosteus aculeatus.* The Sticklebacks abound in the ditches about Oxford, where they may be seen, at the proper time of the year, dressed in their beautiful colours, red, with a tinge of green. Only the males, however, are thus decorated, and of them only those who have proved victors in the engagements so constantly fought by these most combative little fish. Each male stickleback takes to himself a portion of the bank where he resides, and wages war against every other fish that passes near his home. So pugnacious is the little anchorite, that if a walking-stick is pushed into the water, he dashes at it, and really sometimes knocks himself so hard against it that the shock is distinctly felt. No amount of splashing will drive him away, and the only method of tempting him to leave his premises is to take a very fat caddis-worm out of its case and throw it into the water. At it he goes at once, but finds that several other pink and green gentlemen have also determined on appropriating it, and there is a grand struggle. Six or seven little fish all get hold of the caddis at the same time, struggling and pulling in every direction, sometimes hauling it a foot or two one way, and then a foot or two the other, until at last one contrives to pull it away from the others and dashes off, carrying it in his mouth, closely followed by his competitors, who soon catch him and enact the struggle over again, until the caddis is completely torn in pieces. There are six species of sticklebacks inhabiting England, but the habits of all are very similar.—J. G. W.

I forgot to mention, that I once saw in Christ Church College quadrangle, in Oxford, on a very sunny warm morning, a house-martin flying about and settling on the parapet, so late as the 20th of November.

THE LONG-EARED BAT.

At present, I know only two species of bats, the common *vespertilio murinus*, and the *vespertilio auribus*.

(1) *Plecotus communis*. I kept a tame long-eared bat for some time. During the day it was very quiet, and hung in an apparently shapeless mass from the top of its cage, but was brisk enough at night. It was never heard to utter the peculiar shrill shriek of the bats. The bats' cry may be exactly imitated by scraping two keys together, an art which I first saw made use of by a mischievous urchin, who beguiled a long sermon by deluding the congregation into an opinion that a bat was flying about in the church, where they are not unfrequently found. Not long ago we were summoned into the church by a boy who announced that a bat had been captured. On going into the church we found the captor standing by a slab tombstone, pressing his handkerchief against a moulding, under which was the bat. On removing the handkerchief no bat was found, but on examination a hole was discovered under the moulding communicating with a cavity behind the tomb, into which the bat had made its escape, and in which it probably lived.

When the organ at the Cathedral at Oxford was being repaired, on taking down the large pipes, numbers of dead bats were found at the bottom. They must have slipped in, and have been kept there by the contraction of the pipe which would not suffer their wings to play.

There are now in the British Museum fourteen species of British Bats, viz. The Smaller Horse-shoe Bat (*Rhinolophus hipposideros*), found in Devonshire; the Larger Horse-shoe Bat (*Rhinolophus ferrum equinum*), found at Torquay; the Barbastelle (*Barbastellus communis*), also found in Devonshire; the Common Long-eared Bat (*Plecotus communis*), found in all parts of England; the Whiskered Bat (*Vespertilio mystacinus*), found in Cambridgeshire by the Rev. Leonard Jenyns;

I was much entertained last summer with a tame bat, which would take flies out of a person's hand. If you gave it anything to eat, it brought its wings round before the mouth, hovering and hiding its head in the manner of birds of prey when they feed. The adroitness it showed in shearing off the wings of the flies, which were always rejected, was worthy of observation, and pleased me much. Insects seemed to be most acceptable, though it did not refuse raw

SKULL OF BAT.

HAIR OF BAT.

flesh when offered; so that the notion, that bats go down chimneys and gnaw men's bacon, seems no improbable story. While I amused myself with this wonderful quadruped, I saw it several times confute the vulgar opinion, that bats, when down on a flat surface, cannot get on the wing again, by rising with great ease from the floor. It ran, I observed, with more despatch than I was aware of; but in a most ridiculous and grotesque manner.

Daubenton's Bat (*Vespertilio Daubentonii*), taken in Aberdeen Cathedral, Scotland; the Large-eared Bat (*Myotis murinus*); Bechstein's Bat (*Myotis Bechstenii*), taken in the New Forest, Hampshire; Natterer's Bat (*Myotis Natteri:*), taken near London; the Serotine (*Scotophilus serotinus*), taken near London; the Parti-coloured Bat (*Scotophilus discolor*), found at Plymouth; the Hairy-armed Bat (*Scotophilus Leisleri*); the Common Bat (*Scotophilus murinus*), found all over England; and lastly, the Noctule (*Noctulinia altivolans*), found at Tuitagel Castle, at Windsor, and in Cambridgeshire. Twenty-three species are known, nineteen of which are figured in Bell's British Quadrupeds.

As the names of many of the Bats imply, they seldom venture out during the day, but pursue their insect prey in the dusk. When they do, as sometimes happens, emerge into the full light of day, they appear completely bewildered by the unaccustomed glare, and fly about in a very vague manner. About three years since, as I was walking down Holywell Street, Oxford, about eleven in the forenoon, a bat was observed flying wildly about, and evidently completely blinded by the light. It came straight up the street, flying about nine feet from the ground. As it flew along I made an instinctive leap, and cut at the bat with a whalebone cane, when, to my great astonishment, down came the bat, not only struck down, but actually cut in two down the middle as if done with a knife, the two halves falling in different directions. Not a drop of blood escaped. I picked up the bat, or rather the two halves of the bat, and brought it home, but the swarms of vermin that issued from it, soon caused its ejectment.—J. G. W.

Bats drink on the wing, like swallows, by sipping the surface, as they play over pools and streams. They love to frequent waters, not only for the sake of drinking, but on account of insects, which are found over them in the greatest plenty. As I was going, some years ago, pretty late, in a boat from Richmond to Sunbury, on a warm summer's evening, I think I saw myriads of bats between the two places; the air swarmed with them all along the Thames, so that hundreds were in sight at a time.[1]

(1) A few years ago as I was walking in the fields near Hampstead Heath in the dusk of a summer's evening, the air was filled with bats, mostly the Long-eared Bat, but I imagine that several specimens of the Great Horse-shoe Bat were among them. As I had no gun, and the evening was rather dark, it was impossible to be quite certain. They seemed to swarm like gnats, and their continual shrieking was quite oppressive.

The bats form the only instance of flying mammalia, for the swoop of the flying squirrel, or flying-dragon, does not deserve to be called flight. To attain this object, the sternum or breast bone, and the ribs and shoulder bones are modified in their structure as well as the bones of the fore-leg or arm. The principal development is that of the fingers, which are so elongated, that if the fingers of a man were lengthened in proportion, each would be four feet long. The bones of the fore-arm are likewise lengthened. The fingers are connected by a beautifully thin and delicate membrane, very sensitive, supplied with a perfect network of veins and arteries, and devoid of hair. So sensitive is this wing-membrane, that when Spallanzani had stretched threads about his room, he found that bats whom he had previously deprived of sight, threaded their way among the strings with perfect certainty, although there was hardly room for the animal to pass.

The teeth of the bat are like those of all insectivora, sharp and pointed, so that when a bat is eating, it is obliged to make a series of pecks, just as a cat does. The reader will see how well adapted the teeth are for retaining the struggling prey.

The hair of the bat, although to the naked eye it exactly resembles that of the mouse, yet essentially differs from it when examined through a good microscope. The hair that is represented in the accompanying cut is one taken from the common Long-eared Bat, and magnified about two hundred diameters by transmitted light. The protuberances are found to be caused by external scales which can be removed, and for that reason the hair of the bat is often very uneven.

When the bats rest, they suspend themselves head downwards, holding on by their hinder claws. They then fold their wings in a very strange way, and if they are of the long-eared species, they pack their ears under their wings also, and turn themselves into a most grotesquely shapeless mass of brown fur and black membrane, diversified with hooks in unexpected places. The hook of the wings is attached to the end of the thumb, which is not connected to the fingers by a membrane, but is left free. By means of these hooks the animal is enabled to walk, if the sidling plunges which it makes can be dignified with that name.—J. G. W.

LETTER XII.

November 4, 1767.

Sir,—It gave me no small satisfaction to hear that the *falco*[1] turned out an uncommon one. I must confess I should have been better pleased to have heard that I sent you a bird you had never seen before; but that I find would be a difficult task.

I have procured some of the mice mentioned in my former letters,—a young one, and a female with young, both of

THE HARVEST MOUSE.

which I have preserved in brandy. From the colour, shape, size and manner of nesting, I make no doubt but that the species is nondescript. They are much smaller, and more slender, than the *mus domesticus medius* of Ray, and have more of the squirrel or dormouse colour. Their belly is white;

(1) This hawk proved to be the *falco peregrinus*—a variety.

a straight line along their sides divides the shades of their back and belly. They never enter into houses; are carried into ricks and barns with the sheaves; abound in harvest; and build their nests amidst the straws of the corn above the ground, and sometimes in thistles. They breed as many as eight in a litter, in a little round nest composed of the blades of grass or wheat.

One of these nests I procured this autumn, most artificially platted, and composed of blades of wheat; perfectly round, and about the size of a cricket-ball; with the aperture so ingeniously closed, that there was no discovering to what part it belonged. It was so compact and well filled that it would roll across the table without being discomposed, though it contained eight little mice that were naked and blind. As this nest was perfectly full, how could the dam come at her litter respectively, so as to administer a teat to each? Perhaps she opens different places for that purpose, adjusting them again when the business is over; but she could not possibly be contained herself in the ball with her young, which, moreover, would be daily increasing in bulk. This wonderful procreant cradle, an elegant instance of the efforts of instinct, was found in a wheat-field suspended in the head of a thistle.[1]

A gentleman, curious in birds, wrote me word that his servant had shot one last January, in that severe weather, which he believed would puzzle me. I called to see it this summer, not knowing what to expect: but the moment I took it in hand, I pronounced it the male *garrulus bohemicus*, or German silk-tail, from the five peculiar crimson tags, or points, which it carries at the ends of five of the short remiges. It cannot, I suppose, with any propriety, be called an English bird; and yet I see by Ray's *Philosophical Letters* that great

---

(1) *Mus minimus.* A nest of this elegant little animal was brought during my residence in Wiltshire to me by some mowers, who seemed to consider it no uncommon sight. It was about the size of a cricket-ball, as described by White, but was not so firm and compact as drawings make it appear, as the grass of which it was composed was rather loosely put together. It was perfectly circular and empty, but there was not the least appearance of an opening to be discovered. The mouse has been ascertained to be fond of flies and other small insects. It is found in all the northern counties of England.—J. G. W.

flocks of them, feeding on haws, appeared in this kingdom in the winter of 1685.[1]

The mention of haws puts me in mind that there is a total failure of that wild fruit, so conducive to the support of many of the winged nation. For the same severe weather, late in the spring, which cut off all the produce of the more tender and curious trees, destroyed also that of the more hardy and common.

Some birds, haunting with the missel-thrushes, and feeding on the berries of the yew-tree, which answered to the description of the *merula torquata*, or ring-ousel, were lately

BOHEMIAN CHATTERER.

seen in this neighbourhood. I employed some people to procure me a specimen, but without success. (See Letter VIII.)

(1) *Ampelis garrulus.* America has been recently found to be a habitat of this remarkably elegant bird, whose graceful crest, and sealing-wax like appendages to the wing, point it out so surely, that few that reach this country ever leave it unharmed. A very fine specimen was shot at Oxford in 1846, and taken at once to the Ashmolean Museum, where I had an opportunity of inspecting it before it was stuffed.

The young of this bird want the waxen appendages during their first year. While the bird sings it agitates its crest, but does not produce that swelling in the throat so

Query.—Might not Canary birds be naturalised to this climate, provided their eggs were put in the spring into the nests of their congeners, as goldfinches, greenfinches, &c. ? Before winter, perhaps, they might be hardened, and able to shift for themselves.

THE CANARY BIRD.

About ten years ago, I used to spend some weeks yearly at Sunbury, which is one of those pleasant villages lying on the Thames, near Hampton Court. In the autumn I could not help being much amused with those myriads of the swallow kind which assemble in those parts. But what struck me most was, that from the time they began to congregate, forsaking the chimneys and houses, they roosted every night in the osier beds of the aits of that river. Now, this resorting towards that element, at that season of the year, seems to give some countenance to the northern opinion (strange as it is) of their retiring under water. A

conspicuous in the canary and other singing birds. It has been occasionally kept in a cage, but in captivity it loses its vivacity in proportion as it gains in indiscriminate appetite, for although at liberty it eats scarcely anything except berries, it hardly rejects any vegetable substance while in captivity. Its notes are not unlike those of the Thrush, but are weaker and more undecided.—J. G. W.

Swedish naturalist is so much persuaded of that fact, that he talks, in his *Calendar of Flora*, as familiarly of the swallow's going under water in the beginning of September, as he would of his poultry going to roost a little before sunset.

An observing gentleman in London writes me word, that he saw a house-martin on the 23d of last October, flying in and out of its nest in the Borough ; and I myself, on the 29th of last October, as I was travelling through Oxford, saw four or five swallows hovering round and settling on the roof of the County Hospital.[1]

Now, is it likely that these poor little birds, which perhaps had not been hatched but a few weeks, should, at that late season of the year, and from so midland a county, attempt a voyage to Goree or Senegal, almost as far as the equator?[2]

I acquiesce entirely in your opinion that, though most of the swallow kind may migrate, yet some do stay behind and hide with us during the winter.

As to the short-winged, soft-billed birds which come trooping in such numbers in the spring, I am at a loss even what to suspect about them. I watched them narrowly this year, and saw them abound till about Michaelmas, when they appeared no longer. Subsist they cannot openly among us, and yet elude the eyes of the inquisitive; and as to their hiding, no man pretends to have found any of them in a torpid state in the winter. But with regard to their migration, what difficulties attend that supposition ! that such feeble bad fliers, who the summer long never flit but from hedge to hedge, should be able to traverse vast seas and continents, in order to enjoy milder seasons amidst the regions of Africa.

(1) Swallows generally remain in Oxford to a very late period, and are often seen hanging about the public buildings, especially the Radcliffe Library and the old Clarendon, after their friends have gone. The Radcliffe Library is quite beaded with swallows for a week or two before they start.—J. G. W.

(2) See Adanson's *Voyage to Senegal.*

LETTER XIII.

ON THE CHAFFINCH, LINNET, BUNTING, WHEATEAR, AND
HARVEST MOUSE.

SELBORNE, *Jan.* 22, 1768.

SIR,—As in one of your former letters you expressed the
more satisfaction from my correspondence on account of my
living in the most southerly county ; so now I may return
the compliment, and expect to have my curiosity gratified by
your living much more to the north.

For many years past, I have observed, that towards
Christmas vast flocks of chaffinches have appeared in the
fields—many more, I used to think, than could be hatched
in any one neighbourhood. But, when I came to observe
them more narrowly, I was amazed to find that they seemed

THE CHAFFINCH.[1]

to me to be almost all hens. I communicated my suspicions
to some intelligent neighbours, who, after taking pains

(1) *Fringilla cœlebs.* The Chaffinch is called "cœlebs," or "bachelor," because
the females migrate about November, leaving the gentlemen at home. At all events,
if the ladies do remain behind, they keep to themselves, and live only in the society

about the matter, declared that they also thought them mostly all females; at least fifty to one. This extraordinary occurrence brought to my mind the remark of Linnæus, that " before winter, all their hen chaffinches migrate through Holland into Italy." Now, I want to know from some curious person in the north, whether there are any large flocks of these finches with them in the winter, and of which sex they mostly consist? For, from such intelligence one might be able to judge whether our female flocks migrate from the other end of the island, or whether they come over to us from the continent.

THE LINNET [1]

We have, in the winter, vast flocks of the common linnets, more, I think, than can be bred in any one district. These, I observe, when the spring advances, assemble on some tree in the sunshine, and join all in a gentle sort of chirping, as if they were about to break up their winter quarters, and betake themselves to their proper summer homes. It is well known, at least, that the swallows and the fieldfares do congregate with a gentle twittering before they make their respective departures.

of females. Captain Brown suggests that many of the so-called females are the young males, who have not yet attained their full plumage. The nest of the Chaffinch is always placed in the fork of a branch, and is so ingeniously coated on the exterior with mosses and lichens similar in colour to the bark of the tree, that it is not distinguishable except by a practised eye.—J. G. W.

(1) *Fringilla cannabina.*—J. G. W.

You may depend on it that the bunting, *emberiza miliaria*, does not leave this country in the winter. In January, 1767,

THE BUNTING [1]

I saw several dozens of them, in the midst of a severe frost, among the bushes on the downs near Andover: in our woodland enclosed districts it is a rare bird.

Wagtails, both white and yellow, are with us all the winter. Quails crowd to our southern coast, and are often killed in numbers by people that go on purpose.

Mr. Stillingfleet, in his *Tracts*, says, that "if the wheatear [2] (*œnanthe*) does not quit England, it certainly shifts places; for, about harvest, they are not to be found where there was before great plenty of them." This well accounts for the vast quantities that are caught about that time on the south downs near Lewes, where they are esteemed a delicacy. There have been shepherds, I have been credibly informed, that have made many pounds in a season by catching them

(1) *Emberiza miliaria.* The upper mandible of the true Buntings is furnished with a hard sharp knob, to which is opposed on the lower mandible a sharp cutting angle.—J. G. W.

(2) *Sylvia œnanthe.* This bird is sometimes called the English Ortolan, on account of the delicacy of its flesh. The male bird sings both when on the wing, and when seated on the ground. The nest is often placed some way down a rabbit burrow; or if that is unattainable, under a large clod, or behind some rock.—J. G. W.

in traps. And though such multitudes are taken, I never saw (and I am well acquainted with those parts) above two or three at a time; for they are never gregarious. They may perhaps migrate in general; and for that purpose, draw

THE WHEATEAR.

towards the coast of Sussex in autumn; but that they do not all withdraw I am sure, because I see a few stragglers in many counties, at all times of the year, especially about warrens and stone quarries.

I have no acquaintance at present among the gentlemen of the navy, but have written to a friend who was a sea chaplain in the late war, desiring him to look into his minutes, with respect to birds that settled on their rigging during their voyage up or down the Channel. What Hasselquist says on that subject is remarkable; there were little short-winged birds frequently coming on board the ship all the way from our Channel quite up to the Levant, especially before squally weather.

What you suggest with regard to Spain is highly probable. The winters of Andalusia are so mild, that, in all likelihood, the soft-billed birds that leave us at that season may find insects sufficient to support them there.

Some young man, possessed of fortune, health, and leisure, should make an autumnal voyage into that kingdom, and should spend a year there, investigating the natural history of that vast country. Mr. Willoughby[1] passed through that kingdom on such an errand; but he seems to have skirted along in a superficial manner and an ill humour, being much disgusted at the rude dissolute manners of the people.

I have no friend left now at Sunbury to apply to about the swallows roosting on the aits of the Thames; nor can I hear any more about those birds which I suspected were *merulæ torquatæ*.

As to the small mice, I have farther to remark, that though they hang their nests for breeding up amidst the straws of the standing corn, above the ground, yet I find that, in the winter, they burrow deep in the earth, and make warm beds of grass; but their grand rendezvous seems to be in corn-ricks, into which they are carried at harvest. A neighbour housed an oat-rick lately, under the thatch of which were assembled near a hundred, most of which were taken; and some I saw. I measured them, and found that, from nose to tail, they were just two inches and a quarter, and their tails just two inches long. Two of them, in a scale, weighed down just one copper halfpenny, which is about the third of an ounce avoirdupois; so that I suppose they are the smallest quadrupeds in this island. A full-grown *mus medius domesticus* weighs, I find, one ounce lumping weight, which is more than six times as much as the mouse above, and measures, from nose to rump, four inches and a quarter, and the same in its tail. We have had a very severe frost and deep snow this month. My thermometer was one day fourteen degrees and a half below the freezing point, within doors. The tender evergreens were injured pretty much. It was very providential that the air was still, and the ground well covered with snow, else vegetation in general must have suffered prodigiously. There is reason to believe that some days were more severe than any since the year 1739-40.

(1) See Ray's *Travels*, p. 466.

LETTER XIV.

ON THE SPIRACULA OF ANIMALS.

SELBORNE, *March* 12, 1768.

DEAR SIR,—If some curious gentleman would procure the head of a fallow deer and have it dissected, he would find it furnished with two spiracula, or breathing-places,[1] besides the nostrils; probably analogous to the *puncta lachrymalia,* "lachrymal ducts," in the human head. When deer are thirsty, they plunge their noses, like some horses, very deep under water, while in the act of drinking, and continue them in that situation for a considerable time; but, to obviate any inconveniency, they can open two vents, one at the inner corner of each eye, having a communication with the nose. Here seems to be an extraordinary provision of nature worthy our attention, and which has not, that I know of, been noticed by any naturalist. For it looks as if these creatures would not be suffocated, though both their mouths and nostrils were stopped. This curious formation of the head may be of singular service to beasts of chase, by affording them free respiration; and no doubt these additional nostrils are thrown open when they are hard run.[2] Mr. Ray observed that at Malta, the owners slit up the nostrils of such asses as were hard worked; for they, being naturally strait or small, did not admit air sufficient to serve them when they travelled or laboured in that hot climate. And we know that grooms, and gentlemen of the turf, think large nostrils necessary, and a perfection, in hunters and running horses.[3]

Oppian, the Greek poet, by the following line, seems to have had some notion that stags have four spiracula:—

Τετράδυμοι ῥινὲς, πίσυρες πνοίῃσι δίαυλοι.
"Quadrupartite nostrils, four respiratory passages."

OPP. *Cyn.* lib. ii. 1. 181.

---

(1) See Part II. Letter XXI.

(2) In answer to this account, Mr. Pennant sent me the following curious and pertinent reply:—" I was much surprised to find in the antelope something analogous to what you mention as so remarkable in deer. This animal also has a long slit beneath each eye, which can be opened and shut at pleasure. On holding an orange to one, the creature made as much use of those orifices as of his nostrils, applying them to the fruit, and seeming to smell it through them."

(3) These singular organs may probably assist the animal in recognition, but their use is not very clearly defined.—J. G. W.

Writers, copying from one another, make Aristotle say, that goats breathe at their ears, whereas he asserts just the contrary :—Ἀλκμαίων γὰρ οὐκ ἀληθῆ λέγει, φάμενος ἀναπνεῖν τὰς αἶγας κατὰ τὰ ὦτά. "Alcmæon does not advance what is true, when he avers that goats breathe through their ears."— *History of Animals*, book i. chap. xi.

LETTER XV.

ON THE WEASEL, WHITE ROOK, BULLFINCH, STONE CURLEW, ETC.

SELBORNE, *March* 30, 1768.

DEAR SIR,—Some intelligent country people have a notion that we have, in these parts, a species of the genus *mustelinum*, besides the weasel, stoat, ferret, and polecat ; a little reddish

THE WEASEL.

beast, not much bigger than a field mouse, but much longer, which they call a *cane*. This piece of intelligence can be little depended on ; but farther inquiry may be made.[1]

A gentleman in this neighbourhood had two milk-white rooks in one nest. A booby of a carter, finding them before they were able to fly, threw them down and destroyed them, to the regret of the owner, who would have been glad to have preserved such a curiosity in his rookery. I saw the birds myself nailed against the end of a barn, and was surprised to find that their bills, legs, feet, and claws, were milk-white.[2]

(1) The common weasel, *Mustela vulgaris*. The countrymen in this case were wrong, as they generally are when they attempt to describe objects of natural history.—J. G. W.

(2) White varieties of coloured birds are not at all uncommon. There is a white peacock now near Oxford; the keeper at Ashdown in Wiltshire had a tame white

A shepherd saw, as he thought, some white larks on a down above my house this winter : were not these the *emberiza nivalis*, the snow-flake of the Brit. Zool. ? No doubt they were.

A few years ago, I saw a cock bullfinch in a cage, which had been caught in the fields after it was come to its full

THE BULLFINCH.

colours. In about a year it began to look dingy, and blackening every succeeding year, it became coal-black at the end of four. Its chief food was hempseed.[1] Such influence has

pheasant, and in the Ashmolean Museum at Oxford there is a white wren. If there is any decided marking on the feathers, as in the peacock, they are just indicated in the white variety by a faint neutral tint colouring.—J. G. W.

(1) It is a well-known fact, that a constant diet of hempseed will always turn bullfinches black.—J. G. W.

Pyrrhula rubicilla. The Bullfinch presents the curious fact, that the naturally weak notes of a bird may be so strengthened by cultivation, that they become rich and full. The piping Bullfinches, as they are called, are trained by means of an instrument called a bird organ, on which the melody to be learned is played and repeated until the bird has completely learned it. The instruction must be continued through the moulting season, or the bird will almost certainly forget the lessons which have been taught with so much trouble. The greater part of the piping Bullfinches sold in this country are brought from Germany.

A curious anecdote of a bullfinch came lately under my notice. A lady on opening her window, saw a bullfinch seated upon the sill. She was rather astonished to see

food on the colour of animals ! The pied and mottled colours of domesticated animals are supposed to be owing to high, various, and unusual food.

I had remarked for years, that the root of the cuckoo-pint (*arum*) was frequently scratched out of the dry banks of hedges, and in severe snowy weather. After observing, with some exactness, myself, and getting others to do the same, we found it was the thrush kind that scratched it out. The root of the *arum* is remarkably warm and pungent.[1]

Our flocks of female chaffinches have not yet forsaken us. The blackbirds and thrushes are very much thinned down by that fierce weather in January.

In the middle of February, I discovered in my tall hedges a little bird that raised my curiosity ; it was of that yellow-green colour that belongs to the *salicaria* kind, and, I think, was soft-billed. It was no *parus*, and was too long and too big for the golden-crowned wren, appearing most like the largest willow-wren. It hung sometimes with its back downwards, but never continuing one moment in the same place. I shot at it, but it was so desultory that I missed my aim.

I wonder that the stone curlew, *charadrius œdicnemus*, should be mentioned by the writers as a rare bird; it abounds in all the champaign parts of Hampshire and Sussex, and breeds, I think, all the summer, having young ones, I know, very late in the autumn. Already they begin clamouring in the evening. They cannot, I think, with any propriety be called, as they are by Mr. Ray, " *circa aquas versantes;* " for with us (by day at least) they haunt only the most dry, open, upland fields and sheepwalks, far removed from water :

that the bird did not fly away, and on putting out her hand, the bird suffered itself to be taken up and carried into the room. When placed upon the table, it sat quite still, but looked as if it were suffering from illness. On examining the bird, a seed was discovered sticking fast in its throat. A needle soon removed the obstruction, and the bird recovered all its liveliness. Its life, however, was not prolonged by the relief, for while a cage was being prepared for its reception, it escaped from the hands of its benefactress, flew to the window, and dashed itself with such violence against one of the panes that it fell dead on the floor.—J. G. W.

(1) *Arum maculatum*, often called lords and ladies by children. The root of this plant when properly dried and powdered, yields a farinaceous substance very closely resembling arrowroot, and is in some places prepared largely under the title of British Arrowroot.—J. G. W.

what they may do in the night, I cannot say. Worms are their usual food, but they also eat toads and frogs.

I can show you some good specimens of my new mice. Linnæus, perhaps, would call the species *mus minimus*.

LETTER XVI.

ON THE STONE CURLEW, WILLOW WREN, GRASSHOPPER LARK, FLY-CATCHER, AND NUTHATCH.

SELBORNE, *April* 18, 1768.

DEAR SIR,—The history of the stone curlew, *charadrius œdicnemus*, is as follows : It lays its eggs, usually two, never more than three, on the bare ground, without any nest, in

STONE CURLEW [1]

the field, so the countryman, in stirring his fallows, often destroys them. The young run immediately from the egg like partridges, &c., and are withdrawn to some flinty field by the dam, where they skulk among the stones, which are

(1) *Œdicnemus crepitans.*—J. G. W.

their best security; for their feathers are so exactly of the
colour of our grey spotted flints, that the most exact observer, unless he catches the eye of the young bird, may be
eluded. The eggs are short and round, of a dirty white,
spotted with dark bloody blotches. Though I might not be
able, just when I pleased, to procure you a bird, yet I could
show you them almost any day; and any evening you may
hear them round the village; for they make a clamour which
may be heard a mile. *Œdicnemus* is a most apt and expressive name for them, since their legs seem swollen like
those of a gouty man. After harvest, I have shot them
before the pointers in turnip fields.

YELLOW WILLOW WREN

I make no doubt but there are three species of the willow-
wrens; two I know perfectly, but have not been able yet
to procure the third.[2] No two birds can differ more in their
notes, and that constantly, than those two that I am acquainted with; for the one has a joyous, easy, laughing
note, the other a harsh loud chirp. The former is every way
larger, and three-quarters of an inch longer, and weighs two

(1) *Sylvia trochilus.* This bird is also called the Wood-wren.—J. G. W.
(2) *Sylvia trochilus,* the Willow-warbler; *Sylvia sibilatrix,* the Wood-warbler
and *Sylvia rufa,* the Chiff-chaff.—J. G. W.

drachms and a half, while the latter weighs but two ; so that the songster is one-fifth heavier than the chirper. The chirper (being the first summer bird of passage that is heard, the wryneck sometimes excepted) begins his notes in the middle of March, and continues them through the spring and summer, till the end of August, as appears by my journals. The legs of the larger of these two are flesh-coloured ; of the less, black.

The grasshopper lark began his sibilous note in my fields last Saturday. Nothing can be more amusing than the whisper of this little bird, which seems to be close by, though at an hundred yards' distance ; and, when close at your ear,

GRASSHOPPER WARBLER [1]

is scarce any louder than when a great way off. Had I not been a little acquainted with insects, and known that the grasshopper kind is not yet hatched, I should have hardly believed but that it had been a *locusta*, whispering in the bushes. The country people laugh when you tell them that it is the note of a bird. It is a most artful creature, skulking in the thickest part of a bush, and will sing at a yard distance, provided it be concealed. I was obliged to get a person to go on the other side of the hedge where it haunted ;

(1) *Sylvia locustella.* The nest of this bird is very carefully hidden, and contains from five to seven eggs, white, spotted with red. The bird is found in most parts of England, and has been heard near Edinburgh. The note is so like that of a grasshopper that I had lately some difficulty in persuading a friend that the note could be that of a bird. The country people all abjure the bird, and declare that the sound proceeds from the *field mouse.*—J. G. W.

and then it would run, creeping like a mouse before us for an hundred yards together, through the bottom of the thorns; yet it would not come into fair sight; but in a morning early, and when undisturbed, it sings on the top of a twig, gaping and shivering with its wings. Mr. Ray himself had no knowledge of this bird, but received his account from Mr. Johnson, who apparently confounds it with the *reguli non cristati*, from which it is very distinct. See Ray's *Philos. Letters*, p. 108.

The fly-catcher (*stoparola*) has not yet appeared: it usually breeds in my vine. The redstart begins to sing: its note is short and imperfect, but is continued till about the middle of June. The willow-wrens (the smaller sort) are horrid pests in a garden, destroying the peas, cherries, currants, &c., and are so tame that a gun will not scare them.[1]

A List of the Summer Birds of Passage discovered in this neighbourhood, ranged somewhat in the order in which they appear.

LINNÆI NOMINA.

Smallest Willow-wren	*Motacilla trochilus.*
Wryneck	*Iynx torquilla.*
House-swallow	*Hirundo rustica.*
Martin	*Hirundo urbica.*
Sand-martin	*Hirundo riparia.*
Cuckoo	*Cuculus canorus.*
Nightingale	*Motacilla luscinia.*
Blackcap	*Motacilla atricapilla.*
White-throat	*Motacilla sylvia.*
Middle Willow-wren	*Motacilla trochilus.*
Swift	*Hirundo apus.*
Stone Curlew?	*Charadrius œdicnemus?*
Turtle-dove?	*Turtur aldrovandi?*
Grasshopper Lark	*Alauda trivialis.*
Landrail	*Rallus crex.*
Largest Willow-wren	*Motacilla trochilus.*
Redstart	*Motacilla phœnicurus.*
Goatsucker, or Fern-owl	*Caprimulgus Europæus.*
Fly-catcher	*Muscicapa grisola.*

My countrymen talk much of a bird that makes a clatter with its bill against a dead bough or some old pales, calling it a jar-bird. I procured one to be shot in the very fact; it

(1) In another edition of this work, the willow-wren is exonerated from this charge, and the damage attributed to the young of the garden warbler.—J. G. W.

proved to be the *sitta Europæa* (the nuthatch). Mr. Ray
says, that the less spotted woodpecker does the same. This
noise may be heard a furlong or more.

NUTHATCH.[1]

Now is the only time to ascertain the short-winged summer
birds : for, when the leaf is out, there is no making any remarks on such a restless tribe ; and, when once the young
begin to appear, it is all confusion ; there is no distinction of
genus, species, or sex.

In breeding time, snipes play over the moors, piping and
humming ; they always hum as they are descending. Is
not their hum ventriloquous, like that of the turkey? Some
suspect that it is made by their wings.

This morning I saw the golden crowned wren,[3] whose

(1) *Sitta Europæa.* The nest of the Nuthatch is made in the hollow of a tree, and
it is very remarkable, that if the external hole should prove too large, the bird reduces
its size by plastering it up with mud. Yarrell suggests that it does so on account of the
battles which sometimes take place between the Tits and the Nuthatch, who may find
it easier to defend a small hole than a large one. In order to give itself a sufficient
swing to break the nuts, the foot is very large and exceedingly powerful. Although
it runs over the bark of trees after the manner of the Woodpecker, it does not, like
that bird, support itself by the tail feathers, but trusts entirely to the grasping power
of its feet.—J. G. W.

(2) Mr. Yarrell is of opinion that this peculiar humming note is produced by the
wings.—J. G. W.

(3) The nest of this tiny bird is very pretty, and is carefully guarded from the
effects of rain and wind by being hung under the shelter of a thick umbrageous
branch The nest is often built in a fir-tree, and a very pretty specimen in my
procession is placed under the shelter of a branch of fir-cones, that form a canopy
over it. The Golden-crested Wren is common at Oxford, and I have often caught
them in Derbyshire by means of bird-lime.—J. G. W.

crown glitters like burnished gold. It often hangs like a titmouse with its back downwards.

GOLDEN-CRESTED WREN [2]

LETTER XVII.

ON THE FROG, TOAD, WATER-NEWT, VIPER, ETC.

SELBORNE, *June* 18, 1768.

DEAR SIR,—On Wednesday last arrived your agreeable letter of June the 10th. It gives me great satisfaction to find that you pursue these studies still with such vigour, and are in such forwardness with regard to reptiles and fishes.

The reptiles, few as they are, I am not acquainted with so well as I could wish, with regard to their natural history. There is a degree of dubiousness and obscurity attending the propagation of this class of animals something analogous to that of the *cryptogamia* in the sexual system of plants; and the case is the same with regard to some of the fishes, as the eel, &c.

It is strange that the matter with regard to the venom of toads has not been yet settled.[2] That they are not

(1) *Regulus cristatus.*—J. G. W.

(2) An acid fluid is secreted by the tubercles on the toad's skin, but I do not believe that it is actually poisonous. I lately saw an account of savages poisoning their arrows with the secretion obtained from a toad, by placing it near the fire. The account, however, is rather dubious, as it is well known how careful savages are to preserve the secret of their poisons.—J. G. W.

noxious to some animals is plain; for ducks, buzzards, owls, stone curlews, and snakes, eat them, to my knowledge,

TOAD. [1]

with impunity. And I well remember the time, but was not an eye-witness to the fact (though numbers of persons were), when a quack at this village ate a toad to make the country people stare; afterwards he drank oil.

I have been informed also, from undoubted authority, that some ladies (ladies, you will say, of peculiar taste) took a fancy to a toad,[2] which they nourished, summer after summer, for many years, till he grew to a monstrous size, with the

(1) *Bufo vulgaris.*—J. G. W.

(2) Mr. Smee relates an anecdote of a toad, which proves that animal to be possessed of a faculty of which it certainly would appear to be devoid, from its usual habits.

"A gentleman of my acquaintance kept a toad, and used to give it something to eat at dinner time, and it always knew the time regularly, and was always ready at the window to receive his dinner. From family circumstances the time of dining was changed from four to two, and the toad was not there to receive its food. The next day it kept a sharp look-out, and ever afterwards came down at the new time; so that, in point of fact, the creature judges of the interval which existed between one day and another."

I may be pardoned here for introducing a passage from an old book of Natural History, in order to show what difficulties such men as Linnæus had to overcome in disentangling the study of Natural History from the numerous errors with which it was encumbered by the older writers. I may observe that this passage is not more absurd than the remainder of the book.

"It fortuned as the said Earl travailed in Bedfordshire, neer unto a Market town called *Owbourn,* some of his company espyed a Toad fighting with a Spyder, under

maggots, which turn to flesh-flies. The reptile used to come forth every evening from a hole under the garden-steps; and

a hedge in a bottom, by the highway-side, whereat they stood still, until the Earl their Lord and Master came also to behold the same; and there he saw how the Spyder still kept her standing, and the Toad divers times went back from the Spyder, and did eat a piece of an herb, which to his judgement was like a Plantain. At the last, the Earl having seen the Toad do it often, and still return to the combate against the Spyder, he commanded one of his men to go, and with his dagger to cut off the herb, which he performed and brought it away. Presently after the Toad returned to seek it, and not finding it according to her expectation, swelled and broke in pieces: for having received poyson from the Spyder in the combate, nature taught her the vertue of that herb, to expell and drive it out, but wanting the herb, the poyson did instantly work and destroy her. And this (as I am informed) was often times related by the Earl of Bedford himself upon sundry occasions, and therefore I am bolder to insert it into this story.

"I do the more easily believe it because of another like story related by Erasmus in his Book of Friendship, happening likewise in *England*, in manner as followeth. There was a Monk, who had in his chamber divers bundles of green rushes, where-withall he used to strow his chamber at his pleasure. It happened on a day after dinner, that he fell asleep upon one of those bundles of rushes, with his face upward, and while he there slept, a great Toad came and sate upon his lips, bestriding him in such manner, as his whole mouth was covered. Now when his fellows saw it, they were at their wits end, for to pull away the toad was an unavoidable death, but to suffer her to stand still upon his mouth was a thing more cruel than death: and therefore one of them espying a Spyder's web in the window wherein was a great Spyder, he did advise that the Monk should be carryed to that window, and laid with his face upward, right underneath the Spyder's web, which was presently accomplished. And as soon as the Spyder saw her adversary the Toad, she presently wove her thred and descended down upon the Toad, at the first meeting whereof the spyder wounded the toad so that it swelled, and at the second meeting, it swelled more, but at the third time the Spyder kild the Toad, and so became grateful to her host which did nourish her in his Chamber; for at the third time the Toad leaped off from the man's mouth and swelled to death; but the man was preserved whole and alive. And thus much may suffice for the antipathy of Nature betwixt the Toad and the Spider."

Mr. St. John, with whose favourable opinion of toads I perfectly agree, gives in his Highland Sports a very interesting account of some of the habits of these creatures:—

"For my own part, I can see nothing more disgusting in animals usually called reptiles, such as lizards and toads, than in any other living creatures. A toad is a most useful member of society, and deserves the freedom of all floricultural societies, as well as entire immunity from all the pains and penalties which he undergoes at the hands of the ignorant and vulgar. In hot-beds and hot-houses he is extremely useful, and many gardeners take great care of toads in these places, where they do good service by destroying beetles and other insects. In the flower beds, too, they are of similar use. Of quiet and domestic habits, the toad seldom seems to wander far from his seat or form under a loose stone, or at the foot of a fruit-tree or box-edging. There are several habitués of this species in my garden, whom I always see in their respective places during the middle of the day. In the evening they issue out in search of their prey. I found a toad one day caught by the leg in a horse-hair snare which had been placed for birds. The animal, notwithstanding the usual placid and phlegmatic demeanour of its race, seemed to be in a perfect fury, struggling and scratching at everything within his reach, apparently much more in

was taken up, after supper, on the table to be fed. But at last a tame raven, kenning him as he put forth his head, gave him such a severe stroke with his horny beak as put out one eye. After this accident, the creature languished for some time, and died.

I need not remind a gentleman of your extensive reading, of the excellent account there is from Mr. Derham, in Ray's *Wisdom of God in the Creation*, p. 365, concerning the migration of frogs from their breeding-ponds. In this account he at once subverts that foolish opinion, of their dropping from the clouds in rain; showing that it is from the grateful coolness and moisture of those showers that they are tempted to set out on their travels, which they defer till those fall. Frogs are as yet in their tadpole state; but in a few weeks

---

anger than fear. Like many other individuals of quiet exterior, toads are liable to great fits of passion and anger, as is seen in the pools during April, when five or six will contend for the good graces of their sultanas with a fury and pertinacity that is quite wonderful, fighting and struggling for hours together. And where a road intervenes between two ditches I have seen the battle carried on even in the dry dust, till the rival toads, in spite of their natural aquatic propensities, became perfectly dry and covered with sand, and in this powdered state will they continue fighting, regardless of the heat, which shrivels up their skin, or of passers-by, who may tread on them, and maim them, but cannot stop their fighting. There is more character and energy in a toad than is supposed. After the young ones have acquired their perfect shape, they appear to leave the water, and frequently the roads and paths are so covered with minute but well-formed toadlings, that it is impossible to put your foot down without crushing some of them."

I have received a communication from a friend in the country respecting a tame toad. The animal had lived in the family for several years. It was accustomed to sup on a lump of sugar.

I heard lately rather a peculiar method of catching insects by means of toads. A gentleman had skinned a toad preparatory to stuffing it, when he was seized with a curiosity to discover what the reptile had been eating for breakfast. He accordingly emptied its stomach into a basin of water, where he left the contents for three or four days. On examination, he found that the contents of the stomach consisted almost entirely of beetles, some of them very scarce. He pinned out the beetles, and left them for a day or two to harden. On looking at them again, he found the greater part of them alive and sprawling about. One cannot but wonder at the strange tenacity of life in these insects. They had been for some time in the stomach of a living animal, they had then been for several days immersed in water, and lastly they had been impaled on a pin for some days, the pin bearing about the same relation to them as a large broomstick would to us. Yet in spite of all this, they were yet alive, and if released from the pin would probably have recovered.

The gentleman who discovered these insects in the toad, reflected that as the creature had been killed in the early morning, it had probably been devouring those beetles which only come out at night. He accordingly subjected several toads to the same process, and seldom failed to find some rare beetles in their stomachs.—J. G. W.

our lanes, paths, fields, will swarm for a few days with myriads of those emigrants, no larger than my little finger nail. Swammerdam gives a most accurate account of the method and situation in which the male impregnates the spawn of the

FROG[1]

female. How wonderful is the economy of Providence with regard to the limbs of so vile a reptile! While it is an *aquatic*, it has a fish-like tail and no legs: as soon as the legs sprout, the tail drops off as useless, and the animal betakes itself to the land!

Merret, I trust, is widely mistaken when he advances that the *rana arborea* is an English reptile: it abounds in Germany and Switzerland.[2]

It is to be remembered that the *salamandra aquatica* of Ray (the water-newt, or eft) will frequently bite at the angler's bait, and is often caught on his hook. I used to take it for granted that the *salamandra aquatica* was hatched, lived, and died in the water. But John Ellis, Esq. F.R.S. (the coralline Ellis), asserts, in a letter to the Royal Society, dated June the 5th, 1766, in his account of the *mud iguana*, an

---

(1) *Rana temporaria.*—J. G. W.
(2) I perfectly recollect when I was a boy seeing a colony of green frogs inhabiting an apple-tree in a garden at Marston near Oxford. They lived in a hole about five feet from the ground.—J. G. W.

amphibious *bipes* from South Carolina, that the water-eft, or newt, is only the larva of the land eft, as tadpoles are of frogs.

WATER-NEWT. [1]

Lest I should be suspected to misunderstand his meaning, I shall give it in his own words.   Speaking of the *opercula*, or coverings to the gills of the *mud iguana*, he proceeds to say, that " The form of these pennated coverings approaches very near to what I have some time ago observed in the *larva*, or *aquatic* state of our English *lacerta*, known by the name of eft, or newt, which serve them for coverings to their gills, and for fins to swim with while in this state ; and which they

(1) *Triton cristatus.* Female, see p. 81. The Newt belongs to the Batrachians, and not to the Lizards, as is generally imagined. The term "cristatus" or "crested" is given to it on account of the beautiful crimson-tipped crest that extends along the whole back and tail of the male newt, and which when the creature is in the water is constantly waving in a most elegant manner. With regard to Mr. White's dilemma of the hatching of this creature, I can state that its changes take place in the water, as I have watched its development from the egg. After its change to the perfect state, it no longer breathes by means of gills, but is forced to rise to the surface of the water in order to take a fresh supply of air. Its lungs are so large, that one breath is sufficient for a long sojourn under the water. Moreover, the blood of reptiles being only partially oxygenated, they can exist for some time, even when the supply of air is entirely cut off. (See p. 80.)

The female has a very singular habit of tying up her eggs singly in the blades of grass that fringe the water side. The food of the newt consists principally of tadpoles and worms, which it eats with a very curious snap, always taking the worm by the middle. I have heard that this creature is accused of cannibalism, and have frequently seen it attack the smaller newts, but I never saw it eat them.

Many people, especially those who see most of them, have a firm belief in the

lose, as well as the fins of their tails, when they change their state, and become land animals, as I have observed, by keeping them alive for some time myself."

Linnæus, in his *Systema Naturæ*, hints at what Mr. Ellis advances, more than once.

Providence has been so indulgent to us as to allow of but one venomous reptile of the serpent kind in these kingdoms, and that is the viper. As you propose the good of mankind to be an object of your publications, you will not omit to mention common salad oil as a sovereign remedy against the bite of the viper.[1] As to the blind worm (*anguis fragilis*, so called because it snaps in sunder with a small blow) I have found, on examination, that it is perfectly innocuous. A neighbouring yeoman (to whom I am indebted for some good hints) killed and opened a female viper about the 27th of May: he found her filled with a chain of eleven eggs, about the size of those of a blackbird; but none of them were advanced so far towards a state of maturity as to contain any rudiments of young. Though they are oviparous, yet

poisonous character of the newt. There is still a tradition extant in a Wiltshire village where newts abounded, that a girl was putting her hand into the pool, when an "effet" bit her arm, and then spat fire into the wound, so that the girl lost her arm. So firmly do the people believe this, that when I took several newts in my hands, they seemed more disposed to attribute my invulnerability to some hidden charm than to the harmlessness of the creatures themselves. Some of these newts were placed in a trough where the cows drank. After a few days a calf belonging to one of the cows died, of course on account of the "effets." It was in vain to urge that the calf had been shut up in the cow-house, and had never even seen the newts:—there were the effets, and there was the dead calf, and consequently the calf was killed by the effets. So all my newts were summarily ejected.

Some newts that I kept for some time in a glass vessel, changed their skins within a day or two of their change of residence, and I noticed that every newt placed in the vessel always cast its skin. Perhaps the water may have had something to do with it, as the water in the glass vessel was taken from a pump. The skin of the feet came off like little gloves, but I could never see how it was pulled off.

The newt is very tenacious of life. I have always on hand a bottle of spirits of wine poisoned with corrosive sublimate. This is so deadly that it kills, almost instantly, even the large beetles and caterpillars. Some newts, however, which I put into the bottle, fancying that it would be the most merciful way of killing them, lived for nearly a quarter of an hour in it. They continued to form a circle and keep their heads out of the spirit, and I was forced to keep the bottle constantly in motion to keep them under, or they would possibly have lived longer. The singular part of the proceeding was, that the spirit turned their beautiful brilliant eyes to a dead white, and they appeared completely blinded. The effect was such that I determined never to put a living newt into spirit again.—J. G. W.

(1) The viper-catchers apply the oil hot.—J. G. W.

BLIND WORM.[1]

they are viviparous also, hatching their young within their
bellies, and then bringing them forth.   Whereas snakes lay
chains of eggs every summer in my melon beds, in spite of
all that my people can do to prevent them; which eggs do not
hatch till the spring following, as I have often experienced.
Several intelligent folks assure me that they have seen the
viper open her mouth and admit her helpless young down
her throat on sudden surprises, just as the female opossum
does her brood into the pouch under her belly, upon the like
emergencies; and yet the London viper-catchers insist on it
to Mr. Barrington, that no such thing ever happens.[2]   The
serpent kind eat, I believe, but once in a year; or, rather,
but only just at one season of the year.   Country people
talk much of a water-snake, but, I am pretty sure, without
any reason; for the common snake (*coluber natrix*) delights
much to sport in the water, perhaps with a view to procure
frogs and other food.[3]

(1) *Anguis fragilis.*   The blind worm is not a snake, but a lizard of the family
of the Skinks.   The tail of the common lizard also snaps asunder at a very slight
blow.—J. G. W.

(2) The eggs of the snake are not quite so large as those of the blackbird; they
are white, and are covered with a soft membrane.—J. G. W.

(3) The snake delights in the water.   I have captured them in a stone quarry
filled with water, in which also were multitudes of newts.   My scholars made great
pets of their snakes, and used to carry them in their jacket pockets.   It was rather
a singular circumstance, that although when first caught they always emitted their
very unpleasant odour, yet after a day or two they would permit any amount of hand-

VIPER.[1]

I cannot well guess how you are to make out your twelve species of reptiles, unless it be by the various species, or rather varieties, of our *lacerti*, of which Ray enumerates five. I have not had opportunity of ascertaining these, but remember well to have seen, formerly, several beautiful green *lacerti* on the sunny sand-banks near Farnham, in Surrey; and Ray admits there are such in Ireland.

ling without giving out the slightest disagreeable scent. The boys were accustomed to take their snakes on the downs for exercise, and sometimes gave them a swim in the quarry. They had a very large number of snakes, and from practice became quite expert in the chase, particularly an Indian boy, who would plunge into the midst of a nettle-covered ditch on hearing the smallest rustle, and triumphantly bring out a snake, which was transferred to his pocket, and added to the stock at home. The water-snake mentioned by the country people is the common ringed snake. There are only two English snakes, the common snake and the viper.

The snake feeds on frogs, newts, mice, &c. My scholars were accustomed to feed their snakes by the very simple process of making the snake open its mouth, and then pushing a newt or frog down its throat. I once saw a snake struggling with an enormous frog which it had caught by its hinder feet, and was slowly dragging down its throat. When it saw me, the snake moved off, and the frog stopped the screams that it had been uttering at intervals, and hopped away.—J. G. W.

(1) *Pelias Berus.*—J. G. W.

LETTER XVIII.

ON THE STICKLEBACKS, LOACH, TOAD-CURE AND WATER-EFT

STICKLEBACK

SELBORNE, *July* 27, 1768.

DEAR SIR,—I received your obliging and communicative letter of June the 28th, while I was on a visit at a gentleman's house, where I had neither books to turn to, nor leisure to sit down to return you an answer to many queries, which I wanted to resolve in the best manner that I am able.

A person by my order has searched our brooks, but could find no such fish as the *gasterosteus pungitius;* [1] he found *gasterosieus aculeatus* in plenty. This morning, in a basket, I packed a little earthen pot full of wet moss, and in it some sticklebacks. male and female, the females big with spawn ; some lamperns; some bull-heads; but I could procure no minnows. This basket will be in Fleet-street by eight this

(1) The Ten-spined Stickleback; *Gasterosteus aculeatus* has three spines.—J. G. W

evening; so I hope Mazel will have them fresh and fair to-morrow morning. I gave some directions in a letter, to what particulars the engraver should be attentive.

Finding, while I was on a visit, that I was within a reasonable distance of Ambresbury, I sent a servant over to that town, and procured several living specimens of loaches, which he brought safe and brisk, in a glass decanter. They were taken in the gulleys that were cut for watering the meadows. From these fishes (which measured from two to four inches in length) I took the following description:—

' The loach, in its general aspect, has a pellucid appearance; its back is mottled with irregular collections of small black dots, not reaching much below the *linea lateralis*, as are the back and tail fins; a black line runs from each eye down to the nose; its belly is of a silvery white; the upper jaw pro-

LOACH [1]

jects beyond the lower, and is surrounded with six feelers, three on each side; its pectoral fins are large, its ventral much smaller; the fin behind its anus small; its dorsal fin

(1) *Cobitis barbatula.* The Loach is sometimes called the Groundling, in reference to its habits, and, in some places, goes by the name of Beardie, in allusion to the barbules hanging from its lips.—J. G W.

large, containing eight spines; its tail, where it joins to the
tail fin, remarkably broad, without any taperness, so as to be
characteristic of this genus; the tail fin is broad, and square
at the end. From the breadth and muscular strength of the
tail, it appears to be an active nimble fish."

In my visit I was not very far from Hungerford, and did
not forget to make some inquiries concerning the wonderful
method of curing cancers by means of toads. Several intelligent persons, both gentry and clergy, do, I find, give a
great deal of credit to what was asserted in the papers ; and
I myself dined with a clergyman who seemed to be persuaded
that what is related is matter of fact; but, when I came to
attend to his account, I thought. I discerned circumstances
which did not a little invalidate the woman's story of the
manner in which she came by her skill. She says of herself,
that, "labouring under a virulent cancer, she went to some
church where there was a vast crowd; on going into a pew,
she was accosted by a strange clergyman, who, after expressing compassion for her situation, told her, that if she would
make such an application of living toads as is mentioned,[1] she
would be well." Now, is it likely that this unknown gentleman should express so much tenderness for this single suf-
ferer, and not feel any for the many thousands that daily
languish under this terrible disorder ? Would he not have
made use of this invaluable nostrum for his own emolument ?
or, at least, by some means of publication or other, have
found a method of making it public for the good of mankind?
In short, this woman (as it appears to me) having set up for
a cancer doctress, finds it expedient to amuse the country
with this dark and mysterious relation.

The water eft has not, that I can discern, the least appearance of any gills,[2] for want of which it is continually rising
to the surface of the water to take in fresh air. I opened a
big-bellied one, indeed, and found it full of spawn. Not that
this circumstance at all invalidates the assertion that they are
larvæ ; for the *larvæ* of insects are full of eggs, which they

---

(1) The toads were supposed, from their general affinity to poison, to suck out the
poison of the cancer. The idea is still prevalent in some places.—J. G. W.

(2) It has, when young, branchial tufts or gills, but loses them when it attains to
its perfect form.—J. G. W.

exclude the instant they enter their last state. The water-
eft is continually climbing over the brims of the vessel,
within which we keep it in water, and wandering away; and

WATER EFT. [1]

people every summer see numbers crawling out of the pools
where they are hatched, up the dry banks. There are
varieties of them differing in colour; and some have fins up
their tail and back, and some have not.[2]

LETTER XIX.

ON THE WILLOW-LARK, AND WILLOW-WREN.

SELBORNE, *Aug.* 17, 1768.

DEAR SIR.—I have now, past dispute, made out three distinct species of the willow-wrens (*motacillæ trochili*), which
constantly and invariably use distinct notes. But, at the same
time, I am obliged to confess that I know nothing of your
willow-lark. In my letter of April the 18th, I had told you

(1) *Triton cristatus.* Male.—J. G. W.
(2) See note on p. 74.—J. G. W.

peremptorily that I knew your willow-lark, but had not seen
it then ; but, when I came to procure it, it proved in all respects a very *motacilla trochilus;* only that it is a size larger
than the two other, and the yellow-green of the whole upper
part of the body is more vivid, and the belly of a clearer
white. I have specimens of the three sorts now lying before
me ; and can discern that there are three gradations of sizes,
and that the least has black legs, and the other two, flesh-
coloured ones. The yellowest bird is considerably the largest,
and has its quill feathers and secondary feathers tipped with
white, which the others have not. This last haunts only the
tops of trees in high beechen woods, and makes a sibilous
grasshopper-like noise now and then, at short intervals,
shivering a little with its wings when it sings; and is, I make
no doubt now, the *regulus non cristatus* of Ray; which he
says, " *cantat voce stridulâ locustœ.*" Yet this great ornithologist never suspected that there were three species.[1]

(1) See note on p. 65.—J. G. W.

LETTER XX.

ON THE SANDPIPER, BUTCHER-BIRD, AND RING-OUSEL.

SELBORNE, *Oct.* 8, 1768.

IT is, I find, in zoology as it is in botany; all nature is so full, that that district produces the greatest variety which is the most examined. Several birds, which are said to belong to the north only, are, it seems, often in the south. I have discovered this summer three species of birds with us, which writers mention as only to be seen in the northern counties. The first that was brought me (on the 14th of May) was the

SANDPIPER. [1]

sandpiper (*tringa hypoleucus*): it was a cock bird, and haunted the banks of some ponds near the village; and, as it had a companion, doubtless intended to have bred near that water. Besides, the owner has told me since, that on recollection he has seen some of the same birds round his ponds in former summers.

(1) *Tringoïdes hypoleuca* —J. G. W.

The next bird that I procured (on the 21st of May) was a male red-backed butcher-bird (*lanius collurio*). My neighbour, who shot it, says that it might easily have escaped his notice, had not the outcries and chattering of the whitethroats

RED-BACKED BUTCHER BIRD [1]

and other small birds drawn his attention to the bush where it was : its craw was filled with the legs and wings of beetles.

The next rare birds (which were procured for me last week) were some ring-ousels (*turdi torquati*).[2]

This week twelvemonths a gentleman from London being with us, was amusing himself with a gun, and found, he told us, on an old yew hedge where there were berries, some birds like blackbirds, with rings of white round their necks; a neighbouring farmer also at the same time observed the same; but, as no specimens were procured, little notice was taken. I mentioned this circumstance to you in my letter of Novem-

(1) *Lanius collurio.* The propensity of the Shrikes for impaling such creatures as they have killed is well known. The Red-backed Shrike feeds largely on insects, and we used to discover its nest by the insects which it had impaled. The insect most commonly found in this position was *Staphylinus erythropterus*, one of the rove-beetles. The nest is very easy to find, for in addition to the signs already noted, when an intruder approaches the nest, the parent birds make such a noise, and flutter about with such eagerness, that the position is at once pointed out.

The shrikes possess, in common with the falcons and owls, the power of ejecting the indigestible portions of their insect food in the form of pellets.—J. G. W.

(2) *Turdus torquatus.* The nest of this bird is like that of the blackbird, placed either on or very near the ground, and in Scotland is said to be built among the heath. The eggs somewhat resemble those of the common song thrush. In France the ring-ousels are said to make great havoc among the grapes.—J. G. W.

ber the 4th, 1767 (you, however, paid but small regard to what I said, as I had not seen these birds myself): but last week the aforesaid farmer, seeing a large flock, twenty or

RING-OUSEL.

thirty, of these birds, shot two cocks and two hens; and says, on recollection, that he remembers to have observed these birds again last spring, about Lady-day, as it were on their return to the north. Now perhaps these ousels are not the ousels of the north of England, but belong to the more northern parts of Europe; and may retire before the excessive rigour of the frosts in those parts, and return to breed in spring when the cold abates. If this be the case, here is discovered a new bird of winter passage, concerning whose migrations the writers are silent; but if these birds should prove the ousels of the north of England, then here is a migration disclosed within our own kingdom, never before remarked. It does not yet appear whether they retire beyond the bounds of our island to the south; but it is most probable that they usually do, or else one cannot suppose that they would have continued so long unnoticed in the southern counties. The ousel is larger than a blackbird, and feeds on haws; but last autumn (when there were no haws) it fed on yew-berries: in the spring it feeds on ivy-berries, which ripen only at that season, in March and April.

I must not omit to tell you (as you have been so lately on the study of reptiles) that my people, every now and then of late, draw up, with a bucket of water from my well, which is

LIZARD [1]

sixty-three feet deep, a large black warty lizard, with a fin tail and yellow belly. How they first came down at that depth, and how they were ever to have got out thence without help, is more than I am able to say.

My thanks are due to you for your trouble and care in the examination of a buck's head. As far as your discoveries reach at present, they seem much to corroborate my suspicions; and I hope Mr. —— may find reason to give his decision in my favour; and then, I think, we may advance this extraordinary provision of nature as a new instance of the wisdom of God in the creation.

As yet I have not quite done with my history of the *œdic-nemus*, or stone curlew; for I shall desire a gentleman in Sussex (near whose house these birds congregate in vast flocks in the autumn) to observe nicely when they leave him (if they do leave him), and when they return again in the spring: I was with this gentleman lately, and saw several single birds.

(1) *Zootoca vivipara.*—J. G. W.

LETTER XXI.

ON THE STONE CURLEW, JACKDAW, AND RING-OUSEL.

SELBORNE, *Nov.* 28, 1768.

DEAR SIR,—With regard to the *œdicnemus*, or stone curlew, I intend to write very soon to my friend near Chichester, in whose neighbourhood these birds seem most to abound; and shall urge him to take particular notice when they begin to congregate, and afterwards to watch them most narrowly, whether they do not withdraw themselves during the dead of the winter. When I have obtained information with respect to this circumstance, I shall have finished my history of the stone curlew, which, I hope, will prove to your satisfaction, as it will be, I trust, very near the truth. This gentleman, as he occupies a large farm of his own, and is abroad early and late, will be a very proper spy upon the motions of these birds; and besides, as I have prevailed on him to buy the *Naturalist's Journal* (with which he is much delighted), I shall expect that he will be very exact in his dates. It is very extraordinary, as you observe, that a bird so common with us should never straggle to you.

And here will be the properest place to mention, while I think of it, an anecdote which the above-mentioned gentleman told me when I was last at his house; which was, that in a warren joining to his outlet, many daws (*corvi monedulæ*) build every year in the rabbit-burrows under ground.[1] The

---

(1) *Corvus monedula.* A tame jackdaw of my acquaintance was a very amusing bird, and quite domesticated himself; in fact, he evidently considered himself as the principal person in company, and that the sole duty of his master was to wait on him. He was extremely fond of washing himself, and even in the coldest days of winter he would come and ask for water, which was always given to him in a large leaden basin. Into this he would hop, after walking round it once and tasting the water; and then he plunged his head and shoulders beneath the water, and spun round and round in the basin, scattering the water with his wings on every side. I never saw him alarmed except on two occasions, once when I caught him in an entomological net, and once when he was travelling and put his head out of the railway carriage window. The trees apparently rushing by struck him with intense horror, and he dived back into his basket, thrust his bill between the interstices, and remained per-

way he and his brothers used to take their nests, while they were boys, was by listening at the mouths of the holes, and if they heard the young ones cry, they twisted the nest out with a forked stick. Some water-fowls (viz. the puffins) breed, I know, in this manner; but I should never have suspected the daws of building in holes on the flat ground.

JACKDAW.

Another very unlikely spot is made use of by daws as a place to breed in, and that is Stonehenge. These birds deposit their nests in the interstices between the upright and the impost stones of that amazing work of antiquity; which circumstance alone speaks the prodigious height of the upright stones, that they should be tall enough to secure those nests from the annoyance of shepherd boys, who are always idling round that place.

fectly silent for at least half an hour, a most uncommon circumstance with him. At last he feebly cried "Jack," but did not stir, and it was not until he had drunk some water that he recovered his usual loquacity. He used to eat oats in a very clever way; with his foot he held down each grain in succession, and with one blow and twist of his beak, he completely shelled it. He was very fond of large insects, and of mice. The cockchaffers and beetles he seized, and with one bite across the thorax, killed them. He then picked off the head, legs, and wings, and only ate the remaining parts. But with a wasp or bee the case was very different. He then hopped round it several times, and at last made a great peck at the bee, and threw it up in the air. After a little while he ventured upon another peck, but never ventured to carry the bee away until it was reduced to a shapeless mass.—J. G. W.

One of my neighbours last Saturday (November the 26th) saw a martin in a sheltered bottom; the sun shone warm, and the bird was hawking briskly after flies. I am now perfectly satisfied that they do not all leave this island in the winter.

You judge very right, I think, in speaking with reserve and caution concerning the cures done by toads; for, let people advance what they will on such subjects, yet there is such a propensity in mankind towards deceiving and being deceived, that one cannot safely relate anything from common report, especially in print, without expressing some degree of doubt and suspicion.

Your approbation with regard to my new discovery of the migration of the ring-ousel, gives me satisfaction; and I find you concur with me in suspecting that they are foreign birds which visit us. You will be sure, I hope, not to omit to make inquiry whether your ring-ousels leave your rocks in the autumn. What puzzles me most, is the very short stay they make with us, for in about three weeks they are all gone. I shall be very curious to remark whether they will call on us at their return in the spring, as they did last year.

I want to be better informed with regard to ichthyology. If fortune had settled me near the sea-side, or near some great river, my natural propensity would soon have urged me to have made myself acquainted with their productions; but as I have lived mostly in inland parts, and in an upland district, my knowledge of fishes extends little farther than to those common sorts which our brooks and lakes produce.

LETTER XXII.

ON THE JACKDAW, GREEN LIZARD, HERON, GOAT-SUCKER, AND BATS.

SELBORNE, *Jan.* 2, 1769.

DEAR SIR,—As to the peculiarity of jackdaws building with us under ground, in rabbit-burrows, you have, in part, hit upon the reason; for, in reality, there are hardly any towers or steeples in all this country. And perhaps, Norfolk excepted, Hampshire and Sussex are as meanly furnished with churches as almost any counties in the kingdom. We have many livings of two or three hundred pounds a-year, whose houses of worship make little better appearance than dove-cots. When I first saw Northamptonshire, Cambridgeshire, and Huntingdonshire, and the Fens of Lincolnshire, I was amazed at the number of spires which presented themselves in every point of view. As an admirer of prospects, I have reason to lament this want in my own country, for such objects are very necessary ingredients in an elegant land-scape.

What you mention with respect to reclaimed toads raises my curiosity. An ancient author, though no naturalist, has well remarked, that "Every kind of beasts, and of birds, and of serpents, and things in the sea, is tamed, and hath been tamed of mankind." [1]

It is a satisfaction to me to find that a green lizard has actually been procured for you in Devonshire, because it corroborates my discovery, which I made many years ago, of the same sort, on a sunny sand-bank near Farnham, in Surrey. I am well acquainted with the south hams of Devonshire, and can suppose that district, from its southerly situation, to be a proper habitation for such animals in their best colours. [2]

Since the ring-ousels of your vast mountains do certainly not forsake them against winter, our suspicions that those which visit this neighbourhood about Michaelmas are not English birds, but driven from the more northern parts of Europe by the frosts, are still more reasonable; and it will be worth your pains to endeavour to trace from whence they come, and to inquire why they make so very short a stay.

In your account of your error with regard to the two species of herons, you incidentally gave me great entertain-ment in your description of the heronry at Cressy-hall, which is a curiosity I never could manage to see. Fourscore nests of such a bird on one tree, is a rarity which I would ride half as many miles to have a sight of. [3] Pray be sure to tell

---

(1) St. James, chap. iii. 7.

(2) The green lizard is not uncommon on sunny sand-banks near Oxford.—J. G. W.

(3) *Ardea cinerea.* The heron or hern is still not uncommon in England, and may be often seen near water. In the summer of this year, 1853, while passing

me in your next whose seat Cressy-hall is, and near what town it lies.[1] I have often thought that those vast extents of

HERON.

Canterbury on the South Eastern Railway, I saw four herons standing close together on the banks of the Stour. They of course took to flight as the train passed by, and were accompanied by three magpies which started from the same spot. The common crow is very apt to attack the heron as it flies along. I have seen an unfortunate heron passing by Lord Barrington's estate near Shrivenham, completely mobbed by crows, who continued their annoyance until the birds were out of sight. There are upwards of thirty regular heronries in England.

This bird will carry away an amazing quantity of shot if fired at. I once succeeded in surprising a heron on the banks of the Isis, as it was standing on one leg, and watching the water. I was in a boat at the time, having a gun under a cushion for the benefit of the water-fowl. The heron's wings looked so impenetrable, that before firing, I roused the bird, and caused it to open its wings ready for flight, when the entire charge came against its side with a sound audible in the boat. The heron, however, flew off apparently unconcerned, and continued its flight far over the fields with undiminished strength.

Mr. Yarrell, in his "British Birds," gives the following singular account of the tragical death of a heron. " A heron was seen one evening going to a piece of water to feed; the spot was visited the next morning, when it was discovered that the heron had struck its sharp beak through the head of an eel, piercing both eyes. The eel thus held, had coiled itself so tightly round the neck of the heron as to impede the bird's respiration, and both were dead."—J. G. W.

(1) Cressy-hall is near Spalding in Lincolnshire.

fens have never been sufficiently explored. If half-a-dozen gentlemen, furnished with a good strength of water spaniels, were to beat them over for a week, they would certainly find more species.

There is no bird, I believe, whose manners I have studied more than that of the *caprimulgus* (the goat-sucker), as it is

THE GOATSUCKER [1]

a wonderful and curious creature; but I have always found, that though sometimes it may chatter as it flies, as I know it

(1) *Caprimulgus Europæus.* This bird forms one of the tribe Fissirostres, so called from the shape of their beak, which appears as if it had been slit up beyond the eyes, giving the bird a very frog-like expression. The Goat-sucker, or Night-jar, makes no nest, but lays it eggs on the bare ground, taking advantage of some natural depression. Both those singular American birds, the Whip-poor-will, and the Chuck-will's widow, belong to the Night-jars. Of course the English language must feel it a great honour that a native American bird should prefer the tongue of the "Britisher," to that of the Delaware or Sioux.

The flight of the night-jar is generally confined to one spot, and it is usually seen, as related by White, restricting itself to a single tree, round which it is constantly circling. When perching, the goat-sucker, contrary to the habits of most birds, sits lengthwise on the branches, crouching so close to the bark that it cannot be distinguished from the tree itself.—J. G. W.

does, yet in general it utters its jarring note sitting on a bough; and I have for many a half hour watched it as it sat with its under mandible quivering, and particularly this summer. It perches usually on a bare twig, with its head lower than its tail, in an attitude well expressed by your draughtsman in the folio *British Zoology*. This bird is most punctual in beginning its song exactly at the close of day; so exactly, that I have known it strike up more than once or twice just at the report of the Portsmouth evening gun, which we can hear when the weather is still. It appears to me past all doubt, that its notes are formed by organic impulse, by the powers of the parts of its windpipe formed for sound, just as cats pur. You will credit me, I hope, when I assure you, that, as my neighbours were assembled in an hermitage on the side of a steep hill where we drink tea, one of these churn-owls came and settled on the cross of that little straw edifice, and began to chatter, and continued his note for many minutes; and we were all struck with wonder to find that the organs of that little animal, when put in motion, gave a sensible vibration to the whole building! This bird also sometimes makes a small squeak, repeated four or five times; and I have observed that to happen when the cock has been pursuing the hen in a toying manner through the boughs of a tree.

It would not be at all strange if your bat, which you have procured, should prove a new one, since five species have been found in a neighbouring kingdom. The great sort that I mentioned is certainly a nondescript: I saw but one this summer, and that I had no opportunity of taking.

Your account of the Indian grass was entertaining. I am no angler myself; but inquiring of those that are, what they supposed that part of their tackle to be made of, they replied, "Of the intestines of a silkworm."

Though I must not pretend to great skill in entomology, yet I cannot say that I am ignorant of that kind of knowledge: I may now and then perhaps be able to furnish you with a little information.

The vast rain ceased with us much about the same time as with you, and since, we have had delicate weather. Mr. Barker, who has measured the rain for more than thirty years, says, in a late letter, that more rain has fallen this year than in any he ever attended to; though, from July, 1763, to January, 1764, more fell than in any seven months of this year.

LETTER XXIII.

ON THE LIZARD, AND THE MIGRATION OF SWALLOWS.

SELBORNE, *Feb.* 28, 1769.

DEAR SIR,—It is not improbable that the Guernsey lizard and our green lizards may be specifically the same; all that I know is, that when, some years ago, many Guernsey lizards were turned loose in Pembroke College garden, in the University of Oxford, they lived a great while, and seemed to enjoy themselves very well; but never bred. Whether this circumstance will prove anything either way, I shall not pretend to say.

I return you thanks for your account of Cressy-hall; but recollect, not without regret, that in June, 1746, I was visiting for a week together at Spalding, without ever being told that such a curiosity was just at hand. Pray send me word in your next what sort of tree it is that contains such a quantity of herons' nests; and whether the heronry consists of a whole grove or wood, or only of a few trees.

It gave me satisfaction to find we accorded so well about the *caprimulgus;* all I contended for was to prove that it often chatters sitting as well as flying, and therefore the noise was voluntary and from organic impulse, and not from the resistance of the air against the hollow of its mouth and throat.

If ever I saw anything like actual migration, it was last Michaelmas-day. I was travelling, and out early in the morning : at first there was a vast fog; but, by the time that I was got seven or eight miles from home towards the coast, the sun broke out into a delicate warm day. We were then on a large heath, or common, and I could discern, as the mist began to break away, great numbers of swallows (*hirundines rusticæ*) clustering on the stunted shrubs and bushes, as if they had roosted there all night. As soon as the air became clear and pleasant, they all were on the wing at once; and, by a placid and easy flight, proceeded on southward, towards the sea: after this I did not see any more flocks, only now and then a straggler.

I cannot agree with those persons who assert, that the swallow kind disappear some and some, gradually, as they come; for the bulk of them seem to withdraw at once; only

SWALLOW [1]

some stragglers stay behind a long while, and do never, there is the greatest reason to believe, leave this island. Swallows seem to lay themselves up, and to come forth in a warm day, as bats do continually of a warm evening, after they have disappeared for weeks. For a very respectable gentleman assured me that, as he was walking with some friends under Merton-wall on a remarkably hot noon, either in the last week in December, or the first week in January, he espied three or four swallows huddled together on the moulding of one of the windows of that college. I have frequently remarked that swallows are seen later at Oxford than elsewhere: is it owing to the vast massy buildings of that place, to the many waters round it, or to what else?

When I used to rise in a morning last autumn, and see the swallows and martins clustering on the chimneys and thatch of the neighbouring cottages, I could not help being touched with a secret delight mixed with some degree of

(1) *Chelidon urbica.*—J. G. W.

mortification: with delight, to observe with how much ardour and punctuality those poor little birds obeyed the strong impulse towards migration, or hiding, imprinted on their minds by their great Creator; and with some degree of mortification, when I reflected that, after all our pains and inquiries, we are not yet quite certain to what regions they do migrate; and are still further embarrassed to find that some actually do not migrate at all.

These reflections made so strong an impression on my imagination, that they became productive of a composition that may perhaps amuse you for a quarter of an hour when next I have the honour of writing to you.

LETTER XXIV.

ON THE RING-OUSEL, LOCUSTELLA, ETC.

SELBORNE, *May* 29, 1769.

DEAR SIR.—The *scarabæus fullo*[1] I know very well, having seen it in collections; but have never been able to discover one wild in its natural state. Mr. Banks told me he thought it might be found on the sea-coast.

On the 13th of April, I went to the sheep-down, where the ring-ousels have been observed to make their appearance at spring and fall, in their way, perhaps, to the north or south; and was much pleased to see three birds about the usual spot. We shot a cock and a hen; they were plump and in high condition. The hen had but very small rudiments of eggs within her, which proves they are late breeders; whereas those species of the thrush kind that remain with us the whole year, have fledged young before that time. In their crops was nothing very distinguishable, but somewhat that seemed like blades of vegetables nearly digested. In autumn they feed on haws and yew-berries, and in the spring on ivy-berries. I dressed one of these birds, and found it juicy and well-flavoured. It is remarkable that they make but a few days' stay in their spring visit, but

---

(1) *Melolontha fullo.* This beautiful insect is never seen out of Kent. It is therefore, often called the Kentish chafer.—J. G. W.

rest near a fortnight at Michaelmas. These birds, from the observations of three springs and two autumns, are most punctual in their return; and exhibit a new migration unnoticed by the writers, who supposed they never were to be seen in any of the southern counties.

One of my neighbours lately brought me a new salicaria, which, at first, I suspected might have proved your willow-lark;[1] but on a nicer examination, it answered much better to the description of that species which you shot at Reversby, in Lincolnshire. My bird I describe thus:—"It is a size less than the grasshopper-lark; the head, back, and coverts of the wings of a dusky brown, without the dark spots of the grasshopper-lark: over each eye is a milk-white stroke; the chin and throat are white, and the under parts of a yellowish white; the rump is tawny, and the feathers of the tail sharp pointed; the bill is dusky and sharp, and the legs are dusky, the hinder claw long and crooked." The person that shot it says, that it sung so like a reed sparrow, that he took it for one; and that it sings all night: but this account merits farther inquiry. For my part, I suspect it is a second sort of *locustella*, hinted at by Dr. Derham in Ray's *Letters: see* p. 74. He also procured me a grasshopper-lark.

The question that you put with regard to those genera of animals that are peculiar to America, viz. how they came there, and whence? is too puzzling for me to answer; and yet so obvious as often to have struck me with wonder. If one looks into the writers on that subject, little satisfaction is to be found. Ingenious men will readily advance plausible arguments to support whatever theory they shall choose to maintain; but then the misfortune is, every one's hypothesis is each as good as another's, since they are all founded on conjecture. The late writers of this sort, in whom may be seen all the arguments of those that have gone before, as I remember, stock America from the western coast of Africa and the south of Europe; and then break down the isthmus that bridged over the Atlantic. But this is making use of a violent piece of machinery: it is a difficulty worthy of the interposition of a god! "*Incredulus odi*," "Disbelieving I detest."

(1) For this salicaria, see Letter XXV.

H

THE NATURALIST'S SUMMER EVENING WALK.

—— equidem credo, quia sit divinitus illis
Ingenium.   VIRG. *Georg.*

The instructive arts that in their labours shine,
I deem inspired by energy divine.

WHEN day declining sheds a milder gleam,
What time the May-fly[1] haunts the pool or stream;
When the still owl skims round the grassy mead,
What time the timorous hare limps forth to feed;
Then be the time to steal adown the vale,
And listen to the vagrant cuckoo's[2] tale;
To hear the clamorous curlew[3] call his mate,
Or the soft quail his tender pain relate;
To see the swallow sweep the darkening plain,
Belated, to support her infant train;
To mark the swift, in rapid giddy ring,
Dash round the steeple, unsubdued of wing:
Amusive birds! say where your hid retreat,
When the frost rages and the tempests beat?
Whence your return, by such nice instinct led,
When Spring, soft season, lifts her bloomy head?
Such baffled searches mock man's prying pride,
The GOD of NATURE is your secret guide!

   While deepening shades obscure the face of day,
To yonder bench, leaf shelter'd, let us stray,
Till blended objects fail the swimming sight,
And all the fading landscape sinks in night;
To hear the drowsy dorr come brushing by
With buzzing wing, or the shrill cricket[4] cry;

(1) The angler's May-fly, the *ephemera vulgata*, Linn., comes forth from its aurelia state, and emerges out of the water about six in the evening, and dies about eleven at night, determining the date of its fly state in about five or six hours. They usually begin to appear about the 4th of June, and continue in succession for near a fortnight.—See Swammerdam, Derham, Scopoli, &c.

(2) Vagrant cuckoo; so called, because, being tied down by no incubation or attendance about the nutrition of its young, it wanders without control.

(3) *Charadrius œdicnemus.*

(4) *Gryllus campestris.*

THE NATURALIST'S SUMMER EVENING WALK.

"While o'er the cliff the awaken'd churn-owl hung,
Through the still gloom protracts his chattering song."—P. 99.

To see the feeding bat glance through the wood;
To catch the distant falling of the flood;
While o'er the cliff th' awaken'd churn-owl hung,
Through the still gloom protracts his chattering song;
While, high in air, and poised upon his wings,
Unseen, the soft enamour'd woodlark[1] sings:
These, NATURE'S works, the curious mind employ,
Inspire a soothing melancholy joy:
As fancy warms, a pleasing kind of pain
Steals o'er the cheek, and thrills the creeping vein!

    Each rural sight, each sound, each smell combine;
The tinkling sheep-bell, or the breath of kine;
The new-mown hay that scents the swelling breeze,
Or cottage chimney smoking through the trees.

    The chilling night-dews fall:—away, retire;
For see, the glow-worm lights her amorous fire![2]
Thus, ere night's veil had half obscured the sky,
Th' impatient damsel hung her lamp on high:
True to the signal, by love's meteor led,
Leander hasten'd to his Hero's bed.[3]

LETTER XXV.

DIRECTION OF MIGRATION—THE SALICARIA—ODOUR OF REPTILES.

SELBORNE, *Aug.* 30, 1769.

DEAR SIR,—It gives me satisfaction to find that my account
of the ousel migration pleases you.   You put a very shrewd
question when you ask me how I know that their autumnal
migration is southward.   Were not candour and openness
the very life of natural history, I should pass over this query
just as a sly commentator does over a crabbed passage in a
classic; but common ingenuousness obliges me to confess,

----

(1) In hot summer nights, woodlarks soar to a prodigious height, and hang singing in the air.
(2) The light of the female glow-worm (as she often crawls up the stalk of a
grass to make herself more conspicuous) is a signal to the male, which is a slender
dusky *scarabæus*.
(3) See the story of Hero and Leander.

not without some degree of shame, that I only reasoned in that case from analogy. For, as all other autumnal birds migrate from the northward to us, to partake of our milder winters, and return to the northward again, when the rigorous cold abates, so I concluded that the ring-ousels did the same, as well as their congeners, the fieldfares; and especially as ring-ousels are known to haunt cold mountainous countries: but I have good reason to suspect since, that they may come to us from the westward; because I hear from very good authority, that they breed on Dartmoor; and that they forsake that wild district about the time that our visitors appear, and do not return till late in the spring.

I have taken a great deal of pains about your salicaria and mine, with a white stroke over its eye, and a tawny rump. I have surveyed it alive and dead, and have procured several specimens; and am perfectly persuaded myself (and trust you will soon be convinced of the same) that it is no more nor less than the *passer arundinaceus minor* of Ray.[1] This bird, by some means or other, seems to be entirely omitted in the *British Zoology;* and one reason probably was, because it is so strangely classed by Ray, who ranges it among his *pici affines.* It ought, no doubt, to have gone among his *aviculæ caudâ unicolore,* and among your slender-billed small birds of the same division. Linnæus might, with great propriety, have put it into his genus of *motacilla;* and the *motacilla salicaria* of his *fauna suecica* seems to come the nearest to it. It is no uncommon bird, haunting the sides of ponds and rivers, where there is covert, and the reeds and sedges of moors. The country people in some places call it the *sedge-bird.*[2] It sings incessantly night and day, during the breeding time, imitating the note of a sparrow, a swallow, a skylark; and has a strange hurry-ing manner in its song. My specimens correspond most minutely to the description of your *fen salicaria* shot near Revesby. Mr. Ray has given an excellent characteristic of it when he says, *Rostrum et pedes in hâc aviculâ multò*

(1) See Letter XXIV.
(2) *Sylvia salicaria.* The Sedge-Warbler is not uncommon by the banks of rivers where the sedges grow abundantly. It must not be confounded with the Reed Warbler.—J. G. W.

majores sunt quàm pro corporis ratione. The beak and feet
of this bird are too large for the proportions of the rest of
the body.

I have got you the egg of an *œdicnemus*, or stone curlew,
which was picked up in a fallow on the naked ground: there
were two ; but the finder inadvertently crushed one with
his foot before he saw them.

When I wrote to you last year on reptiles, I wish I had
not forgot to mention the faculty that snakes have of stinking in self-defence. I knew a gentleman who kept a tame
snake, which was in its person as sweet as any animal, while
in good humour and unalarmed ; but, as soon as a stranger,
or a dog or cat, came in, it fell to hissing, and filled the
room with such nauseous effluvia, as rendered it hardly sup-
portable. Thus the squnck, or stonck[1] of Ray's *Synop.
Quadr.*, is an innocuous and sweet animal; but, when pressed
hard by dogs and men, it can eject such a most pestilent
and fetid smell and excrement, that nothing can be more
horrible.

A gentleman sent me lately a fine specimen of the *lanius
minor cinerascens cum maculâ in scapulis albâ, Raii;* Ray's
lesser butcher-bird, ash-coloured, with a white spot at the
insertion of the wings; which is a bird that, at the time of
your publishing your two first volumes of *British Zoology,*
I find you had not seen. You have described it well from
Edward's drawing.

(1) *Mephitis varians.* The horribly powerful odour emitted by this animal when
offended, cannot be borne by any animal. The Skunk is even safe from the puma
and jaguar, neither of whom can withstand the effluvia. It is an American animal
and belongs to the Weasel tribe.—J. G. W.

LETTER XXVI.

SELBORNE, *Dec.* 8, 1769.

DEAR SIR,—I was much gratified by your communicative
letter on your return from Scotland, where you spent, I find,
some considerable time, and gave yourself good room t
examine the natural curiosities of that extensive kingdom,
both those of the islands, as well as those of the Highlands.
The usual bane of such expeditions is hurry; because men
seldom allot themselves half the time they should do; but,
fixing on a day for their return, post from place to place,
rather as if they were on a journey that required despatch,
than as philosophers investigating the works of nature.
You must have made, no doubt, many discoveries, and laid
up a good fund of materials for a future edition of the
British Zoology, and will have no reason to repent that you
have bestowed so much pains on a part of Great Britain that
perhaps was never so well examined before.

It has always been matter of wonder to me, that fieldfares,
which are so congenerous to thrushes and blackbirds, should
never choose to breed in England: but that they should not
think even the Highlands cold, and northerly, and sequestered enough, is a circumstance still more strange and
wonderful. The ring-ousel, you find, stays in Scotland the
whole year round; so that we have reason to conclude that
those migrators that visit us for a short space every autumn,
do not come from thence.

And here, I think, will be the proper place to mention,
that those birds were most punctual again in their migration
this autumn, appearing, as before, about the 30th of September; but their flocks were larger than common, and
their stay protracted somewhat beyond the usual time. I
they came to spend the whole winter with us, as some o
their congeners do, and then left us, as they do, in spring, I
should not be so much struck with the occurrence, since i
would be similar to that of the other winter birds of passage

but when I see them for a fortnight at Michaelmas, and
again for about a week in the middle of April, I am seized
with wonder, and long to be informed whence these travellers come, and whither they go, since they seem to use our
hills merely as an inn, or baiting place.

Your account of the greater brambling, or snow-fleck, is
very amusing; and strange it is that such a short-winged bird
should delight in such perilous voyages over the northern
ocean! Some country people in the winter time have every
now and then told me that they have seen two or three white
larks on our downs; but, on considering the matter, I begin
to suspect that these are some stragglers of the birds we are
talking of, which sometimes, perhaps, may rove so far to the
southward.

It pleases me to find that white hares are so frequent on

WHITE HARE.[1]

the Scottish mountains, and especially as you inform me that
it is a distinct species; for the quadrupeds of Britain are so
few, that every new species is a great acquisition.

(1) *Lepus variabilis*, the Alpine-hare. This animal is rather smaller than the
common hare, but larger than the rabbit. In the winter months, its fur becomes

The eagle-owl, could it be proved to belong to us, is so majestic a bird, that it would grace our *fauna* much. I never was informed before where wild geese are known to breed.

EAGLE OWL.[1]

You admit, I find, that I have proved your *fen salicaria* to be the lesser reed-sparrow of Ray; and I think you may be secure that I am right; for I took very particular pains to clear up that matter, and had some fair specimens; but as they were not well preserved they are decayed already. You will, no doubt, insert it in its proper place in your next edition. Your additional plates will much improve your work.

De Buffon, I know, has described the water shrew-mouse;[2]

perfectly white, like that of the ermine, but the black tips of the ears still retain their original colour. The summer coat of the Alpine-hare is a light grey.—J. G. W

(1) *Bubo maximus.* This bird is sometimes called the Great-eared Owl.—J. G. W

(2) *Crossopus fodiens.* I have often seen this beautiful little animal at Little Hinton in Wiltshire. It had formed its habitation in the banks of a very little stream and had made regular runs like those of the water-rat. When diving, its coat appears as if speckled with silver, owing to the bubbles of air which adhere to its fur It lives on the grubs of the various aquatic insects inhabiting the muddy banks from which its long flexible snout soon extracts them.—J. G. W.

but still I am pleased to find you have discovered it in Lincolnshire, for the reason I have given in the article of the white hare.

SHREW-MOUSE.

As a neighbour was lately ploughing in a dry chalky field, far removed from any water, he turned out a water-rat, that was curiously laid up in an hybernaculum, artificially formed of grass and leaves. At one end of the burrow lay about a gallon of potatoes, regularly stowed, on which it was to have supported itself for the winter. But the difficulty with me is how this *amphibius mus* came to fix its winter station at such a distance from the water. Was it determined in its choice of that place by the mere accident of finding the potatoes which were planted there? or is it the constant practice of the aquatic rat to forsake the neighbourhood of the water in the colder months?

Though I delight very little in analogous reasoning, knowing how fallacious it is with respect to natural history; yet in the following instance I cannot help being inclined to think it may conduce towards the explanation of a difficulty that I have mentioned before with respect to the invariable early retreat of the *hirundo apus*, or swift, so many weeks before its congeners; and that not only with us, but also in Andalusia, where they begin to retire about the beginning of August.

The great large bat[1] (which, by the by, is at present a nondescript in England, and what I have never been able yet to procure) retires or migrates very early in the summer: it

(1) *Noctulinia altivolans.* Gray.—J. G. W.

also ranges very high for its food, feeding in a different region of the air ; and that is the reason I never could procure one. Now, this is exactly the case with the swifts ; for they take their food in a more exalted region than the other species, and are very seldom seen hawking for flies near the ground, or over the surface of the water. From hence I would conclude, that these *hirundines*, and the larger bats, are supported by some sorts of high-flying gnats, scarabs, or *phalœnœ*, that are of short continuance, and that the short stay of these strangers is regulated by the defect of their food.

By my journal it appears that curlews clamoured on to October the thirty-first, since which I have not seen or heard any. Swallows were observed on to November the third.

––––––

LETTER XXVII.

HEDGEHOGS—THE FIELDFARE.

SELBORNE, *Feb.* 22, 1770.

DEAR SIR,—Hedgehogs [1] abound in my gardens and fields The manner in which they eat the roots of the plaintain in my grass walks is very curious ; with their upper mandible,

–––

(1) *Erinaceus Europæus.* The Hedgehog is the only English quadruped whose body is defended by spikes. The method by which the spikes are retained in the skin may be seen in the accompanying sketch. The spike is as it were thrust through the skin and retained by the little head, like a pin or a nail. The curve is such that while the animal is in its ordinary position, all its quills lie smoothly, pointing backwards; but when it is curled up, the quills are immediately drawn upwards, each pointing a different way. When the creature is rolled up, the array of quills present so elastic a surface, that it does not hesitate to throw itself from a height, trusting to the elasticity of its prickly covering to break its fall.

The muscle spoken of by White, which enables the hedgehog to roll itself up, is a very singular one indeed. It extends over the whole back and sides, being very thin on the back, and very thick and strong round the sides, so that when the creature has contracted the muscles, it is as it were tied together by a thick rope of muscle.

With regard to its food, there have been many contradictory remarks. Of course the cow-sucking story is sufficiently absurd but then there is a report which has often been absurd and laughed at, that the hedgehog will roll itself upon apples, and carry them away. Now a friend, but a short time since, told me that he had actually *seen* a hedgehog deliberately roll itself up on some fallen apples, and walk away with them. I confess that I could not believe the story for some time, but I always

which is much longer than their lower, they bore under the plant, and so eat the root off upwards, leaving the tuft of leaves untouched. In this respect they are serviceable, as they destroy a very troublesome weed: but they deface the walks in some measure by digging little round holes. It

HEDGEHOG.

appears, by the dung that they drop upon the turf, that beetles are no inconsiderable part of their food. In June last, I procured a litter of four or five young hedgehogs, which appeared to be about five or six days old; they, I find, like puppies, are born blind, and could not see when they came to my hands. No doubt their spines are soft and flexible at the time of their birth, or else the poor dam would have but a bad time of it in the critical moment of parturition: but it is plain that they soon harden; for these little pigs had

SPINE OF HEDGEHOG.

such stiff prickles on their back and sides, as would easily have fetched blood, had they not been handled with caution. Their spines are quite white at this age; and they have little hang-

found him consistent in his report, that he had actually seen it. Dr. Buckland's experiment with a hedgehog and a snake, is too generally known to need insertion.

It is singular enough, that while hedgehogs have been wrongfully accused of sucking cows, a delinquent should exist, who has not even been suspected. This is the common pig, which when small has been repeatedly seen to suck the cows as they lie down to rest, the cows offering no opposition whatever.

The flesh of the hedgehog is said to be very good eating. The approved method of dressing it, is to open and clean it, leaving the skin on. It is then enveloped in a thick mass of clay, and placed on the fire. When it is sufficiently baked, the clay is taken off, and brings the skin with it.—J. G. W.

ing ears, which I do not remember to be discernible in the old ones. They can, in part, at this age, draw their skin down over their faces; but are not able to contract themselves into a ball, as they do, for the sake of defence, when full grown. The reason, I suppose, is, because the curious muscle, that enables the creature to roll itself up in a ball, was not then arrived at its full tone and firmness. Hedgehogs make a deep and warm hybernaculum with leaves and moss, in which they conceal themselves for the winter; but I never could find that they stored in any winter provision, as some quadrupeds certainly do.

I have discovered an anecdote with respect to the fieldfare (*turdus pilaris*), which, I think, is particular enough: this

FIELDFARE.

bird, though it sits on trees, in the day-time, and procures the greatest part of its food from white-thorn hedges; yea, moreover, builds on very high trees, as may be seen by the *fauna suecica;* yet always appears with us to roost on the ground. They are seen to come in flocks just before it is dark, and to settle and nestle among the heath in our forest. And besides, the larkers, in dragging their nets by night, frequently catch them in the wheat-stubbles; while the bat fowlers, who take many red-wings in the hedges, never entangle any of this species. Why these birds, in the matter of roosting, should differ from all their congeners, and from themselves, also, with respect to their proceedings by day, is a fact for which I am by no means able to account.[1]

I have somewhat to inform you of concerning the moose-deer; but, in general, foreign animals fall seldom in my way; my little intelligence is confined to the narrow sphere of my own observations at home.

LETTER XXVIII.

MOOSE-DEER.

SELBORNE, *March*, 1770.

ON Michaelmas-day, 1768, I managed to get a sight of the female moose belonging to the Duke of Richmond, at Goodwood; but was greatly disappointed, when I arrived at the spot, to find that it had died, after having appeared in a languishing way for some time, on the morning before. However, understanding that it was not stripped, I proceeded to examine this rare quadruped: I found it in an old greenhouse, slung under the belly and chin by ropes, and in a standing posture; but, though it had been dead for so short a time, it was in so putrid a state that the stench was hardly sup-

---

(1) *Turdus pilaris.* The Fieldfare has been known occasionally to build in this country, but such examples are very rare, and may be looked upon as accidents. Some years they are very plentiful and comparatively tame, while in others they are scarce, and almost impossible to approach. In a year of the latter description, I was shooting, or according to the transatlantic idiom, gunning, and had for some time been pursuing a flock of fieldfares. At last, being quite in despair at their caution, just as they were topping a hedge, hopelessly out of range, I fired at them, and to my great amazement saw one fall on the opposite side of the hedge. On going to pick it up, it could nowhere be found, and after a fruitless search the conviction forced itself upon my mind that the bird had not fallen at all. So I proceeded on my journey. On returning by the same place, about two hours afterwards, it was very natural that my eyes should once more wander over the spot. In doing so, they were arrested by another pair of eyes, sparkling at the bottom of a rather large, but unfinished rat hole. On ejecting the occupant I was much pleased to find my missing fieldfare, who had lain perfectly quiet as long as it saw itself unobserved, but used the most strenuous efforts to escape after it had been discovered. The nest of the Fieldfare is placed at various heights from the ground, mostly against the trunk of the fir-trees, which abound in Norway, a country which seems to be a particular stronghold of the fieldfare.—J. G. W.

portable. The grand distinction between this deer, and any
other species that I have ever met with, consisted in the strange
length of its legs; on which it was tilted up much in the
manner of the birds of the *grallæ* order. I measured it as
they do a horse, and found that, from the ground to the
wither, it was just five feet four inches; which height answers
exactly to sixteen hands, a growth that few horses arrive at;
but then, with this length of legs, its neck was remarkably
short, no more than twelve inches; so that, by straddling
with one foot forward and the other backward, it grazed on
the plain ground, with the greatest difficulty, between its legs:
the ears were vast and lopping, and as long as the neck; the

MOOSE-DEER. [1]

head was about twenty inches long, and ass-like; and had
such a redundancy of upper lip as I never saw before, with
huge nostrils. This lip, travellers say, is esteemed a dainty

(1) *Alces palmatus.*—J. G. W.

dish in North America. It is very reasonable to suppose, that this creature supports itself chiefly by browsing off trees, and by wading after water plants; towards which way of livelihood the length of legs and great lips must contribute much. I have read somewhere, that it delights in eating the *nymphœa*, or water-lily. From the fore-feet to the belly, behind the shoulder, it measured three feet and eight inches; the length of the legs, before and behind, consisted a great deal in the *tibia*, which was strangely long; but, in my haste to get out of the stench. I forgot to measure that joint exactly. Its scut seemed to be about an inch long; the colour was a grizzly black; the mane about four inches long; the fore-hoofs were upright and shapely, the hind flat and splayed. The spring before, it was only two years old, so that, most probably, it was not then come to its growth. What a vast tall beast must a full-grown stag be! I have been told some arrive at ten feet and a half! This poor creature had at first a female companion of the same species, which died the spring before. In the same garden was a young stag, or red-deer, between whom and this moose it was hoped that there might have been a breed; but their inequality of height must have always been a bar to any commerce of the amorous kind. I should have been glad to have examined the teeth, tongue, lips, hoofs, &c. minutely; but the putrefaction precluded all farther curiosity. This animal, the keeper told me, seemed to enjoy itself best in the extreme frost of the former winter. In the house, they showed me the horn of a male moose, which had no front antlers, but only a broad palm, with some snags on the edge. The noble owner of the dead moose proposed to make a skeleton of her bones.

Please to let me hear if my female moose corresponds with that you saw; and whether you think still that the American moose and European elk are the same creature.

LETTER XXIX.

MIGRATION—PAIRING OF BIRDS—CATS—THE OTTER.

SELBORNE, *May* 12, 1770.

DEAR SIR,—Last month we had such a series of cold turbulent weather, such a constant succession of frost, and snow,
and hail, and tempest, that the regular migration, or appearance of the summer birds was much interrupted. Some did
not show themselves (at least were not heard till weeks
after their usual time, as the black-cap and white-throat;
and some have not been heard yet, as the grasshopper-lark
and largest willow-wren. As to the fly-catcher, I have not
seen it; it is, indeed, one of the latest, but should appear
about this time; and yet, amidst all this meteorous strife
and war of the elements, two swallows discovered themselves
as long ago as the eleventh of April, in frost and snow; but
they withdrew quickly, and were not visible again for many
days. House-martins, which are always more backward than
swallows, were not observed till May came in.

HOUSE-SPARROW.

Among the *monogamous* birds, several are to be found
after pairing time, single and of each sex; but whether thi

state of celibacy is matter of choice or necessity is not so easily discoverable. When the house-sparrows deprive my martins of their nests, as soon as I cause one to be shot, the other, be it cock or hen, presently procures a mate, and so for several times following.

I have known a dove-house infested by a pair of white owls, which made great havoc among the young pigeons: one

WHITE OWL [1]

of the owls was shot as soon as possible; but the survivor readily found a mate, and the mischief went on. After some time the new pair were both destroyed, and the annoyance ceased.

Another instance I remember of a sportsman, whose zeal

(1) *Nyctea nivea.* This bird is properly a native of the north of Europe, but as it has once or twice been found in England, I have admitted it among the illustrations. The white owl mentioned by White as destroying the pigeons, is the common barn owl, *Strix flammea.*—J. G. W.

for the increase of his game being greater than his humanity,
after pairing time, he always shot the cock bird of every
couple of partridges upon his grounds: supposing that the
rivalry of many males interrupted the breed. He used to
say, that though he had widowed the same hen several times,
yet he found she was still provided with a fresh paramour,
that did not take her away from her usual haunt.

Again: I knew a lover of setting, an old sportsman, who
has often told me that, soon after harvest, he has frequently
taken small coveys of partridges consisting of cock birds
alone: these he pleasantly used to call old bachelors.

There is a propensity belonging to common house cats
that is very remarkable: I mean their violent fondness for

CAT. [1]

fish, which appears to be their most favourite food; and yet
nature, in this instance, seems to have planted in them an
appetite that, unassisted, they know not how to gratify: for
of all quadrupeds, cats are the least disposed towards water
and will not, when they can avoid it, deign to wet a foot
much less to plunge into that element.

Quadrupeds that prey on fish are amphibious; such is the

(1) *Felis domestica.*—J. G. W.

otter, which by nature is so well formed for diving, that it makes great havoc among the inhabitants of the waters. Not supposing that we had any of those beasts in our shallow brooks, I was much pleased to see a male otter brought to me, weighing twenty-one pounds, that had been shot on the bank of our stream, below the Priory, where the rivulet divides the parish of Selborne from Harteley-wood.

OTTER

LETTER XXX.

MOOSE—AMERICAN BIRDS.

SELBORNE, *Aug.* 1, 1770.

DEAR SIR,—The French, I think, in general are strangely prolix in their natural history. What Linnæus says with

(1) *Lutra vulgaris.* The otter appears to play the same part in the water as the weasels do upon the land, and the semblance extends even to the method of devouring their prey, for as the weasels only suck the blood and eat the brain, so the otter, if it has plenty of food, daintily bites off the flakes of flesh at the back of the fish's neck, and leaves the remainder upon the bank for the benefit of less fastidious animals, among whom man may be reckoned. Mr. St. John tells us of an old woman who bitterly lamented the extirpation of the otters from the neighbouring river, as they generally left a fine salmon on the bank every day, which, as she philosophically remarked, was none the worse for the bit that the poor beasts had taken for themselves.

It is a pity that, instead of persecuting the otter, we do not tame it, and make it catch fish for our use as well as for its own food. This has often been done, and indeed is a common practice in India, where Bishop Heber saw many fastened by a straw collar and long rope to stakes by the river side. The Hindoo fishermen appeared to consider an otter as necessary to their fishing, as a shepherd finds his dog necessary for the management of his flock.

The movements of this animal when in the water are exceedingly graceful : it appears to glide about without the slightest effort, and certainly with the very lightest possible disturbance of the water. Although essentially an aquatic animal, it has been known, in seasons when the supply of fish has been scanty, to usurp the habits of its congeners the weasels, and to attack the smaller farm animals, such as poultry, or young pigs.—J. G. W.

respect to insects holds good in every other branch: " *Verbositas præsentis sœculi, calamitas artis.*"

Pray how do you approve of Scopoli's new work? As I admire his *Entomologia,* I long to see it.

I forgot to mention in my last letter (and had not room to insert in the former), that the male moose, in rutting time, swims from island to island, in the lakes and rivers of North America, in pursuit of the females. My friend, the chaplain, saw one killed in the water, as it was on that errand, in the river St. Lawrence: it was a monstrous beast, he told me; but he did not take the dimensions.

When I was last in town, our friend Mr. Barrington most obligingly carried me to see many curious sights. As you were then writing to him about horns, he carried me to see many strange and wonderful specimens. There is, I remember, at Lord Pembroke's, at Wilton, a horn-room furnished with more than thirty different pairs; but I have not seen that house lately.

Mr. Barrington showed me many astonishing collections of stuffed and living birds from all quarters of the world. After I had studied over the latter for a time, I remarked that every species almost, that came from distant regions, such as South America, the coast of Guinea, &c., were thick-billed birds, of the *loxia* and *fringilla* genera; and no *motacillæ* or *muscicapidæ* were to be met with. When I came to consider, the reason was obvious enough; for the hard-billed birds subsist on seeds which are easily carried on board, while the soft-billed birds, which are supported by worms and insects, or, what is a succedaneum for them, fresh raw meat, can meet with neither in long and tedious voyages. It is from this defect of food that our collections (curious as they are) are defective, and we are deprived of some of the most delicate and lively genera.[1]

(1) White was correct in this surmise, for there are plenty of *motacillæ*, or rather *sylviadæ* and *muscicapidæ*, in South America. See Audubon, &c.—J. G. W.

LETTER XXXI.

SELBORNE, *Sept.* 14, 1770.

DEAR SIR,—You saw, I find, the ring-ousels again among their native crags; and are farther assured that they continue resident in those cold regions the whole year. From whence then do our ring-ousels migrate so regularly every September, and make their appearance again, as if in their return, every April? They are more early this year than common, for some were seen at the usual hill on the fourth of this month.

An observing Devonshire gentleman tells me, that they frequent some parts of Dartmoor, and breed there, but leave those haunts about the end of September, or beginning of October, and return again about the end of March.

Another intelligent person assures me, that they breed in great abundance all over the Peak of Derby, and are called there tor-ousels, withdraw in October and November, and return in spring. This information seems to throw some light on my new migration.

Scopoli's new work [1] (which I have just procured) has its merits, in ascertaining many of the birds of the Tyrol and Carniola. Monographers, come from whence they may, have, I think, fair pretence to challenge some regard and approbation from the lovers of natural history; for, as no man can alone investigate all the works of nature, these partial writers may, each in his department, be more accurate in their discoveries, and freer from errors, than more general writers, and so by degrees may pave the way to an universal correct natural history. Not that Scopoli is so circumstantial and attentive to the life and conversation of his birds as I could wish: he advances some false facts; as when he says of the *hirundo urbica*, that, "*pullos extra nidum non nutrit.*" This assertion I know to be wrong, from repeated observation this summer; for house-martins do feed their young

(1) " Annus primus Historico-Naturalis."

flying, though, it must be acknowledged, not so commonly
as the house-swallow; and the feat is done in so quick a
manner as not to be perceptible to indifferent observers.
He also advances some (I was going to say) improbable
facts; as when he says of the woodcock, that, "*pullos rostro*

WOODCOCK.

portat fugiens ab hoste,"—flying from the enemy it carries
its young in its beak.' But candour forbids me to say
absolutely that any fact is false, because I have never been
witness to such a fact. I have only to remark, that the
long unwieldy bill of the woodcock is, perhaps, the worst
adapted of any among the winged creation for such a feat of
natural affection.

(1) *Scolopax rusticola.* Jesse states that the singular mode of conveyance here
mentioned is really practised by the woodcock.—J. G. W.

LETTER XXXII.

SELBORNE, *Oct.* 29, 1770.

DEAR SIR,—After an ineffectual search in Linnæus, Brisson, &c., I begin to suspect that I discern my brother's *hirundo hyberna* in Scopoli's new-discovered *hirundo rupestris*, p. 167. His description of "*Supra murina, subtus albida; tectrices maculâ ovali albâ in latere interno; pedes nudi nigri; rostrum nigrum; remiges obscuriores quam plumæ dorsales; rectrices remigibus concolores; caudâ emarginatâ nec forcipatâ,*" agrees very well with the bird in question; but when he comes to advance that it is "*statura hirundinis urbicæ,*" and that "*definitio hirundinis ripariæ Linnæi huic quoque convenit,*" he, in some measure, invalidates all he has said; at least, he shows at once that he compares them to these species merely from memory; for I have compared the birds themselves, and find they differ widely in every circumstance of shape, size, and colour. However, as you will have a specimen, I shall be glad to hear what your judgment is in the matter.

Whether my brother is forestalled in his nondescript or not, he will have the credit of first discovering that they spend their winters under the warm and sheltery shores of Gibraltar and Barbary.

Scopoli's characters of his Ordines and Genera are clear, just, and expressive, and much in the spirit of Linnæus. These few remarks are the result of my first perusal of Scopoli's *Annus Primus*.

The bane of our science is the comparing one animal to the other by memory. For want of caution in this particular, Scopoli falls into errors. He is not so full with regard to the manners of his indigenous birds as might be wished, as you justly observe: his Latin is easy, elegant, and expressive, and very superior to Kramer's.[1]

(1) See his *Elenchus Vegetabilium et Animalium per Austriam Inferiorem*, &c.,— Summary of Vegetables and Animals in Lower Austria."

LETTER XXXIII.

ALPINE SWALLOW—STONE CURLEW—BUSTARD.

SELBORNE, *Nov.* 26, 1770.

DEAR SIR,—I was much pleased to see among the collection of birds from Gibraltar. some of those short-winged English summer birds of passage, concerning whose departure we have made so much inquiry.   Now, if these birds are found in Andalusia to migrate to and from Barbary, it may easily be supposed that those that come to us may migrate back to the continent, and spend their winters in some of the warmer parts of Europe.   This is certain, that many soft-billed birds that come to Gibraltar appear there only in spring and autumn, seeming to advance in pairs towards the northward for the sake of breeding during the summer months, and retiring in parties and broods towards the south at the

ALPINE SWIFT. [1]

decline of the year; so that the rock of Gibraltar is the great rendezvous and place of observation, from whence the

(1) *Cypselus melba.* The Alpine, or White-bellied Swift.—J. G. W.

take their departure each way towards Europe or Africa. It is therefore no mean discovery, I think, to find that our small short-winged summer birds of passage are to be seen, spring and autumn, on the very skirts of Europe;—it is a presumptive proof of their emigrations.

Scopoli seems to me to have found the *hirundo melba* (the great Gibraltar swift) in Tyrol, without knowing it. For what is the *hirundo alpina*, but the aforementioned bird in other words? Says he, " *Omnia prioris* (meaning the swift) *sed pectus album; paulo major priore.*" "All the marks of the former but the white breast; a little larger than the former." I do not suppose this to be a new species. It is true also of the melba, that "*nidificat in excelsis Alpium rupibus,*"—"it builds its nest in the lofty cliffs of the Alps." Vid. *Annum Primum.*

My Sussex friend, a man of observation and good sense, but no naturalist, to whom I applied on account of the stone curlew (*œdicnemus*), sends me the following account:—"In looking over my Naturalist's Journal for the month of April, I find the stone curlews are first mentioned on the 17th and 18th, which date seems to me rather late. They live with us all the spring and summer, and at the beginning of autumn prepare to take leave, by getting together in flocks. They seem to me a bird of passage that may travel into some dry hilly country south of us, probably Spain, because of the abundance of sheep-walks in that country; for they spend their summers with us in such districts. This conjecture I hazard, as I have never met with any one that has seen them in England in the winter. I believe they are not fond of going near the water, but feed on earth-worms, that are common on sheep-walks and downs. They breed on fallows and lay-fields abounding with grey mossy flints, which much resemble their young in colour, among which they skulk and conceal themselves. They make no nest, but lay their eggs on the bare ground, producing in common but two at a time. There is reason to think their young run soon after they are hatched, and that the old ones do not feed them, but only lead them about at the time of feeding, which, for the most part, is in the night." Thus far my friend.

In the manners of this bird, you see, there is something

very analogous to the bustard, whom it also somewhat resembles in aspect and make, and in the structure of its feet.

BUSTARD [1]

For a long time I have desired my relation to look out for these birds in Andalusia; and now he writes me word that, for the first time, he saw one dead in the market on the 3d of September.

When the *œdicnemus* flies, it stretches out its legs straight behind, like a heron.

LETTER XXXIV.

NOXIOUS INSECTS.

SELBORNE, *March* 30, 1771.

DEAR SIR,—There is an insect [2] with us, especially on chalky districts, which is very troublesome and teasing all the latter end of the summer, getting into people's skins, especially

(1) *Otus tarda.*—J. G. W.
(2) This little creature is not an insect, but belongs to the same class of creatures as the cheese-mite, &c.—J. G. W.

those of women and children, and raising tumours which itch intolerably. This animal (which we call an harvest-bug) is

HARVEST-BUG. [1]

very minute, scarce discernible to the naked eye, of a bright scarlet colour, and of the genus of *acarus*. They are to be met with in gardens, on kidney-beans, or any legumens, but prevail only in the hot months of summer. Warreners, as some have assured me, are much infested by them on chalky downs, where these insects swarm sometimes to so infinite a degree as to discolour their nets, and to give them a reddish cast ; while the men are so bitten as to be thrown into fevers.

There is a small, long, shining fly in these parts, very troublesome to the housewife, by getting into the chimneys, and laying its eggs in the bacon while it is drying. These eggs produce maggots, called jumpers, which, harbouring in the gammons and best part of the hogs, eat down to the bone, and make great waste. This fly I suspect to be a variety of the *musca putris* of Linnæus. It is to be seen in the summer in farm kitchens, on the bacon-racks, and about he mantel-pieces and on the ceilings.[2]

The insect that infests turnips, and many crops in the garden, (destroying often whole fields, while in their seedling leaves,) is an animal that wants to be better known. The

(1) *Leptus autumnalis.*—J. G. W.

(2) *Piophila casei,* Fall. The leaps executed by the larvæ of this fly, are managed in a manner somewhat similar to that adopted by salmon, namely, by the creature bending itself into a circle, and then suddenly springing straight again.—J. G. W.

country people here call it the turnip fly and black dolphin; but I know it to be one of the *coleoptera*, the *"chrysomela oleracea, saltatoria, femoribus posticis crassissimis."* [1]—" The cabbage chrysomela, moving by a leap, with very thick hind-legs." In very hot summers they abound to an amazing degree, and as you walk in a field, or in a garden, make a pattering like rain, by jumping on the leaves of the turnips or cabbages.

There is an œstrus, known in these parts to every ploughboy, which, because it is omitted by Linnæus, is also passed over by late writers; and that is the *curvicauda* of old Mouffet, mentioned by Derham in his *Physico-Theology*, p. 250: an insect worthy of remark, for depositing its eggs, as it flies, in so dexterous a manner on the single hairs of the legs and flanks of grass-horses.[2] But then, Derham is mistaken when he advances that this œstrus is the parent of that wonderful star-tailed maggot which he mentions afterwards; for more modern entomologists have discovered that singular production to be derived from the egg of the *musca chamæleon.* See Geoffroy, t. xvii. f. 4.[3]

A full history of noxious insects, hurtful in the field, garden, and house, suggesting all the known and likely means of destroying them, would be allowed by the public to be a most useful and important work. What knowledge there is of this sort lies scattered, and wants to be collected : great improvements would soon follow of course. A knowledge of the properties, economy, propagation, and, in short, of the life and conversation, of these animals, is a necessary step to lead us to some method of preventing their depredations.

As far as I am a judge, nothing would recommend entomology more than some neat plates that should well express the generic distinctions of insects according to Linnæus ; for I am well assured, that many people would study insects, could they set out with a more adequate notion of those distinctions than can be conveyed at first by words alone.

(1) *Haltica nemorum*, a most destructive little insect, as it attacks the leaves when they first spring, and utterly ruins the plant.—J. G. W.

(2) *Gasterophilus equi.*—J. G. W.

(3) The star-tailed maggot is the larva of *Eristalis tenax*, the common drone-fly, which sits in the sunshine on leaves and walls, and is often mistaken for the drone bee.—J. G. W.

LETTER XXXV.

TRAIN OF PEACOCK—OX CALCULUS.

SELBORNE, 1770.

DEAR SIR,—Happening to make a visit to my neighbour's peacocks, I could not help observing, that the trains of those

PEACOCK. [1]

magnificent birds appear by no means to be their tails, those long feathers growing not from their *uropygium*, but all up

(1) *Pavo cristatus.*—J. G. W.

their backs. A range of short, brown, stiff feathers, about six inches long, fixed in the *uropygium*, is the real tail, and serves as the *fulcrum* to prop the train, which is long and top-heavy, when set on end. When the train is up, nothing appears of the bird before but its head and neck; but this would not be the case were these long feathers fixed only in the rump, as may be seen by the turkey-cock, when in a strutting attitude. By a strong muscular vibration, these birds can make the shafts of their long feathers clatter like the swords of a sword-dancer; they then trample very quick with their feet, and run backwards towards the females.

I should tell you that I have got an uncommon *calculus œgogropila*, taken out of the stomach of a fat ox. It is perfectly round, and about the size of a large Seville orange; such are, I think, usually flat.

LETTER XXXVI.

BATS.

September, 1771.

DEAR SIR,—The summer through, I have seen but two of that large species of bat which I call *vespertilio altivolans*, from its manner of feeding high in the air. I procured one of them, and found it to be a male, and made no doubt, as they accompanied together, that the other was a female; but happening, in an evening or two, to procure the other likewise, I was somewhat disappointed when it appeared to be also of the same sex. This circumstance, and the great scarcity of this sort, at least in these parts, occasions some suspicions in my mind whether it is really a species, or whether it may not be the male part of the more known species, one of which may supply many females, as is known to be the case in sheep, and some other quadrupeds. But this doubt can only be cleared by a further examination, and some attention to the sex, of more specimens.

In the extent of their wings, they measured fourteen inches and a half, and four inches and a half from the nose to the tip of the tail: their heads were large, their nostrils bilobated, their shoulders broad and muscular, and their whole bodies fleshy and plump. Nothing could be more sleek and soft than their fur, which was of a bright chestnut colour; their maws were full of food, but so macerated, that the quality could not be distinguished; their livers, kidneys, and hearts were large, and their bowels covered with fat. They weighed each, when entire, full one ounce and one drachm. Within the ear, there was somewhat of a peculiar structure that I did not understand perfectly; but refer it to the observation of the curious anatomist.[1] These creatures send forth a very rancid and offensive smell.

(1) This peculiar structure is, in all probability, the "tragus," a most singular organ, forming part of the ear of the bats. Its size and shape are very variable, and afford valuable distinctions in determining the species. In the Long-eared Bat, for example, it is very long and pointed, while in the Horse-shoe Bat it is hardly perceptible, although the ears are of considerable size. Indeed, it appears to be an example of the extraordinary development of the skin in the Bat tribe. In those Bats which, like the Horse-shoe Bat, possess a cutaneous development on the nose, the tragus is very small. The Horse-shoe Bat is so called, on account of the shape of this nasal crest, whose use is but conjectured. Mr. Bell appears to think that it may be an aid to the sense of smell, and that it comes into use after the bat has retired to its hiding-place for the day. The Horse-shoe Bat especially takes a delight in dark caverns beyond all reach of the light of day, and may probably require some additional aid to enable it to find its way about in its dark home. But this is only conjectural, as the habits of the bats are, from their nature, very difficult to watch, and are consequently but little known. It is much to be wished that some naturalist would take up these singular creatures as his particular study.—J. G. W.

LETTER XXXVII.

GOATSUCKER—SWALLOWS—RING-OUSEL.

SELBORNE, 1771.

DEAR SIR,—On the twelfth of July, I had a fair opportunity of contemplating the motions of the *caprimulgus*, or fern-owl, as it was playing round a large oak that swarmed with *scarabœi solstitiales*, or fern chafers. The powers of its wing were wonderful, exceeding, if possible, the various evolutions and quick turns of the swallow genus. But the circumstance that pleased me most was, that I saw it distinctly more than once put out its short leg when on the wing, and, by a bend of the head, deliver somewhat into its mouth. If it takes any part of its prey with its foot, as I have now the greatest reason to suppose it does these chafers, I no longer wonder at the use of its middle toe, which is curiously furnished with a serrated claw.

Swallows and martins, the bulk of them, I mean, have forsaken us sooner this year than usual; for, on September the 22d, they rendezvoused in a neighbour's walnut-tree, where it seemed probable they had taken up their lodgings for the night. At the dawn of the day, which was foggy, they rose all together in infinite numbers, occasioning such a rushing from the strokes of their wings against the hazy air as might be heard to a considerable distance; since that, no flock has appeared, only a few stragglers.

Some swifts stayed late. till the 22d of August; a rare instance, for they usually withdraw within the first week.

On September the 24th, three or four ring-ousels appeared in my fields for the first time this season. How punctual are these visitors in their autumnal and spring migrations!

LETTER XXXVIII.

HOUSE-MARTINS—RING-OUSELS.

SELBORNE, *March* 15, 1773.

DEAR SIR,—By my journal for last autumn, it appears that the house-martins bred very late, and stayed very late in these parts; for on the 1st of October I saw young martins in their nests, nearly fledged; and again, on the 21st of October, we had at the next house a nest full of young martins, just ready to fly, and the old ones were hawking for insects with great alertness. The next morning the brood forsook their nest, and were flying round the village. From this day I never saw one of the swallow kind till November the 3d; when twenty, or perhaps thirty, house-martins were playing all day long by the side of the Hanging-wood, and over my fields. Did these small weak birds, some of which were nestlings twelve days ago, shift their quarters at this late season of the year to the other side of the northern tropic? Or rather, is it not more probable that the next church, ruin, chalk-cliff, steep covert, or perhaps sand-bank, lake, or pool, (as a more northern naturalist would say,) may become their hybernaculum, and afford them a ready and obvious retreat? We now begin to expect our vernal migration of ring-ousels every week. Persons worthy of credit assure me that ring-ousels were seen at Christmas, 1770, in the forest of ere, on the southern verge of this county. Hence we may conclude that their migrations are only internal, and not extended to the continent southward, if they do at first come all from the northern parts of this island only, and not from the north of Europe. Come from whence they will, it is plain, from the fearless disregard that they show for men and guns, that they have been little accustomed to places of much resort. Navigators mention that in the Isle of Ascension, and other desolate districts, birds are so little acquainted with the human form, that they settle on men's shoulders, and have no more dread of a sailor than they would have of a goat that was grazing. A young man at Lewes, in Sussex, assured me that about seven years ago ring-ousels abounded so about that town in the autumn, that he killed sixteen himself in one afternoon: he added further, that some had appeared since in every autumn; but he could not find that any had been observed before the season in which he shot so many. I myself have found these birds in little parties in the autumn, cantoned all along the Sussex downs, wherever there were shrubs and bushes, from Chichester to Lewes; particularly in the autumn of 1770.

LETTER XXXIX.

OSPREY—BUTCHER-BIRD—CHOUGHS—TITLARK—SNIPES—MISCELLANEOUS
OBSERVATIONS.

SELBORNE, *Nov.* 9, 1773.

DEAR SIR,—As you desire me to send you such observations as may occur, I take the liberty of making the following remarks, that you may, according as you think me right or

OSPREY.

wrong, admit or reject what I here advance, in your intended new edition of the *British Zoology*.

The osprey was shot about a year ago at Frinsham Pond

a great lake, at about six miles from hence, while it was sitting on the handle of a plough and devouring a fish ; it used to precipitate itself into the water, and so take its prey by surprise.[1]

A great ash-coloured butcher-bird [2] was shot last winter in Tisted Park, and a red-backed butcher-bird at Selborne. They are scarce birds in this country.

GREAT GREY SHRIKE.

(1) *Pandion Haliaëtus.* The chief points to be remarked in this bird are the mobility of the outer toe and the roughness of the soles of its feet. When the bird seizes a fish, it grasps it across the body, forcing its four sharp curved talons firmly into its flesh. This hold is never relaxed until the fish is entirely eaten, the Osprey picking the flesh from between its claws. It is very rare in England, but abounds in America.—J. G. W.

(2) *British Zoology,* vol. i. p. 161.—WHITE. *Lanius excubitor.* The name *Excubitor* is given to this bird from its habit of posting itself as a watchman or sentinel on the look-out for hawks and other birds of prey, of whom it entertains a most especial dread. This propensity of making a great outcry whenever it sees a hawk, is made use of by the continental fowlers in the capture of the Peregrine falcon. The fowler places his nets properly baited with a pigeon close to a little turf hut erected for the purpose, and fastens a shrike near it by a string, which gives it sufficient scope to run about. Near the shrike, he makes a miniature hut of two pieces of turf. His trap thus set, the fowler sits quietly in his hut making or mending his nets, knowing full well that the shrike will keep a better watch than himself. No sooner does a hawk appear at any distance than the shrike begins to be alarmed ; as it approaches, he screams aloud with fear, and just as the hawk is near the pigeon, he runs under his turf hut, which is the signal for the fowler to prepare for pulling a string communicating with the net, and which draws the net over both hawk and pigeon.— J. G. W.

Crows[1] go in pairs the whole year round.

Cornish choughs abound and breed on Beechy Head, and on all the cliffs of the Sussex coast.

CHOUGH[2]

The common wild pigeon,[3] or stock-dove, is a bird of passage in the south of England, seldom appearing till towards the end of November,—is usually the latest winter bird of passage. Before our beechen woods were so much destroyed, we had myriads of them, reaching in strings for a mile together, as they went out in a morning to feed. They leave us early in spring. Where do they breed?

The people of Hampshire and Sussex call the missel-bird[4] the storm-cock, because it sings early in the spring, in blowing showery weather. Its song often commences with the year: with us it builds much in orchards.

A gentleman assures me he has taken the nests of ring-ousels[5] on Dartmoor; they build in banks on the sides of streams.

(1) *British Zoology*, vol. i. p. 167.
(2) *Ibid.* p. 198.—WHITE. *Coracia gracula.*—J. G. W.
(3) *British Zoology*, p. 216.   (4) *Ibid.* p. 224.
(5) *Ibid.* p. 229.

MISSEL-THRUSH [1]

Titlarks [2] not only sing sweetly as they sit on trees, but also as they play and toy about on the wing; and particularly

TITLARK [3]

while they are descending, and sometimes as they stand on the ground.

Adamson's [4] testimony seems to me to be a very poor

(1) *Turdus viscivorus.*—J. G. W.   (2) *British Zoology*, vol. ii. p. 237.
(3) *Anthus pratensis.*—J. G. W.   (4) *Ibid.* vol. ii. p. 242.

evidence that European swallows migrate during our winter
to Senegal ; he does not talk at all like an ornithologist, and
probably saw only the swallows of that country, which I know
build within Governor O'Hara's hall against the roof. Had
he known European swallows, would he not have mentioned
the species ?

The house-swallow washes by dropping into the water as it
flies ;[1] this species appears commonly about a week before
the house-martin, and about ten or twelve days before the
swift.

In 1772, there were young house-martins[2] in their nest
till October the 23d.

The swift[3] appears about ten or twelve days later than the
house-swallow ; viz. about the 24th or 26th of April.

Whin-chats[4] and stone-chatters[5] stay with us the whole
year.

Some wheatears[6] continue with us the winter through.

Wagtails, all sorts, remain with us all the winter.

Bullfinches,[7] when fed on hempseed, often become wholly
black.

We have vast flocks of female chaffinches[8] all the winter,
with hardly any males among them.

When you say that, in breeding time, the cock snipes[9]
make a bleating noise,[10] and a drumming (perhaps I should

---

(1) From this habit has arisen the idea that swallows retire beneath the water in
the winter, and rise in the spring. — J. G. W.

(2) *British Zoology*, vol. ii. p. 224.   (3) *Ibid.* p. 245.

(4) *Saxicola rubetra*, Bechst.—J. G. W.

(5) *Saxicola rubicola*, Bechst.—J. G. W.   *British Zoology*, vol. ii. pp. 270, 271.
—WHITE.

(6) *Ibid.* p. 269.   (7) *Ibid.* p. 300. See note on p. 62.—J. G. W.

(8) *Ibid.* p. 306.   (9) *Ibid.* p. 358.

(10) *Numenius scolopacinus.* There are four species of Snipes inhabiting England,
the Common snipe, the Jack-snipe, Sabine's snipe, and the Brown snipe. The
common snipe is very plentiful near Oxford, particularly along the river-side. A
rather curious circumstance occurred to me some years since while out snipe shooting
with a friend. He got four or five shots at snipes, and although an excellent marksman in general, missed them all and returned home without one bird, while I got no
shots at all, and yet bagged three fine snipes. Two I picked up as they lay on the
ground, wounded by a sportsman who had preceded us by half an hour, and the
third I took away from a hawk who had just struck it down, and was employed in
tearing open the snipe's breast, when it was alarmed by my companion's gun, and
flew off, leaving its dying prey behind it.

The jack-snipe is much rarer than the common snipe, and has only once or twice

have rather said a humming), I suspect we mean the same thing. However, while they are playing about on the wing, they certainly make a loud piping with their mouths; but

SNIPE.

whether that bleating or humming is ventriloquous, or proceeds from the motion of their wings, I cannot say; but this I know, that, when this noise happens, the bird is always descending, and his wings are violently agitated.

Soon after the lapwings[1] have done breeding, they congregate, and leaving the moors and marshes, betake themselves to downs and sheep-walks.[2]

been known to breed in this country. It has a curious habit of remaining near the same place, always returning even after being shot at. In Stanley's Birds there is an account of a jack-snipe that was regularly hunted for a whole winter by a gentleman who was a very bad marksman, and who, when he at last succeeded in killing it, bitterly lamented the loss of a bird that had afforded him amusement for so many weeks. Mr. Yarrell mentions an instance of a jack-snipe remaining in this country until June, when it was so weak, that it was captured by a man, who actually knocked it down with his hat.

Sabine's snipe is remarkable for the total absence of white from its plumage. Only one or two specimens have been discovered in this country. The brown snipe, although very rare in England, is exceedingly common in America, where it is called either the red-breasted snipe, the brown snipe, or the grey snipe, according to the variations of its plumage at different periods of the year.—J. G. W.

(1) *British Zoology*, vol. ii. p. 360.

(2) The lapwing or peewit is well known for the art which it displays in attracting an intruder away from its nest. This art is however sometimes turned against itself, for by long practice men are enabled to judge from the appearance of the

LAPWING. [1]

Two years ago[2] last spring, the little auk was found alive
and unhurt, but fluttering and unable to rise, in a lane a few
miles from Alresford, where there is a great lake; it was kept
a while, but died.

I saw young teals taken alive in the ponds of Wolmer

TEAL [3]

bird where her nest is, and can walk up to the very spot. The eggs are sold a
a high price in London, under the title of plover's eggs. The lapwing makes n
nest, but places a few pieces of hay and straw in a hollow, and lays its four larg
dark eggs upon them.—J. G. W.

(1) *Vanellus cristatus.*—J. G. W.   (2) *British Zoology*, vol. ii. p. 409.
(3) *Querquedula crecca.*—J. G. W.   *British Zoology*, vol. ii. p. 475.

Forest in the beginning of July last, along with flappers, or young wild ducks.

Speaking of the swift,[1] that page says, "*its drink the dew;*" whereas it should be, "it drinks on the wing;" for all the swallow kind sip their water as they sweep over the face of pools or rivers: like Virgil's bees, they drink flying—"*flumina summa libant,*" "they sip the surface of the stream." In this method of drinking, perhaps this genus may be peculiar.

Of the sedge-bird[2] be pleased to say, it sings most part of the night; its notes are hurrying, but not unpleasing, and imitative of several birds, as the sparrow, swallow, skylark. When it happens to be silent in the night, by throwing a stone or clod into the bushes where it sits, you immediately set it a-singing, or, in other words, though it slumbers sometimes, yet, as soon as it is awakened, it reassumes its song.

LETTER XL.

SWALLOWS — NIGHTINGALES — KESTREL — EELS — BLACKBIRD—REDBREAST—
GREAT TITMOUSE—CROW—CARP—WHITETHROAT, ETC.

SELBORNE, *Sept.* 2, 1774.

DEAR SIR,—Before your letter arrived, and of my own accord, I had been remarking and comparing the tails of the male and female swallow, and this ere any young broods appeared; so that there was no danger of confounding the dams with their *pulli;* and, besides, as they were then always in pairs, and busied in the employ of nidification, there could be no room for mistaking the sexes, nor the individuals of different chimneys, the one for the other. From all my observations, it constantly appeared that each sex has the long feathers in its tail that give it that forked shape; with this difference, that they are longer in the tail of the male than in that of the female.

Nightingales,[3] when their young first come abroad, and are helpless, make a plaintive and a jarring noise; and also a

---

(1) *British Zoology,* vol. ii. p. 15.   (2) *Ibid.* p. 475.

(3) *Luscinia Philomela.* I cannot resist quoting part of the nightingale's song as written in words, or rather syllables :—

snapping or cracking, pursuing people along the hedges as they walk: these last sounds seem intended for menace and defiance.

NIGHTINGALE.

The grasshopper-lark chirps all night in the height of summer.

Swans turn white the second year, and breed the third.

Weasels[1] prey on moles, as appears by their being sometimes caught in mole-traps.

"Tiuu tiuu tiuu tiuu—Spe tiu zqua—
Tiō tiō tiō tio tio tio tio tix—Qutio qutio qutio qutio—
Zquo zquo zquo zquo—Tzü tzü tzü tzü tzü tzü tzü tzü tzü tzi—
Quorror tiu zqua pipiquisi—Zozozozozozozozozozozozozo zirrhading ! &c. &c."

So does a well-known naturalist endeavour to express the wild and spiritual melody of this most exquisite of British song-birds, the NIGHTINGALE. And in truth it is perhaps as good a description as can be given without the aid of music. Even its own marvellous notes sound comparatively weak unless backed by the accompaniments of night and tranquillity ; for the inimitable song of this Mendelssohn among birds loses great part of its beauty when uttered by day, deadened and confused with other sounds.

In some counties of England it is never found, but in many its nightly strains are frequently heard. The fields and College gardens of Oxford are full of nightingales, whose songs add greatly to the effect of the scene.

An unsuccessful attempt was made some years ago, by Sir John Sinclair, to naturalise the nightingale in Scotland. At a great expense, he procured a large number of the eggs, and borrowing the habits of the cuckoo, deposed a number of robins' eggs to make room for the intruders. They were hatched and flew at the proper season. When the migrating time arrived, they disappeared, and never returned.

The nest of the nightingale is always placed on the ground, often resting against the old stump of a tree, and as its construction is very slight, and the eggs of an uniform olive-brown, it is by no means an easy nest to find.—J. G. W.

(1) *Mustela vulgaris.* This is the least of all the *Mustelidæ.* It is a most courageous animal, and has been even known to attack a man. There is a well

Sparrow-hawks sometimes breed in old crows' nests; and the kestrel in churches and ruins.

KESTREL. [1]

There are supposed to be two sorts of eels in the island of Ely. The threads sometimes discovered in eels are perhaps their young: the generation of eels is very dark and mysterious.[2]

authenticated story of a man who had disturbed a nest of weasels, being resolutely attacked by his diminutive though formidable antagonists. They ran up his body, instinctively endeavouring to reach his throat, nor was it until help came that he could escape from his assailants. As it was, he was quite wearied with the exertion he had made in tearing them away from his person. The weasel hunts by scent like the dog. Not long since a weasel ran into one of the Oxford churches during service, and pursued its course with the utmost deliberation. After looking about for some time, it appeared satisfied, and leaving the church, speedily returned, bearing a frog, which it ate with great coolness.—J. G. W.

(1) *Tinnunculus alaudarius.* This bird is sometimes called the Windhover, on account of its singular habit of remaining hovering in one place for a considerable length of time. This it does while its quick eye sweeps the ground, and no sooner does an unfortunate field-mouse poke its head from under a stone or out of its hiding-place, than the kestrel sweeps down upon it with unerring skill, and carries it off in an instant. The farmers, who are most virulent against this bird, ought to encourage it as much as possible, instead of disgracing their barns by nailing the murdered windhover on them.

These birds are not at all uncommon. There are plenty in Bagley Wood near Oxford, and in Dovedale they live in numbers. Their nest is often made in the deserted habitation of a crow or magpie, and contains four eggs of a very dark reddish brown.—J. G. W.

(2) There are supposed to be four species of British eel, namely, the Sharp-nosed,

EEL. [1]

Hen-harriers breed on the ground, and seem never to settle on trees.

When redstarts shake their tails, they move them horizontally, as dogs do when they fawn: the tail of the wagtail, when in motion, bobs up and down, like that of a jaded horse.

Hedge-sparrows have a remarkable flirt with their wings in breeding time: as soon as frosty mornings come, they make a very piping, plaintive noise.

Many birds which become silent about midsummer reassume their notes again in September; as the thrush, blackbird,[2] woodlark, willow-wren, &c.; hence August is by much the most mute month, the spring, summer, and autumn through. Are birds induced to sing again because the temperament of autumn resembles that of spring?

Linnæus ranges plants geographically: palms inhabit the tropics; grasses the temperate zones; and mosses and lichens the polar circles: no doubt animals may be classed in the same manner with propriety.

---

the Broad-nosed, the Snig, and the Grig. The threads here mentioned are intestinal worms, with which they are much infested.—J. G. W.

(1) *Anguilla acutirostris*—the Sharp-nosed Eel.—J. G. W.

(2) The blackbird (*Turdus merula*) is distributed over the whole of Europe, and has been seen in countries of another continent; Darwin mentions that he saw it inhabiting one of the Azores. Its own magnificent voice is too well known to need description, but it has been frequently heard imitating the notes of other birds, and has carried its polylogistic power so far, as to have been both heard and seen imitating the harsh crowing of the domestic cock, as well as the guttural clatter of the hens.—J. G. W.

BLACKBIRD

House-sparrows[1] build under eaves in the spring; as the weather becomes hotter, they get out for coolness, and nest in plum-trees and apple-trees. These birds have been known sometimes to build in rooks' nests, and sometimes in the forks of boughs under rooks' nests.

As my neighbour was housing a rick, he observed that his dogs devoured all the little red mice that they could catch, but rejected the common mice; and that his cats ate the common mice, refusing the red.

Redbreasts sing all through the spring, summer, and autumn. The reason that they are called autumn songsters is, because in the two first seasons their voices are lost and drowned in the general chorus : in the latter, their song becomes distinguishable. Many songsters of the autumn seem

(1) *Passer domesticus.* There are two kinds of British sparrow, the house-sparrow, and the tree-sparrow. These two birds are very much alike, and are often confounded together.—J. G. W.

to be the young cock redbreasts of that year: notwithstanding the prejudices in their favour, they do much mischief in gardens to the summer fruits.[1]

REDBREAST [2]

The titmouse, which early in February begins to make two quaint notes, like the whetting of a saw, is the marsh titmouse; the great titmouse sings with three cheerful joyous notes, and begins about the same time.

GREAT TITMOUSE [3]

(1) They eat also the berries of the ivy, the honeysuckle, and the *Euonymus Europæus,* or spindle-tree.

(2) *Erythacus rubecula.*—J. G. W.   (3) *Parus major.*—J. G. W.

Wrens sing all the winter through, frost excepted.

House-martins came remarkably late this year, both in Hampshire and Devonshire: is this circumstance for or against either hiding or migration?

Most birds drink, sipping at intervals; but pigeons take a long continued draught, like quadrupeds.

Notwithstanding what I have said in a former letter, no grey crows were ever known to breed on Dartmoor; it was my mistake.

HOODED CROW [1]

The appearance and flying of the *Scarabæus solstitialis*, or fern-chafer,[2] commence with the month of July, and cease about the end of it. These scarabs are the constant food of *caprimulgi*, or fern-owls, through that period. They abound on the chalky downs, and in some sandy districts, but not in the clays.

In the garden of the Black Bear Inn, in the town of Reading, is a stream or canal, running under the stables, and out into the fields on the other side of the road: in this water are many carps, which lie rolling about in sight, being fed by

(1) *Corvus cornix*, Penn. This bird is very common on the western islands of Scotland, but rarely is seen in the south of England.—J G. W.
(2) *Amphimalla solstitialis*, Lat.—J. G. W.

travellers, who amuse themselves by tossing them bread ; but as soon as the weather grows at all severe, these fishes are no

CARP. [1]

longer seen, because they retire under the stables, where they remain till the return of spring. Do they lie in a torpid state ? if they do not, how are they supported ?

WHITETHROAT [2]

The note of the whitethroat, which is continually repeated, and often attended with odd gesticulations on the

(1) *Cyprinus carpio.*—J. G. W.
(2) *Sylvia cinerea.* The song of the whitethroat is very peculiar, and the manner in which the bird sings is rather singular. The male bird frequently sings in the

wing, is harsh and displeasing. These birds seem of pugnacious disposition; for they sing with an erected crest, and attitudes of rivalry and defiance; are shy and wild in breeding-time, avoiding neighbourhoods, and haunting lonely lanes and commons; nay, even the very tops of the Sussex Downs, where there are bushes and covert; but in July and August they bring their broods into gardens and orchards, and make great havoc among the summer fruits.

The black-cap has, in common, a full, sweet, deep, loud, and wild pipe; yet that strain is of short continuance, and his motions are desultory; but, when that bird sits calmly and engages in song in earnest, he pours forth very sweet, but inward melody, and expresses great variety of soft and gentle modulations, superior, perhaps, to those of any of our warblers, the nightingale excepted.

Black-caps mostly haunt orchards and gardens: while they warble, their throats are wonderfully distended.

REDSTART [1]

The song of the redstart is superior, though somewhat like that of the whitethroat; some birds have a few more notes

midst of short flights in the air, always returning to the same twig from which he started. A small ash copse near Little Hinton, in Wiltshire, abounded with these pretty birds, whose rival songs were constantly heard.—J. G. W.

(1) *Ruticilla phœnicura.* I once found the nest of a redstart built in a wall forming a side of a narrow passage in Merton College, through which the servants

than others. Sitting very placidly on the top of a tall tree
in a village, the cock sings from morning to night; he affects
neighbourhoods, and avoids solitude, and loves to build in
orchards and about houses; with us he perches on the vane
of a tall maypole.

The fly-catcher is, of all our summer birds, the most mute
and the most familiar; it also appears the last of any. It
builds in a vine, or a sweet-brier, against the wall of a house,
or in the hole of a wall, or on the end of a beam or plate, and
often close to the post of a door where people are going in
and out all day long. This bird does not make the least
pretension to song, but uses a little inward wailing note, when
it thinks its young in danger from cats or other annoyances:
it breeds but once, and retires early.

Selborne parish alone can and has exhibited at times more
than half the birds that are ever seen in all Sweden: the
former has produced more than one hundred and twenty
species, the latter only two hundred and twenty-one. Let me
add, also, that it has shown near half the species that were
ever known in Great Britain.[1]

On a retrospect, I observe that my long letter carries with
it a quaint and magisterial air, and is very sententious: but
when I recollect that you requested stricture and anecdote,
I hope you will pardon the didactic manner for the sake of
the information it may happen to contain.

LETTER XLI.

SOFT-BILLED BIRDS—TITMICE.

It is matter of curious inquiry to trace out how those species
of soft-billed birds that continue with us the winter through
subsist during the dead months. The imbecility of birds
seems not to be the only reason why they shun the rigour of

---

were constantly passing with their mops, pails, and brooms; the birds did not appear
to be in the least disturbed by them, and had nearly hatched the eggs. The name
of the bird is derived from its habit of flirting up its tail as it flies, displaying the
red upper tail coverts and tail feathers.—J. G. W.

(1) Sweden, 221; Great Britain, 252 species.

our winters ; for the robust wryneck (so much resembling the hardy race of woodpeckers) migrates, while the feeble little golden-crowned wren, that shadow of a bird, braves our severest frost, without availing himself of houses or villages, to which most of our winter birds crowd in distressful seasons, while he keeps aloof in fields and woods ; but perhaps this may be the reason why they may often perish, and why they are almost as rare as any bird we know.

I have no reason to doubt but that the soft-billed birds which winter with us, subsist chiefly on insects in their aurelia state. All the species of wagtails in severe weather haunt shallow streams, near their spring-heads, where they never freeze ; and, by wading, pick out the aurelias of the genus of *phryganeæ*,[1] &c.

Hedge-sparrows frequent sinks and gutters in hard weather, where they pick up crumbs and other sweepings ; and in mild weather they procure worms, which are stirring every month in the year, as any one may see that will only be at the trouble of taking a candle to a grass-plot on any mild winter's night. Redbreasts and wrens, in the winter, haunt outhouses, stables, and barns, where they find spiders and flies that have laid themselves up during the cold season. But the grand support of the soft-billed birds in winter, is that infinite profusion of aureliæ of the *lepidoptera ordo*, which is fastened to the twigs of trees and their trunks, to the pales and walls of gardens and buildings, and is found in every cranny and cleft of rock or rubbish, and even in the ground itself.

Every species of titmouse winters with us. They have what I call a kind of intermediate bill, between the hard and the soft, between the Linnæan genera of *fringilla* and *motacilla*. One species alone spends its whole time in the woods and fields, never retreating for succour, in the severest seasons, to houses and neighbourhoods,—and that is the delicate long-tailed titmouse,[2] which is almost as minute as the golden-crowned wren ; but the blue titmouse, or nun (*parus cœruleus*), the cole-mouse (*parus ater*), the great black-headed

---

(1) See Derham's *Physico-Theology*, p. 235.
(2) *Parus caudatus.*—J. G. W.

titmouse (*fringillago*), and the marsh titmouse (*parus palus-tris*), all resort, at times, to buildings, and in hard weather particularly. The great titmouse, driven by stress of wea-

LONG-TAILED TITMOUSE.

ther, much frequents houses ; and, in deep snows, I have seen this bird, while it hung with its back downwards (to my no small delight and admiration), draw straws lengthwise from out the eaves of thatched houses, in order to pull out the flies that were concealed between them, and that in such numbers that they quite defaced the thatch, and gave it a ragged appearance.

The blue titmouse, or nun, is a great frequenter of houses, and a general devourer. Besides insects, it is very fond of flesh, for it frequently picks bones on dunghills. It is a vast admirer of suet, and haunts butchers' shops.[1] When a boy I have known twenty in a morning caught with snap mouse traps, baited with tallow or suet. It will also pick holes in apples left on the ground, and be well entertained with the

(1) I once kept a tame sparrow-hawk in a garden. He was a shocking coward, and used to run away from the wagtails and robins if they approached him in a threatening manner. The blue titmice soon discovered this weakness, and used to drive him away directly he was fed. and take his meal for themselves.—J. G. W.

seeds on the head of a sunflower. The blue, marsh, and great titmice will, in very severe weather, carry away barley and oat-straws from the sides of ricks.

BLUE TITMOUSE [1]

How the wheatear and whin-chat support themselves in winter, cannot be so easily ascertained, since they spend their time on wild heaths and warrens, the former, especially, where there are stone-quarries. Most probable it is, that their maintenance arises from the aurelia of the *lepidoptera ordo*, which furnish them with a plentiful table in the wilderness.

LETTER XLII.

IRELAND AND SCOTLAND.

SELBORNE, *March* 9, 1775.

DEAR SIR,—Some future faunist, a man of fortune, will, I hope, extend his visits to the kingdom of Ireland; a new field, and a country little known to the naturalist. He will not, it is to be wished, undertake that tour unaccompanied by a botanist, because the mountains have scarcely been sufficiently examined; and the southerly counties of so mild an island may possibly afford some plants little to be expected within the British dominions. A person of a thinking turn of mind will draw many just remarks from the modern

(1) *Parus cæruleus.*—J. G. W.

improvements of that country, both in arts and agriculture, where premiums obtained long before they were heard of with us. The manners of the wild natives, their superstitions, their prejudices, their sordid way of life, will extort from him many useful reflections. He should also take with him an able draughtsman, for he must, by no means, pass over the noble castles and seats, the extensive and picturesque lakes and water-falls, and the lofty, stupendous mountains, so little known, and so engaging to the imagination, when described and exhibited in a lively manner. Such a work would be well received.

As I have seen no modern map of Scotland, I cannot pretend to say how accurate or particular any such may be; but this I know, that the best old maps of that kingdom are very defective.

The great obvious defect that I have remarked in all maps of Scotland that have fallen in my way is, a want of a coloured line, or stroke, that shall exactly define the just limits of that district called the Highlands. Moreover, all the great avenues to that mountainous and romantic country want to be well distinguished. The military roads formed by General Wade are so great and Roman-like an undertaking, that they well merit attention. My old map, Moll's map, takes notice of Fort William, but could not mention the other forts that have been erected long since; therefore a good representation of the chain of forts should not be omitted.

The celebrated zigzag up the Coryarich must not be passed over. Moll takes notice of Hamilton and Drumlanrig, and such capital houses; but a new survey, no doubt, should represent every seat and castle remarkable for any great event or celebrated for its paintings, &c. Lord Breadalbane's seat and beautiful policy are too curious and extraordinary to be omitted.

The seat of the Earl of Eglingtoun, near Glasgow, is worthy of notice. The pine plantations of that nobleman are very grand and extensive indeed.

LETTER XLIII.

HONEY-BUZZARD—SPARROW-HAWK.

A PAIR of honey buzzards, *buteo apivorus, sive vespivorus,*
Raii, built them a large shallow nest, composed of twigs, and
lined with dead beechen leaves, upon a tall slender beech near
the middle of Selborne Hanger, in the summer of 1780.   In
the middle of the month of June, a bold boy climbed this

HONEY-BUZZARD [1]

ree, though standing on so steep and dizzy a situation, and
brought down an egg, the only one in the nest, which had
been set on for some time, and contained the embryo of a
young bird.   The egg was smaller, and not so round as those
of the common buzzard; was dotted at each end with small

(1) *Pernis apivorus.* The genus Pernis is at once distinguished by the feathered
drum round the eyes.   This bird so seldom visits our coasts at all, that an example
of its breeding in England is particularly valuable.   The usual food of this bird is
wasps,—their comb and larva; indeed, if it wished to feed upon bees, its chance of
living in England would be but small.   The bird has been seen to scratch a wasp's
nest out of the earth; a feat which would appear impossible, if one were to judge
from the look of its feet and legs.   Yet it was not only seen to rise from a newly
scratched wasp's nest, but was actually secured on the spot by a steel trap.—J. G. W.

red spots, and surrounded in the middle with a broad bloody
zone.

The hen bird was shot, and answered exactly to Mr. Ray's
description of that species; had a black cere, short thick legs,
and a long tail. When on the wing, this species may be
easily distinguished from the common buzzard by its hawk-
like appearance, small head, wings not so blunt, and longer
tail. This specimen contained in its craw some limbs of
frogs, and many grey snails without shells. The irides
of the eyes of this bird were of a beautiful bright yellow
colour.

About the 10th of July, in the same summer, a pair of
sparrow-hawks bred in an old crow's nest on a low beech in
the same hanger; and as their brood, which was numerous,
began to grow up, became so daring and ravenous, that they

SPARROW-HAWK.

were a terror to all the dames in the village that had chickens
or ducklings under their care. A boy climbed the tree,
and found the young so fledged that they all escaped from
him, but discovered that a good house had been kept. The
larder was well stored with provisions; for he brought down

(i) *Accipiter nisus*. See note on p. 148.—J. G. W.

a young blackbird, jay, and house-martin, all clean picked, and some half devoured. The old birds had been observed to make sad havoc for some days among the new-flown swallows and martins, which, being but lately out of their nests, had not acquired those powers and command of wing that enable them, when more mature, to set such enemies at defiance.

LETTER XLIV.

STOCK-DOVE—RING-DOVE.

SELBORNE, *Nov.* 30, 1780.

DEAR SIR, — Every incident that occasions a renewal of our correspondence will ever be pleasing and agreeable to me.

As to the wild wood-pigeon,[1] the *œnas* or *vinago* of Ray, I am much of your mind, and see no reason for making it the origin of the common house-dove; but suppose those that have advanced that opinion may have been misled by another appellation, often given to the *œnas,* which is that of stock-dove.

Unless the stock-dove in the winter varies greatly in manners from itself in summer, no species seems more unlikely to be domesticated, and to make a house-dove. We very rarely see the latter settle on trees at all, nor does it ever haunt the woods; but the former, as long as it stays with us, from November perhaps to February, lives the same wild life with the ring-dove (*palumbus torquatus*); frequents coppices and groves, supports itself chiefly by mast, and delights to roost in the tallest beeches. Could it be known in what manner stock-doves build, the doubt would be settled with

(1) *Columba palumbus.* The nest of the ring-dove or wood-pigeon is composed of a few loose sticks laid on a spray, so loosely and carelessly that the eggs can be seen from below. The nest is seldom less than sixteen or twenty feet from the ground, and often much higher; but I have found a nest with two eggs placed in a hedge, not a yard from the ground. The ring-dove has a great predilection for the r-tree as a building place, and is often found near the top of a large holly.

This is the largest of our native pigeons, of which we have three; the ring-dove (*Columba palumbus*), the stock-dove (*Columba œnas*), and the turtle-dove (*Turtur uritus*). It is called by other names in different parts of the country, the Cushat and Quest being the most common. — J. G. W.

me at once, provided they construct their nests on trees, like the ring-dove, as I much suspect they do.

You received, you say, last spring, a stock-dove from Sussex; and are informed that they sometimes breed in that

STOCK-DOVE [1]

county. But why did not your correspondent determine the place of its nidification, whether on rocks, cliffs, or trees? If he was not an adroit ornithologist, I should doubt the fact, because people with us perpetually confound the stock-dove with the ring-dove.

For my own part, I readily concur with you in supposing that house-doves are derived from the small blue rock-pigeon for many reasons. In the first place, the wild stock-dove is manifestly larger than the common house-dove, against the usual rule of domestication, which generally enlarges the breed. Again, those two remarkable black spots on the remiges of each wing of the stock-dove, which are so charac

(1) *Columba œnas.*—J. G. W.

teristic of the species, would not, one should think, be totally lost by its being reclaimed; but would often break out among its descendants. But what is worth a hundred arguments, is the instance you give in Sir Roger Mostyn's house-doves in Carnarvonshire; which, though tempted by plenty of food and gentle treatment, can never be prevailed on to inhabit their cote for any time; but, as soon as they begin to breed, betake themselves to the fastnesses of Ormshead, and deposit their young in safety amidst the inaccessible caverns and precipices of that stupendous promontory.

> "Naturam expellas furcâ . . tamen usque recurret."
>
> "Nature, expell'd by force, will still return."

I have consulted a sportsman, now in his seventy-eighth year, who tells me that, fifty or sixty years back, when the beechen woods were much more extensive than at present, the number of wood-pigeons was astonishing; that he has often killed near twenty in a day; and that, with a long wild-fowl piece, he has shot seven or eight at a time on the wing, as they came wheeling over head. He moreover adds—which I was not aware of—that often there were among them little parties of small blue doves, which he calls rockiers. The food of these numberless emigrants was beech-mast and some acorns; and particularly barley, which they collected in the stubbles. But of late years, since the vast increase of turnips, that vegetable has furnished a great part of their support in hard weather; and the holes they pick in these roots greatly damage the crop. From this food their flesh has contracted a rancidness, which occasions them to be rejected by nicer judges of eating, who thought them before a delicate dish. They were shot not only as they were feeding in the fields, and especially in snowy weather, but also at the close of the evening, by men who lay in ambush among the woods and groves to kill them as they came in to roost.[1] These are the principal circumstances relating to this wonderful internal migration, which with us takes place towards the end of November, and ceases early in the spring. Last winter we

(1) Some old sportsmen say, that the main part of these flocks used to withdraw as soon as the heavy Christmas frosts were over.

had, in Selborne High-wood, about a hundred of these doves; but in former times the flocks were so vast, not only with us, but all the district around, that on mornings and evenings they traversed the air, like rooks, in strings, reaching for a

RING-DOVE [1]

mile together. When they thus rendezvoused here by thousands, if they happened to be suddenly roused from their roost-trees on an evening,

> "Their rising all at once was like the sound
> Of thunder heard remote."

It will by no means be foreign to the present purpose to add, that I had a relation in this neighbourhood who made it a practice for a time, whenever he could procure the eggs of a ring-dove, to place them under a pair of doves that were sitting in his own pigeon-house, hoping thereby, if he could bring about a coalition, to enlarge his breed, and teach his

(1) *Columba palumbus.* See note on p. 153.—J. G. W.

own doves to beat out into the woods, and to support themselves by mast. The plan was plausible, but something
always interrupted the success; for though the birds were
usually hatched, and sometimes grew to half their size, yet
none ever arrived at maturity. I myself have seen these
foundlings in their nest displaying a strange ferocity of
nature, so as scarcely to bear to be looked at, and snapping
with their bills by way of menace. In short, they always
died, perhaps for want of proper sustenance; but the owner
thought that by their fierce and wild demeanour they frighted
their foster-mothers, and so were starved.

Virgil, as a familiar occurrence, by way of simile, describes
a dove haunting the cavern of a rock, in such engaging
numbers, that I cannot refrain from quoting a passage; and
John Dryden has rendered it so happily in our language,
that, without further excuse, I shall add his translation
also :—

> "Qualis speluncâ subitò commota columba,
> Cui domus, et dulces latebroso in pumice nidi,
> Fertur in arva volans, plausumque exterrita pennis
> Dat tecto ingentem : mox aëre lapsa quieto
> Radit iter liquidum, celerès neque commovet alas."

> "As when the dove her rocky hold forsakes,
> Roused in a fright, her sounding wings she shakes;
> The cavern rings with clattering; out she flies,
> And leaves her callow care, and cleaves the skies.
> At first she flutters; but at length she springs
> To smoother flight, and shoots upon her wings."

Part II.

TO THE HON. DAINES BARRINGTON.

LETTER I.

SUMMER AND WINTER BIRDS OF PASSAGE.

SELBORNE, *June* 30, 1769.

DEAR SIR,—When I was in town last month, I partly engaged that I would some time do myself the honour to write to you on the subject of natural history; and I am the more ready to fulfil my promise, because I see you are a gentleman of great candour, and one that will make allowances, especially where the writer professes to be an outdoor naturalist, one that takes his observations from the subject itself, and not from the writings of others.

The following is a List of the Summer Birds of Passage which I have discovered in this neighbourhood, ranged somewhat in the order in which they appear.

	RAII NOMINA.	USUALLY APPEARS ABOUT
1. Wryneck,	*Iynx, sive Torquilla.*	The middle of March : harsh note.
2. Smallest Willow-wren,	*Regulus non cristatus.*	March 23 : chirps till September.
3. Swallow,	*Hirundo domestica.*	April 13.
4. Martin,	*Hirundo rustica.*	Ditto.
5. Sand-martin,	*Hirundo riparia.*	Ditto.
6. Blackcap,	*Atricapilla.*	Ditto : a sweet wild note.
7. Nightingale,	*Luscinia.*	Beginning of April
8. Cuckoo,	*Cuculus.*	Middle of April.
9. Middle Willow-wren,	*Regulus non cristatus.*	Ditto : a sweet plaintive note.
10. Whitethroat,	*Ficedula affinis.*	Ditto : mean note ; sings on till September.
11. Redstart,	*Ruticilla.*	Ditto : more agreeable song.

	RAII NOMINA.	USUALLY APPEARS ABOUT
12. Stone-curlew,	*Œdicnemus.*	End of March : loud nocturnal whistle.
13. Turtle-dove,	*Turtur.*	
14. Grasshopper-lark,	*Alauda minima, locustæ voce.*	Middle of April : a small sibilous note, till the end of July.
15. Swift,	*Hirundo apus.*	bout April 27.
16. Less Reed-sparrow,	*Passer arundinaceus minor.*	A sweet polyglot, but hurrying : it has the notes of many birds.
17. Landrail,	*Ortygometra.*	A loud harsh note, crex, crex.
18. Largest Willow-wren,	*Regulus non cristatus.*	*Cantat voce stridulâ locustæ :* end of April, on the tops of high beeches.
19. Goat-sucker, or Fern-owl,	*Caprimulgus.*	Beginning of May : chatters by night with a singular noise.
20. Fly-catcher,	*Stoparola.*	May 12. A very mute bird : this is the latest summer bird of passage.

This assemblage of curious and amusing birds belongs to ten several genera of the Linnæan system ; and are all of the *ordo* of *passeres*, save the *iynx* and *cuculus*, which are *picæ*, and the *charadrius* (*œdicnemus*) and *rallus* (*ortygometra*), which are *grallæ*.

These birds, as they stand numerically, belong to the following Linnæan genera:—

1,	*Iynx.*	13,	*Columba.*
2, 6, 7, 9, 10, 11, 16, 18,	*Motacilla.*	17,	*Rallus.*
3, 4, 5, 15,	*Hirundo.*	19,	*Caprimulgus.*
8,	*Cuculus.*	14,	*Alauda.*
12,	*Charadrius.*	20,	*Muscicapa.*

Most soft-billed birds live on insects, and not on grain and seeds, and therefore at the end of summer they retire ; but the following soft-billed birds, though insect eaters, stay with us the year round :—

	RAII NOMINA.	
Redbreast,	*Rubecula.*	These frequent houses ; and haunt out-buildings in the winter : eat spiders.
Wren,	*Passer troglodytes.*	
Hedge-sparrow,	*Curruca.*	Haunt sinks, for crumbs and other sweepings.
White Wagtail,	*Motacilla alba.*	These frequent shallow rivulets, near the spring heads, where they never freeze : eat the aureliæ of Phryganea. The smallest birds that walk.
Yellow Wagtail,	*Motacilla flava.*	
Grey Wagtail, ·	*Motacilla cinerea.*	
Wheatear,	*Œnanthe.*	Some of these are to be seen with us the winter through.
Whinchat,	*Œnanthe secunda.*	
Stone-chatter,	*Œnanthe tertia.*	
Golden-crowned Wren,	*Regulus cristatus.*	This is the smallest British bird : haunts the tops of tall trees ; stays the winter through.

A List of the Winter Birds of Passage round this neighbourhood, ranged somewhat in the order in which they appear.

RAII NOMINA.

1. Ring-ousel,	*Merula torquata.*	This is a new migration, which I have lately discovered about Michaelmas week, and again about the 14th of March.
2. Red-wing,	*Turdus iliacus.*	About old Michaelmas.
3. Fieldfare,	*Turdus pilaris.*	Though a percher by day, roosts on the ground.
4. Royston-crow,	*Cornix cinerea.*	Most frequently on downs.
5. Woodcock,	*Scolopax.*	Appears about old Michaelmas.
6. Snipe,	*Gallinago minor.*	Some snipes constantly breed with us.
7. Jack-snipe,	*Gallinago minima.*	
8. Wood-pigeon,	*Œnas.*	Seldom appears till late ; not in such plenty as formerly.
9. Wild-swan,	*Cygnus ferus.*	On some large waters.
10. Wild-goose,	*Anser ferus.*	
11. Wild-duck,	*Anas torquata minor.*	
12. Pochard,	*Anas fera fusca.*	
13. Widgeon,	*Penelope.*	On our lakes and streams.
14. Teal, breeds with us in Wolmer Forest,	*Querquedula.*	
15. Crossbeak,	*Coccothraustes.*	These are only wanderers that appear occasionally, and are not observant of any regular migration.
16. Crossbill,	*Loxia.*	
17. Silk-tail,	*Garrulus Bohemicus.*	

These birds, as they stand numerically, belong to the following Linnæan genera:—

1, 2, 3,	*Turdus.*	9, 10, 11, 12, 13, 14.	*Anas.*
4,	*Corvus.*	15, 16,	*Loxia.*
5, 6, 7,	*Scolopax.*	17,	*Ampelis.*
8,	*Columba.*		

Birds that sing in the night are but few:—

Nightingale,	*Luscinia.*	" In shadiest covert hid." MILTON.
Woodlark,	*Alauda arborea.*	Suspended in mid air.
Less Reed-sparrow,	*Passer arundinaceus minor.*	Among reeds and willows.

I should now proceed to such birds as continue to sing after midsummer; but as they are rather numerous, they would exceed the bounds of this paper; besides, as this is now the season for remarking on that subject, I am willing to repeat my observations on some birds, concerning the continuation of whose song I seem at present to have some doubt.

LETTER II.

SINGING BIRDS.

SELBORNE, *Nov.* 2, 1769.

DEAR SIR,—When I did myself the honour to write to you, about the end of last June, on the subject of natural history, I sent you a list of the summer birds of passage which I have observed in this neighbourhood, and also a list of the winter birds of passage; I mentioned, besides, those soft-billed birds that stay with us the winter through in the south of England, and those that are remarkable for singing in the night.

According to my proposal, I shall now proceed to such birds (singing birds, strictly so called) as continue in full song till after midsummer, and shall range them somewhat in the order in which they first begin to open as the spring advances.

RAII NOMINA.

1. Woodlark,	*Alauda arborea.*	In January, and continues to sing through all the summer and autumn.
2. Song-thrush,	*Turdus simpliciter dictus.*	In February, and on to August; reassume their song in autumn.
3. Wren,	*Passer troglodytes.*	All the year, hard frost excepted.
4. Redbreast,	*Rubecula.*	Ditto.
5. Hedge-sparrow,	*Curruca.*	Early in February, to July the 10th.
6. Yellow-hammer,	*Emberiza flava.*	Early in February, and on through July to August the 21st.
7. Skylark,	*Alauda vulgaris.*	In February, and on to October.
8. Swallow,	*Hirundo domestica.*	From April to September.
9. Black-cap,	*Atricapilla.*	Beginning of April to July the 13th.
10. Titlark,	*Alauda pratorum.*	From middle of April to July 16th.
11. Blackbird,	*Merula vulgaris.*	Sometimes in February and March, and so on to July the 23d; reassumes in autumn.
12. Whitethroat,	*Ficedula affinis.*	In April, and on to July the 23d.
13. Goldfinch,	*Carduelis.*	April, and through to September the 16th.
14. Greenfinch,	*Chloris.*	On to July and August the 2d.
15. Less Reed-sparrow,	*Passer arundinaceus minor.*	May, on to beginning of July.
16. Common Linnet,	*Linaria vulgaris.*	Breeds and whistles on till August; reassumes its note when they begin to congregate in October, and again early before the flocks separate.

Birds that cease to be in full song, and are usually silent at or before midsummer:—

RAII NOMINA.

17. Middle Willow-wren,	} *Regulus non cristatus.*	Middle of June; begins in April.
18. Redstart,	*Ruticilla.*	Ditto; begins in May.
19. Chaffinch,	*Fringilla.*	{ Beginning of June; sings first in February.
20. Nightingale,	*Luscinia.*	Middle of June; sings first in April.

Birds that sing for a short time, and very early in the spring:—

21. Missel-bird,	*Turdus viscivorus.*	{ January the 2d, 1770, in February. Is called in Hampshire and Sussex the storm-cock, because its song is supposed to forebode windy wet weather; is the largest singing bird we have.
22. Great Titmouse, or Ox-eye,	} *Fringillago.*	{ In February, March, April; reassumes for a short time in September.

Birds that have somewhat of a note or song, and yet are hardly to be called singing birds:—

23. Golden-crowned Wren,	} *Regulus cristatus.*	{ Its note as minute as its person; frequents the tops of high oaks and firs; the smallest British bird.
24. Marsh Titmouse,	*Parus palustris.*	{ Haunts great woods; two harsh sharp notes.
25. Small Willow-wren,	*Regulus non cristatus.*	Sings in March and on to September.
26. Largest ditto,	Ditto.	{ *Cantat voce stridula locustæ;* from end of April to August.
27. Grasshopper-lark,	{ *Alauda minima voce locustæ.*	{ Chirps all night, from the middle of April to the end of July.
28. Martin,	*Hirundo agrestis.*	{ All the breeding time; from May to September.
29. Bullfinch,	*Pyrrhula.*	
30. Bunting,	*Emberiza alba.*	From the end of January to July.

All singing birds, and those that have any pretensions to song, not only in Britain, but perhaps the world through, come under the Linnæan *ordo* of *passeres.*

The above-mentioned birds, as they stand numerically, belong to the following Linnæan genera:—

1, 7, 10, 27,	*Alauda.*	8, 28,	*Hirundo.*
2, 11, 21,	*Turdus.*	13, 16, 19,	*Fringilla.*
3, 4, 5, 9, 12, 15, 17, 18, 20, 23, 25, 26,	} *Motacilla.*	22, 24,	*Parus.*
6, 30,	*Emberiza.*	14, 29,	*Loxia.*

Birds that sing as they fly are but few:—

	RAII NOMINA.	
Skylark,	*Alauda vulgaris.*	Rising, suspended, and falling.
Titlark,	*Alauda pratorum.*	In its descent; also sitting on trees, and walking on the ground.
Woodlark,	*Alauda arborea.*	Suspended; in hot summer nights all night long.
Blackbird,	*Merula.*	Sometimes from bush to bush.
Whitethroat,	*Ficedula affinis.*	Uses, when singing on the wing, odd jerks and gesticulations.
Swallow,	*Hirundo domestica.*	In soft sunny weather.
Wren,	*Passer troglodytes.*	Sometimes from bush to bush.

Birds that breed most early in these parts:—

Raven,	*Corvus.*	Hatches in February and March.
Song-thrush,	*Turdus.*	In March.
Blackbird,	*Merula.*	In March.
Rook,	*Cornix frugilega.*	Builds the beginning of March.
Woodlark,	*Alauda arborea.*	Hatches in April.
Ring-dove.	*Palumbus torquatus.*	Lays the beginning of April.

All birds that continue in full song till after midsummer, appear to me to breed more than once.

Most kinds of birds seem to me to be wild and shy, somewhat in proportion to their bulk: I mean in this island, where they are much pursued and annoyed; but in Ascension Island, and many other desolate places, mariners have found fowls so unacquainted with a human figure, that they would stand still to be taken, as is the case with boobies, &c. As an example of what is advanced, I remark that the golden-crested wren (the smallest British bird) will stand unconcerned till you come within three or four yards of it; while the bustard (*otis*), the largest British land fowl, does not care to admit a person within so many furlongs.

LETTER III.

BIRDS IN CAGES—INCREASE OF BIRDS IN SUMMER.

SELBORNE, *Jan.* 15, 1770.

DEAR SIR,—It was no small matter of satisfaction to me to find that you were not displeased with my little *methodus*, or systematic table of birds. If there was any merit in the sketch, it must be owing to its punctuality. For many months I carried a list in my pocket of the birds that were to be remarked, and as I rode or walked about my business, I noted each day the continuance or omission of each bird's song, so that I am as sure of the certainty of my facts as a man can be of any transaction whatsoever.

I shall now proceed to answer the several queries which you put in your two obliging letters, in the best manner that I am able. Perhaps Eastwick, and its environs, where you heard so very few birds, is not a woodland country, and therefore not stocked with such songsters. If you will cast your eye on my last letter, you will find that many species continued to warble after the beginning of July.

The titlark and yellow-hammer breed late, the latter very late; and therefore it is no wonder that they protract their song; for I lay it down as a maxim in ornithology, that as long as there is any incubation going on, there is music. As to the redbreast and wren, it is well known to the most incurious observer that they whistle the year round, hard frost excepted; especially the latter.

It was not in my power to procure you a black-cap, or a less reed-sparrow, or sedge-bird, alive. As the first is undoubtedly, and the last, as far as I can yet see, a summer bird of passage, they would require more nice and curious management in a cage than I should be able to give them: they are both distinguished songsters. The note of the former has such a wild sweetness that it always brings to my mind those lines in a song in "As You Like It:"—

> And tune his merry note
> Unto the *wild* bird's throat.

The latter has a surprising variety of notes, resembling the song of several other birds; but then it has also a hurrying manner, not at all to its advantage. It is, notwithstanding, a delicate polyglot.

It is new to me that titlarks in cages sing in the night, perhaps only caged birds do so. I once knew a tame redbreast[1] in a cage that always sang as long as candles were in

---

(1) *Erythacus rubecula.* There is a popular opinion prevalent, that if redbreast are confined in a cage, they speedily die. Like most popular opinions, this is erroneous, as a female redbreast was not only put in a cage but kept there, and was sedulously attended by her mate for a long time. A white variety of this bird has

the room; but in their wild state no one supposes they sing in the night.

I should be almost ready to doubt the fact, that there are to be seen much fewer birds in July than in any former month, notwithstanding so many young are hatched daily. Sure I am, that it is far otherwise with respect to the swallow tribe, which increases prodigiously as the summer advances; and I saw at the time mentioned many hundreds of young wagtails on the banks of the Cherwell, which almost covered the meadows. If the matter appears, as you say, in the other species, may it not be owing to the dams being engaged in incubation, while the young are concealed by the leaves?

Many times have I had the curiosity to open the stomachs of woodcocks and snipes; but nothing ever occurred that helped to explain to me what their subsistence might be; all that I could ever find was a soft mucus, among which lay many pellucid small gravels.

LETTER IV.

THE CUCKOO.

SELBORNE, *Feb.* 19, 1770.

DEAR SIR,—Your observation, that "the cuckoo does not deposit its egg indiscriminately in the nest of the first bird that comes in its way, but probably looks out a nurse in some degree congenerous, with whom to entrust its young,"[1]

---

been occasionally seen. During my early days, a redbreast attached itself to us, and not only favoured us with his company during the frosts of winter, but continued his visits through the summer. On hearing his name "Bob," he would chirp in reply, and come to see if there were any food for him. We have recognised him while walking, and heard him respond to his name. He used to come and sit by while we fed the rabbits, of which we had a large stock, and at last became so impudent as to sit on the edge of the pan from which a rabbit was eating, and assist in the demolition of its dinner.—J. G. W.

(1) *Cuculus canorus.* The egg of this ornithological Skimpole is very small indeed, hardly larger than that of a sparrow. This fact will account for the seeming stupidity of such favoured birds as the wagtail, the robin, or the hedge-sparrow, who permit an intruding egg to be placed in their nest, and hatch and feed the young one. It is said that the cuckoo during its first few days in the foster-parent's nest, ejects the whole of its previous occupants, whether eggs or nestlings, by means of

is perfectly new to me; and struck me so forcibly, that I naturally fell into a train of thought that led me to consider whether the fact were so, and what reason there was for it. When I came to recollect and inquire, I could not find that

CUCKOO

any cuckoo had ever been seen in these parts, except in the nest of the wagtail, the hedge-sparrow, the titlark, the white-throat and the redbreast, all soft-billed insectivorous birds. The excellent Mr. Willoughby mentions the nest of the *palumbus* (ring-dove), and of the *fringilla* (chaffinch), birds that subsist on acorns and grains, and such hard food; but

a singular hollow in its back; but that after a few days the hollow is filled up, and the young cuckoo no longer interferes with any other occupant.

The food of the cuckoo consists principally of the large hairy caterpillars, such as those of the *Arctia caja*. These hairs of the caterpillar come off during the process of digestion, and fix themselves on the interior of the stomach, so that it was for some time imagined that the stomach of the cuckoo was lined with natural hairs, until microscopical examination proved that the hairs were those of a caterpillar.

The cuckoo is sometimes a very bold bird; and will come and settle within a few paces, if not alarmed by any sudden movement. I have often succeeded in decoying a cuckoo by imitating its note, until it perched on a tree close to my station.

This bird has been known to eat young mice. It first beat the mouse against a stone until it was reduced to a soft mass, and then swallowed it.—J. G. W.

then he does not mention them as of his own knowledge; but says afterwards, that he saw himself a wagtail feeding a cuckoo. It appears hardly possible that a soft-billed bird should subsist on the same food with the hard-billed; for the former have thin membranaceous stomachs suited to their soft food; while the latter, the granivorous tribe, have strong muscular gizzards, which, like mills, grind, by the help of small gravels and pebbles, what is swallowed. This proceeding of the cuckoo, of dropping its eggs as it were by chance, is such a monstrous outrage on maternal affection, one of the first great dictates of nature, and such a violence on instinct, that had it only been related of a bird in the Brazils or Peru, it would never have merited our belief. But yet, should it farther appear that this simple bird, when divested of that natural στοργή that seems to raise the kind in general above themselves, and inspire them with extraordinary degrees of cunning and address, may be still endued with a more enlarged faculty of discerning what species are suitable and congenerous nursing mothers for its disregarded eggs and young, and may deposit them only under their care, this would be adding wonder to wonder, and instancing, in a fresh manner, that the methods of Providence are not subjected to any mode or rule, but astonish us in new lights, and in various and changeable appearances.

What was said by a very ancient and sublime writer concerning the defect of natural affection in the ostrich, may be well applied to the bird we are talking of:—

"She is hardened against her young ones, as though they were not hers:

"Because God hath deprived her of wisdom, neither hath he imparted to her understanding."[1]

Query. Does each female cuckoo lay but one egg in a season, or does she drop several in different nests, according as opportunity offers?[2]

(1) Job xxix. 16, 17.   (2) It lays several eggs.—J. G. W.

LETTER V.

YELLOW-HAMMER—REED-SPARROW—EFFECTS OF FROST ON ANIMALS—SILENCE
OF SINGING BIRDS.

SELBORNE, *April* 12, 1770.

DEAR SIR,—I heard many birds of several species sing last
year after midsummer; enough to prove that the summer
solstice is not the period that puts a stop to the music of the

YELLOW-HAMMER.

woods.   The yellow-hammer,[1] no doubt, persists with more
steadiness than any other; but the woodlark, the wren, the
redbreast, the swallow, the whitethroat, the goldfinch,[2] the

(1) *Emberiza citrinella*—Yellow-hammer. Mr. Yarrell remarks that the name of
the bird ought to be Yellow-Ammer; the word "ammer" being the German term for
bunting. The nest of this bird is usually found on or near the ground, often in
a casual hole in a hay-field, and very well concealed. The eggs are five in number,
and quite unmistakeable; being scribbled all over with slender black lines. Sometimes, however, the lines are almost entirely wanting.—J. G. W.
(2) *Fringilla carduelis.* This pretty little bird is mostly found in those counties
where thistles most abound. Wiltshire is full of them. I have repeatedly seen
them in large flocks, flying along the little hedges that skirt the Wiltshire downs.
Its nest is a beautiful structure, usually placed on the extremity of a branch. I was
once beating the boughs of an oak-tree for caterpillars, when a goldfinch's egg fell
into the sheet that was held under the branch. So well, however, was the nest
concealed among the leaves, at the extremity of the branch, that even with the clue

GOLDFINCH.

common linnet, are all undoubted instances of the truth of what I advanced.

If this severe season does not interrupt the regularity of the summer migrations, the black-cap will be here in two or three days. I wish it was in my power to procure you one of those songsters; but I am no bird-catcher; and so little used to birds in a cage, that I fear, if I had one, it would soon die for want of skill in feeding.

Was your reed-sparrow, which you kept in a cage, the thick-billed reed-sparrow of the *Zoology*, p. 320; or was it the less reed-sparrow of Ray, the sedge-bird of Mr. Pennant's last publication, p. 16?

As to the matter of long-billed birds growing fatter in moderate frosts, I have no doubt within myself what should be the reason. The thriving at those times appears to me to arise altogether from the gentle check which the cold throws upon insensible perspiration. The case is just the same with blackbirds, &c.; and farmers and warreners observe, the first, that their hogs fat more kindly at such times; and the latter, that their rabbits are never in such good case

thus given, it could hardly be discovered. One singular point was, that it required a very hard blow indeed to send the eggs out of the nest; and after we had been battering for some time with a leaping-pole, one egg remained in the nest.

The goldfinch is almost the most docile of all the cage birds, and soon attaches itself to its owner.—J. G. W.

as in a gentle frost. But, when frosts are severe and of long continuance, the case is soon altered; for then a want of food soon overbalances the repletion occasioned by a checked perspiration. I have observed, moreover, that some human constitutions are more inclined to plumpness in winter than in summer.

When birds come to suffer by severe frost I find that the first that fail and die are the red-wing fieldfares, and then the song-thrushes.

You wonder, with good reason, that the hedge-sparrows, &c., can be induced at all to sit on the egg of the cuckoo, without being scandalised at the vast disproportioned size of

HEDGE-SPARROW. [1]

the supposititious egg; but the brute creation, I suppose, have very little idea of size, colour, or number.[2] For the common hen, I know, when the fury of incubation is on her, will sit on a single shapeless stone, instead of a nest full of eggs that have been withdrawn; and moreover, a hen turkey, in the same circumstances, would sit on in the empty nest till she perished with hunger.

(1) *Accentor modularius.*—J. G. W.
(2) As I have before mentioned, p. 165, note (1), the egg of the cuckoo is not larger than that of a sparrow.—J. G. W.

I think the matter might be easily determined whether a cuckoo lays one or two eggs, or more, in a season, by opening a female during the laying time. If more than one were come down out of the ovary, and advanced to a good size, doubtless, then, she would that spring lay more than one.

I will endeavour to get a hen, and examine.

Your supposition that there may be some natural obstruction in singing birds while they are mute, and that when this is removed the song recommences, is new and bold. I wish you could discover some good grounds for this suspicion.

I was glad you were pleased with my specimen of the *caprimulgus*, or fern-owl: you were, I find, acquainted with the bird before.

When we meet, I shall be glad to have some conversation with you concerning the proposal you make of my drawing up an account of the animals in this neighbourhood. Your partiality towards my small abilities persuades you, I fear, that I am able to do more than is in my power; for it is no small undertaking for a man, unsupported and alone, to begin a natural history from his own autopsia. Though there is endless room for observation in the field of nature, which is boundless, yet investigation (where a man endeavours to be sure of his facts) can make but slow progress; and all that one could collect in many years would go into a very narrow compass.

Some extracts from your ingenious "Investigations of the difference between the present temperature of the air in Italy," &c., have fallen in my way, and given me great satisfaction. They have removed the objection that always arose in my mind whenever I came to the passages which you quote. Surely the judicious Virgil, when writing a didactic poem for the region of Italy, could never think of describing freezing rivers, unless such severity of weather pretty frequently occurred!

P.S. Swallows appear amidst snows and frost.

LETTER VI.

SELBORNE, *May* 21, 1770.

DEAR SIR,—The severity and turbulence of last month so
interrupted the regular process of summer migration, that
some of the birds do but just begin to show themselves, and
others are apparently thinner than usual; as the whitethroat, the black-cap, the redstart, the fly-catcher. I well
remember that, after the very severe spring in the year
1739-40, summer birds of passage were very scarce. They
come probably hither with a south-east wind, or when it
blows between those points; but, in that unfavourable year,
the winds blew the whole spring and summer through from
the opposite quarters. And yet, amidst all these disadvan-
tages, two swallows, as I mentioned in my last, appeared
this year as early as the 11th of April, amidst frost and
snow; but they withdrew again for a time.

I am not pleased to find that some people seem so little
satisfied with Scopoli's new publication.[1] There is room to
expect great things from the hands of that man, who is a
good naturalist; and one would think that a history of the
birds of so distant and southern a region as Carniola would
be new and interesting. I could wish to see that work, and
hope to get it sent down. Dr. Scopoli is physician to the
wretches that work in the quicksilver mines of that district.

When you talked of keeping a reed-sparrow, and giving it
seeds, I could not help wondering; because the reed-sparrow
which I mentioned to you (*passer arundinaceus minor*, Raii)
is a soft-billed bird, and most probably migrates hence before
winter; whereas the bird you kept (*passer torquatus*, Raii)
abides all the year, and is a thick-billed bird. I question
whether the latter be much of a songster; but in this matter
I want to be better informed. The former has a variety of
hurrying notes, and sings all night. Some part of the song
of the former, I suspect, is attributed to the latter. We
have plenty of the soft-billed sort, which Mr. Pennant had

---

(1) This work he calls his "*Annus Primus Historico-Naturalis.*"

entirely left out of his *British Zoology,* till I reminded him of his omission. See *British Zoology* last published, p. 16.[1]

I have somewhat to advance on the different manners in which different birds fly and walk; but as this is a subject that I have not enough considered, and is of such a nature as not to be contained in a small space, I shall say nothing farther about it at present.[2]

No doubt the reason why the sex of birds in their first plumage is so difficult to be distinguished is, as you say, "because they are not to pair and discharge their parental functions till the ensuing spring." As colours seem to be the chief external sexual distinction in many birds, these colours do not take place till sexual attachments begin to obtain. And the case is the same in quadrupeds, among whom, in their younger days, the sexes differ but little: but, as they advance to maturity, horns and shaggy manes, beards and brawny necks, &c. &c., strongly discriminate the male from the female We may instance still farther in our own species, where a beard and stronger features are usually characteristic of the male sex; but this sexual diversity does not take place in earlier life; for a beautiful youth shall be so like a beautiful girl, that the difference shall not be discernible :—

> " Quem si puellarum insereres choro,
>   Mirè sagaces falleret hospites
>   Discrimen obscurum, solutis
>   Crinibus, ambiguove vultu."—HOR.

LETTER VII.

SCOPOLI'S "ANNUS PRIMUS"—PULVERATRICES—YOUNG FERN-OWL—CUCKOO —CROSSBEAK—WATER-OUSEL—RING-OUSEL—LAND-TORTOISE.

RINGMER, near LEWES, *Oct.* 8, 1770.

DEAR SIR,—I am glad to hear that Kuckalm is to furnish you with the birds of Jamaica. A sight of the *hirundines* of that hot and distant island would be a great entertainment to me.

The *Anni* of Scopoli are now in my possession; and I

(1) See Letter XXV. to Thomas Pennant, Esq.
(2) See Letter XLI. to the Hon. Daines Barrington.

have read the *Annus Primus* with satisfaction; for, though some parts of this work are exceptionable, and he may advance some mistaken observations, yet the ornithology of so distant a country as Carniola is very curious. Men that undertake only one district, are much more likely to advance natural knowledge, than those that grasp at more than they can possibly be acquainted with. Every kingdom, every province, should have its own monographer.

The reason, perhaps, why he mentions nothing of Ray's *Ornithology*, may be the extreme poverty and distance of his country, into which the works of our great naturalists may have never yet found their way. You have doubts, I know, whether this *Ornithology* is genuine, and really the work of Scopoli : as to myself, I think I discover strong tokens of authenticity ; the style corresponds with that of his *Entomology;* and his characters of his Ordines and Genera are many of them new, expressive, and masterly. He has ventured to alter some of the Linnæan genera, with sufficient show of reason.

It might, perhaps, be mere accident that you saw so many swifts and no swallows at Staines, because, in my long observation of those birds, I never could discover the least degree of rivalry or hostility between the species.

Ray remarks, that birds of the *gallinæ* order, as cocks and hens, partridges and pheasants, &c., are *pulveratrices,* such as dust themselves, using that method of cleansing their feathers, and ridding themselves of their vermin. As far as I can observe, many birds that dust themselves never wash; and I once thought that those birds that wash themselves would never dust : but here I find myself mistaken ; for common house-sparrows are great *pulveratrices,* being frequently seen grovelling and wallowing in dusty roads ; and yet they are great washers. Does not the skylark dust ?

Query. Might not Mahomet and his followers take one method of purification from these *pulveratrices?* because I find, from travellers of credit, that if a strict Mussulman is journeying in a sandy desert, where no water is to be found, at stated hours he strips off his clothes, and most scrupulously rubs his body over with sand or dust.

A countryman told me he had found a young fern-owl in

the nest of a small bird on the ground: and that it was fed by the little bird. I went to see this extraordinary pheno-menon, and found that it was a young cuckoo hatched in the nest of a titlark; it was become vastly too big for its nest, appearing

> "———————in tenui re
> Majores pennas nido extendisse."

> "Though by poverty depress'd,
> Spreading its wings beyond the nest;"

and was very fierce and pugnacious, pursuing my finger, as I teased it, for many feet from the nest, and sparring and buffeting with its wings like a game-cock. The dupe of a dam appeared at a distance, hovering about with meat in its mouth, and expressing the greatest solicitude.

In July, I saw several cuckoos swimming over a large pond; and found, after some observation, that they were feeding on the *libellulæ*, or dragon-flies, some of which they caught as they settled on the weeds, and some as they were on the wing. Notwithstanding what Linnæus says, I cannot be induced to believe that they are birds of prey.

CROSSBEAK [1]

This district affords some birds that are hardly ever heard of at Selborne. In the first place, considerable flocks of

(1) *Loxia curvirostra.*—J. G. W.

crossbeaks (*loxiæ curvirostræ*) have appeared this summer in the pine groves belonging to this house;[1] the water-ousel is said to haunt the mouth of the Lewes river, near New-

WATER-OUSEL.

haven,[2] and the Cornish chough builds, I know, all along the chalky cliffs of the Sussex shore.

I was greatly pleased to see little parties of ring-ousels

(1) The apparently awkward form of this bird's bill is admirably adapted fo the life which it leads. Its food consists chiefly of the seeds from fir-cones and these it obtains in a rather singular manner. It first cuts off the cone at th stem with its scissor-like beak, and while one foot holds the cone firmly agains a branch, it inserts both its mandibles beneath the scales, then by separating th mandibles, the scale is raised up while the seed is scooped out by the horny tip c the tongue. The bird has been seen to hold part of a cone in one foot, and peck i to pieces after the manner of the parrots. It can also shell almonds, and posses itself of the kernel, by a method exactly similar to that employed by the idle schoo boy, when he endeavours to beguile "the graver hours that bring constraint," b noiselessly opening nuts with his pocket-knife. The crossbill first pecks a hole i the end of the nut, and then by inserting its beak and giving a sudden wrench, th shell is split open.—J. G. W.

(2) *Hydrobata cinclus.* The conflicting statements that have been made respectin this bird are known to all naturalists, and to many general readers. One party declare that the dipper, or water-ousel, walks about under the water, because they have see it do so; while the other party as strenuously declares that the dipper does not wal under the water, because they have not seen it. So the disputed point is left sti open. Waterton and Mc Gillivray both unite in declaring that its movements unde the water cannot be called walking, but rather come under the designation of scu fling. The latter author, relating the movements of a dipper, which he was observin says, "It appeared to tumble about in a very extraordinary manner, with its hea downwards, as if picking something; and at the same time great exertion was use both by wings and legs." He further proceeds to observe that it is not likely walk under water, inasmuch as it never does so on land, but makes a kind of lea ing hop.—J. G. W.

(my newly-discovered migrators), scattered, at intervals, all along the Sussex downs from Chichester to Lewes. Let them come from whence they will, it looks very suspicious that they are cantoned all along the coast, in order to pass the Channel, when severe weather advances. They visit us again in April, as it should seem, in their return, and are not to be found in the dead of winter. It is remarkable that they are very tame, and seem to have no manner of apprehensions of danger from a person with a gun. There are bustards[1] on the wide downs near Brighthelmstone. No doubt you are acquainted with the Sussex downs. The prospects and rides round Lewes are most lovely.

As I rode along near the coast I kept a very sharp lookout in the lanes and woods, hoping I might, at this time of the year, have discovered some of the summer short-winged birds of passage crowding towards the coast, in order for their departure ; but it was very extraordinary that I never saw a redstart, whitethroat, black-cap, uncrested wren, flycatcher, &c. ; and I remember to have made the same remark in former years, as I usually come to this place annually about this time. The birds most common along the coast, at present, are the stone-chatters, whinchats, buntings, linnets, some few wheatears, titlarks, &c. Swallows and housemartins abound yet, induced to prolong their stay by this soft, still, dry season.

A land tortoise, which has been kept for thirty years in a little walled court belonging to the house where I am now visiting, retires under ground about the middle of November, and comes forth again about the middle of April. When it first appears in the spring, it discovers very little inclination towards food, but in the height of summer grows voracious, and then, as the summer declines, its appetite declines; so

(1) *Otus tarda.* See p. 122. This bird may now be considered almost extinct in England. Formerly it used to be tolerably common, and hunted with great zeal, more particularly as the chase of the bird required such caution and patience, as would sorely try the temper of even a practised deer-stalker. There is a very remarkable anatomical peculiarity connected with this bird. In the neck there is a large membranous pouch opening between the under side of the tongue and the lower mandible. The use of the pouch is not exactly ascertained, but it must perform some important part in the economy of the bird, as it is capable of holding six or seven pints of water. Only the male bird is furnished with this singular pouch. The eggs of the bustard are laid on the bare ground, no nest being made.—J. G. W.

that for the last six weeks in autumn it hardly eats at all. Milky plants, such as lettuces, dandelions, sow-thistles, are its favourite dish. In a neighbouring village one was kept till, by tradition, it was supposed to be an hundred years old, —an instance of vast longevity in such a poor reptile !

LETTER VIII.

CHAFFINCHES—WOODCOCKS—FIELDFARES—NIDIFICATION—RAIN.

SELBORNE, *Dec.* 20, 1770.

DEAR SIR,—The birds that I took for aberdavines were reed-sparrows (*passeres torquati*).

There are, doubtless, many home internal migrations within this kingdom that want to be better understood; witness those vast flocks of hen chaffinches that appear with us in the winter without hardly any cocks among them. Now were there a due proportion of each sex, it would seem very improbable that any one district should produce such numbers of these little birds, and much more when only one half of the species appears; therefore we may conclude that the *fringillæ cœlebes*, for some good purposes, have a peculiar migration of their own, in which the sexes part. Nor should it seem so wonderful that the intercourse of sexes in this species of birds should be interrupted in winter; since in many animals, and particularly in bucks and does, the sexes herd separately, except at the season when commerce is necessary for the continuance of the breed. For this matter of the chaffinches, see *Fauna Suecica*, p. 85, and *Systema Naturæ*, p. 318. I see every winter vast flights of hen chaffinches, but none of cocks.

Your method of accounting for the periodical motions of the British singing birds, or birds of flight, is a very probable one, since the matter of food is a great regulator of the actions and proceedings of the brute creation: there is but one that can be set in competition with it, and that is love. But I cannot quite acquiesce with you in one circumstance when you advance that, " When they have thus feasted, they again separate into small parties of five or six, and get the best fare they can within a certain district, having no inducement to go in quest of fresh-turned earth." Now, if you mean that the business of congregating is quite at an end from the conclusion of wheat-sowing to the season of barley and oats, it is not the case with us; for larks and chaffinches, and particularly linnets, flock and congregate as much in the very dead of winter as when the husbandman is busy with his ploughs and harrows.

Sure there can be no doubt but that woodcocks and fieldfares leave us in the spring, in order to cross the seas, and to retire to some districts more suitable to the purpose of breeding. That the former pair before they retire, and that the hens are forward with egg, I myself, when I was a sportsman, have often experienced. It cannot be denied but that now and then we hear of a woodcock's nest, or young birds, discovered in some part or other of this island; but then they are always mentioned as rarities, and somewhat out of the common course of things; but as to redwings and fieldfares, no sportsman or naturalist has ever yet, that I could hear, pretended to have found the nest or young of those species in any part of these kingdoms. And I the more admire at this instance as extraordinary, since, to all appearance, the same food, in summer as well as in winter, might support them here which maintains their congeners, the blackbirds and thrushes, did they choose to stay the summer through. From hence it appears, that it is not food alone which determines some species of birds with regard to their stay or departure. Fieldfares and redwings disappear sooner or later, according as the warm weather comes on earlier or later; for I well remember, after that dreadful winter, 1739-40, that cold north-east winds continued to blow on through April and May, and that these kinds of birds (what few remained of them) did not depart as usual, but were seen lingering about till the beginning of June.

The best authority that we can have for the nidification of the birds above mentioned, in any district, is the testimony of faunists that have written professedly the natural history of particular countries. Now, as to the fieldfare, Linnæus, in his *Fauna Suecica*, says of it, that "*maximis in arboribus*

nidificat;" "it builds in the largest trees." And of the
redwing, he says in the same place, that "*nidificat in mediis
arbusculis, sive sepibus: ova sex cœruleo-viridia, maculis nigris
variis.*" "It builds in the midst of shrubs or hedges; it
produces six eggs, of a sea-green colour, with varied black
spots." Hence we may be assured that fieldfares and redwings breed in Sweden. Scopoli says, in his *Annus Primus,*
of the woodcock, that "*nupta ad nos venit circa œquinoctium
vernale:*" "when mated, it comes to us about the vernal
equinox;" meaning in Tyrol, of which he is a native. And
afterwards he adds, "*nidificat in paludibus alpinis: ova
ponit* 3—5." "It builds its nest in the Alpine marshes,
and lays from three to five eggs." It·does not appear from
Kramer that woodcocks breed at all in Austria; but he says,
"*Avis hæc septentrionalium provinciarum œstivo tempore
incola est; ubi plerumque nidificat. Appropinquante hyeme
australiores provincias petit: hinc circa plenilunium potissimum mensis Octobris plerumque Austriam transmigrat. Tunc
rursus circa plenilunium potissimum mensis Martii per Austriam matrimonio juncta ad septentrionales provincias redit.*"
"This bird, in summer, inhabits the northern provinces
where it generally nests. On the approach of winter it seeks
more southern provinces; it usually leaves this at the October
full-moon, generally in the direction of Austria. Then it
returns back, after mating, generally about the March full
moon." For the whole passage (which I have abridged), see
Elenchus, &c., p. 351. This seems to be a full proof of the
emigration of woodcocks; though little is proved concerning
the place of their breeding.

P.S.—There fell in the county of Rutland, in three weeks
of this present very wet weather, seven inches and a half of
rain, which is more than has fallen in any three weeks for
these thirty years past, in that part of the world. A mean
quantity in that county for one year is twenty inches and
a half.

LETTER IX.

MIGRATION.

ENFIELD, near ANDOVER, *Feb.* 12, 1771.

DEAR SIR,—You are, I know, no great friend to migration; and the well-attested accounts from various parts of the kingdom, seem to justify you in your suspicions, that at least many of the swallow kind do not leave us in the winter, but lay themselves up, like insects and bats, in a torpid state, and slumber away the more uncomfortable months, till the return of the sun and fine weather awakens them.

But then we must not, I think, deny migration in general; because migration certainly does subsist in some places, as my brother in Andalusia has fully informed me. Of the motions of these birds he has ocular demonstration, for many

BEE-BIRD. [1]

weeks together, both spring and fall; during which periods myriads of the swallow kind traverse the Straits from north to south, and from south to north, according to the season. And these vast migrations consist not only of hirundines, but of bee-birds,[1] hoopoes, *oro pendolos*, or golden thrushes,

(1) *Merops apiaster*—the Bee-eater.—J. G. W.

GOLDEN-THRUSH.[1]

&c. &c., and also of many of our soft-billed summer birds of
passage; and moreover, of birds which never leave us, such
as all the various sorts of hawks and kites.   Old Belon,
two hundred years ago, gives a curious account of the incredible armies of hawks and kites which he saw in the spring
time traversing the Thracian Bosphorus, from Asia to
Europe.   Besides the above-mentioned, he remarks that
the procession is swelled by whole troops of eagles[2] and
vultures.

(1) *Oriolus galbula*, the golden thrush, or oriole. The name oriole is said to
be derived from the call-note of the bird. It is a rare visitant in this country, never
making its appearance except between spring and autumn.—J. G. W.

(2) The eagle stands first in the diurnal birds of prey. All the birds of prey, and
especially the eagles, possess great powers of vision; and can, moreover, alter the
focus of the eye at will, so as to suit various distances. The kestrel, for example,
can, from a lofty elevation in the air, espy so small and inconspicuous an animal
as the field-mouse; and yet never miscalculates its swoop upon the devoted
creature, although it lies quite close to the ground. It is evident, then, that the
focus of the eye must be altered, and if we search for the mode of alteration, we
shall find it beautifully displayed in the eye of the eagle. The orb of the eagle's
eye is supported and surrounded by fifteen very slight bony plates, so curved as to
bend round the eye. These plates are capable of movement one upon the other.
Now, if the eagle wishes to examine an object closely, it contracts the bony circle
which has the effect of compressing the edge of the lens, and consequently forcing
it into a more convex shape. If, on the contrary, it wishes to examine a far-distant

GOLDEN EAGLE [1]

Now, it is no wonder that birds residing in Africa should retreat before the sun as it advances, and retire to milder

object, it merely permits the bony plates to recede, and the lens loses part of its convexity. The same circumstance takes place, although not quite to such an extent, in man. If we have for some time been gazing at a distant landscape, and then suddenly turn our eyes to a book held near the face, we shall find that the eye requires a short time to elapse before it can distinguish the words in the book. In striving to read a book or to examine an object placed too near the eyes, we always find ourselves frowning with great ferocity; thereby contracting the orbit of the eye, and consequently making it more convex. The eye of the eagle is defended from the rays of the sun by an overhanging eyebrow.

There are two British eagles,—the Golden and the White-tailed or Cinereous Eagle, which are readily distinguished from each other by the foot; that of the Golden Eagle being covered with little reticulations as far as the toes, which are defended by three broad scales; whereas the foot of the White-tailed Eagle is entirely covered with broad scales.

In the younger state of the Golden Eagle, the plumage differs from that of the adult bird. The young Golden Eagle has been described as the Ring-tailed Eagle. —J. G. W.

(1) *Aquila chrysaëtos.* J. G. W.

regions, and especially birds of prey, whose blood being
heated with hot animal food, are more impatient of a sultry
climate; but then I cannot help wondering why kites and
hawks, and such hardy birds as are known to defy all the
severity of England, and even of Sweden and all north
Europe, should want to migrate from the south of Europe,
and be dissatisfied with the winters of Andalusia.

It does not appear to me that much stress may be laid on
the difficulty and hazard that birds must run in their migrations, by reason of vast oceans, cross winds, &c.: because, if
we reflect, a bird may travel from England to the Equator
without launching out and exposing itself to boundless seas,
and that by crossing the water at Dover, and again at
Gibraltar. And I with the more confidence advance this
obvious remark, because my brother has always found that
some of his birds, and particularly the swallow kind, are
very sparing of their pains in crossing the Mediterranean;
for when arrived at Gibraltar, they do not,

> "————— Ranged in figure, wedge their way,
> ————————— and set forth
> Their airy caravan, high over seas
> Flying, and over lands with mutual wing
> Easing their flight;"   MILTON.

but scout and hurry along in little detached parties of six or
seven in a company; and, sweeping low, just over the surface of the land and water, direct their course to the opposite
continent at the narrowest passage they can find. They
usually slope across the bay to the south-west, and so pass
over opposite to Tangier, which, it seems, is the narrowest
space.

In former letters, we have considered, whether it was
probable that woodcocks, in moonshiny nights, cross the
German Ocean from Scandinavia. As a proof that birds of
less speed may pass that sea, considerable as it is, I shall
relate the following incident, which, though mentioned to
have happened so many years ago, was strictly matter of
fact:—As some people were shooting in the parish of
Trotten, in the county of Sussex, they killed a duck in that
dreadful winter, 1708-9, with a silver collar about its neck,

(1) I have read a like anecdote of a swan.

on which were engraven the arms of the King of Denmark. This anecdote the Rector of Trotten at that time has often told to a near relation of mine, and to the best of my remembrance, the collar was in the possession of the rector.

At present, I do not know anybody near the sea-side that will take the trouble to remark at what time of the moon woodcocks first come : if I lived near the sea myself, I would soon tell you more of the matter. One thing I used to observe when I was a sportsman, that there were times in which woodcocks were so sluggish and sleepy, that they would drop again when flushed just before the spaniels, nay, just at the muzzle of a gun that had been fired at them : whether this strange laziness was the effect of a recent fatiguing journey, I shall not presume to say.

Nightingales not only never reach Northumberland and Scotland, but also, as I have been always told, Devonshire and Cornwall. In those two last counties, we cannot attribute the failure of them to the want of warmth : the defect in the west is rather a presumptive argument that these birds come over to us from the continent at the narrowest passage, and do not stroll so far westward.

Let me hear from your own observation whether skylarks do not dust. I think they do; and if they do, whether they wash also.

The *alauda pratensis* of Ray was the poor dupe that was educating the booby of a cuckoo mentioned in my letter of October last.[1]

Your letter came too late for me to procure a ring-ousel for Mr. Tunstal during their autumnal visit ; but I will endeavour to get him one when they call on us again in April. I am glad that you and that gentleman saw my Andalusian birds ; I hope they answered your expectation. Royston, and grey crows, are winter birds that come much about the same time with the woodcock : they, like the fieldfare or redwing, have no apparent reason for migration ; for, as they fare in the winter like their congeners, so might they, in all appearance, in the summer.[2] Was not

(1) Letter VII. to the Hon. Daines Barrington.

(2) *Coracia gracula.* See p. 132. The chough is distinguished from the true rows by the peculiar shape of its bill. It is a sea-coast bird, and lives on the most

Tennant, when a boy, mistaken? Did he not find a missel-thrush's nest, and take it for the nest of a fieldfare?

The stock-dove or wood-pigeon, *œnas* Raii, is the last winter bird of passage which appears with us, and is not seen till towards the end of November. About twenty years ago, they abounded in the district of Selborne, and strings of them were seen morning and evening that reached a mile or more; but since the beechen woods have been greatly thinned, they have much decreased in number. The ring-dove, *palumbus* Raii, stays with us the whole year, and breeds several times through the summer.

Before I received your letter of October last, I had just remarked in my journal that the trees were unusually green. This uncommon verdure lasted on late into November, and may be accounted for from a late spring, a cool and moist summer, but more particularly from vast armies of chaffers, or tree-beetles, which, in many places, reduced whole woods to a leafless naked state. These trees shot again at mid-summer, and then retained their foliage till very late in the year.

My musical friend, at whose house I am now visiting, has tried all the owls that are his near neighbours, with a pitch-pipe set at concert pitch, and finds they all hoot in B flat. He will examine the nightingales next spring.

inaccessible rocks that it can find. It feeds principally on the smaller insects, but does not despise some of the larger and softer insects; such as the grasshopper and the cockchaffer, which last he picks to pieces, as has been related of the jackdaw, p. 88. Now and then a specimen has been found inland; possibly blown there, as is sometimes the case with other sea-birds. The common gull, for instance, is not unfrequently found at Oxford. I have seen it sitting on a bridge, over a branch of the Isis, and looking quite out of its element.

It is a curious fact, that the chough has a great objection to walking upon grass; and will never do so, unless from some very strong temptation.—J. G. W.

LETTER X.

HOOTING OF OWLS—MIGRATION—WOODCOCKS.

SELBORNE, *Aug.* 1, 1771.

DEAR SIR,—From what follows, it will appear that neither owls nor cuckoos keep to one note. A friend remarks that many (most) of his owls hoot in B flat; but that one went almost half a note below A. The pipe he tried their notes by was a common half-crown pitch-pipe, such as masters use for tuning of harpsichords; it was the common London pitch.

A neighbour of mine, who is said to have a nice ear, remarks that the owls about this village hoot in three different keys, in G flat or F sharp, in B flat, and A flat. He heard two hooting to each other, the one in A flat, and the other in B flat. *Query:* Do these different notes proceed from different species, or only from various individuals? The same person finds, upon trial, that the note of the cuckoo (of which we have but one species) varies in different individuals; for, about Selborne wood, he found they were mostly in D; he heard two sing together, the one in D, and the other in D sharp, which made a disagreeable concert; he afterwards heard one in D sharp, and, about Wolmer Forest, some in C. As to nightingales, he says, that their notes are so short, and their transitions so rapid, that he cannot well ascertain their key. Perhaps in a cage, and in a room, their notes may be more distinguishable. This person has tried to settle the notes of a swift, and of several other small birds, but cannot bring them to any criterion.

As I have often remarked that redwings are some of the first birds that suffer with us in severe weather, it is no wonder at all that they retreat from Scandinavian winters; and much more the *ordo* of *grallæ,* who all, to a bird, forsake the northern parts of Europe at the approach of winter. "*Grallæ tanquam conjuratæ unanimiter in fugam se conjiciunt; ne earum unicam quidem inter nos habitantem invenire possimus; ut enim æstate in australibus degere nequeunt ob*

REDWING. [1]

defectum lumbricorum, terramque siccam; ita nec in frigidis ob eandem causam." "The grallæ, as if by agreement, take flight, nor can we find one residing here; for as, during summer, the deficiency of earth-worms and the hardness of the ground prevents them from abiding in hot countries; so neither can they dwell in cold climes, for the same reason," says Ekmarck the Swede, in his ingenious little treatise called *Migrationes Avium*, which, by all means, you ought to read, while your thoughts run on the subject of migration.— See *Amœnitates Academicæ*, vol. iv. p. 565.

Birds may be so circumstanced as to be obliged to migrate in one country, and not in another; but the *grallæ* (which procure their food from marshes and boggy ground) must, in winter, forsake the more northerly parts of Europe, or perish for want of food.

I am glad you are making inquiries from Linnæus concerning the woodcock; it is expected of him that he should

(1) *Turdus iliacus.* The redwing arrives in this country rather before the fieldfare. They cannot sustain the cold as well as that bird. The song of the redwing is so delightful, that Linnæus compared it with that of the nightingale.—J. G. W.

be able to account for the motions and manner of life of the animals of his own *Fauna*.

Faunists, as you observe, are too apt to acquiesce in bare descriptions, and a few synonymes: the reason is plain, because all that may be done at home in a man's study; but the investigation of the life and conversation of animals is a concern of much more trouble and difficulty, and is not to be attained but by the active and inquisitive, and by those that reside much in the country.

Foreign systematists are, I observe, much too vague in their specific differences; which are almost universally constituted by one or two particular marks, the rest of the description running in general terms. But our countryman, the excellent Mr. Ray, is the only describer that conveys some precise idea in every term or word, maintaining his superiority over his followers and imitators, in spite of the advantage of fresh discoveries and modern information.

At this distance of years, it is not in my power to recollect at what periods woodcocks used to be sluggish or alert, when I was a sportsman; but, upon my mentioning this circumstance to a friend, he thinks he has observed them to be remarkably listless against snowy, foul weather; if this should be the case, then the inaptitude for flying arises only from an eagerness for food, as sheep are observed to be very intent on grazing against stormy wet evenings.

LETTER XI.

CONGREGATING OF BIRDS.

SELBORNE, *Feb.* 8, 1772.

DEAR SIR,—When I ride about in winter and see such prodigious flocks of various kinds of birds, I cannot help admiring at these congregations, and wishing that it was in my power to account for those appearances, almost peculiar to the season. The two great motives which regulate the proceedings of the brute creation are love and hunger; the former incites animals to perpetuate their kind, the latter induces them to preserve individuals. Whether either of these should seem to be the ruling passion in the matter of congregating, is to be considered. As to love, that is out of the question at a time of the year when that soft passion is not indulged; besides, during the amorous season, such a jealousy prevails between the male birds, that they can hardly bear to be together in the same · hedge or field. Most of the singing and elation of spirits at that time seem to me to be the effect of rivalry and emulation; and it is to this spirit of jealousy that I chiefly attribute the equal dispersion of birds in the spring over the face of the country.

Now as to the business of food. As these animals are actuated by instinct to hunt for necessary food, they should not, one would suppose, crowd together in pursuit of sustenance, at a time when it is most likely to fail; yet such associations do take place in hard weather chiefly, and thicken as the severity increases. As some kind of self-interest and self-defence is, no doubt, the motive for the proceeding, may it not arise from the helplessness of their state in such rigorous seasons; as men crowd together, when under great calamities, though they know not why? Perhaps approximation may dispel some degree of cold: and a crowd may make each individual appear safer from the ravages of birds of prey, and other dangers.

If I admire when I see how much congenerous birds love to congregate, I am the more struck when I see incongruous ones in such strict amity. If we do not much wonder to see a flock of rooks usually attended by a train of daws, yet it is strange that the former should so frequently have a flight of starlings for their satellites. Is it because rooks have a more discerning scent than their attendants, and can lead them to spots more productive of food? Anatomists say that rooks, by reason of two large nerves which run down between the eyes into the upper mandible, have a more delicate feeling in their beaks than other round-billed birds, and can grope for their meat when out of sight. Perhaps, then, their associates attend them on the motive of interest, as greyhounds wait on the motions of their finders, and as lions are said to do on the yelpings of jackals. Lapwings and starlings sometimes associate.

LETTER XII.

March 9, 1772.

DEAR SIR,—As a gentleman and myself were walking, on the 4th of last November, round the sea-banks at Newhaven, near the mouth of the Lewes river, in pursuit of natural knowledge, we were surprised to see three house swallows gliding very swiftly by us. That morning was rather chilly, with the wind at north-west; but the tenor of the weather for some time before had been delicate, and the noons remarkably warm. From this incident, and from repeated accounts which I meet with, I am more and more induced to believe that many of the swallow kind do not depart from this island, but lay themselves up in holes and caverns, and do, insect-like and bat-like, come forth at mild times, and then retire again to their *latebræ*, or lurking-places. Nor make I the least doubt but that, if I lived at Newhaven, Seaford, Brighthelmstone, or any of those towns near the chalk cliffs of the Sussex coast, by proper observations, I should see swallows stirring at periods of the winter, when the noons were soft and inviting, and the sun warm and invigorating. And I am the more of this opinion, from what I have remarked during some of our late springs; and though some swallows did make their appearance about the usual time, viz. the 13th or 14th of April, yet, meeting with an harsh reception, and blustering, cold, north-east winds, they immediately withdrew, absconding for several days, till the weather gave them better encouragement.

LETTER XIII.

TORTOISE.

April 12, 1772.

DEAR SIR,—While I was in Sussex last autumn, my residence was at the village near Lewes, from whence I had formerly the pleasure of writing to you. On the 1st of November,
I remarked that the old tortoise, formerly mentioned, began
first to dig the ground, in order to the forming of its hybernaculum, which it had fixed on just beside a great turf of
hepaticas. It scrapes out the ground with its forefeet, and

TORTOISE.

throws it up over its back with its hind; but the motion of
its legs is ridiculously slow, little exceeding the hour-hand
of a clock, and suitable to the composure of the animal.
Nothing can be more assiduous than this creature, night and
day, in scooping the earth, and forcing its great body into the
cavity; but, as the noons of that season proved unusually

warm and sunny, it was continually interrupted, and called forth by the heat, in the middle of the day ; and though I continued there till the 13th of November, yet the work remained unfinished. Harsher weather and frosty mornings would have quickened its operations. No part of its behaviour ever struck me more than the extreme timidity it always expresses with regard to rain ; for though it has a shell that would secure it against the wheel of a loaded cart, yet does it discover as much solicitude about rain as a lady dressed in all her best attire, shuffling away on the first sprinklings, and running its head up in a corner. If attended to, it becomes an excellent weather-glass ; for as sure as it walks elate, and as it were on tiptoe, feeding with great earnestness in a morning, so sure will it rain before night. It is totally a diurnal animal, and never attempts to stir after it becomes dark. The tortoise, like other reptiles, has an arbitrary stomach, as well as lungs ; and can refrain from eating as well as breathing for a great part of the year. When first awakened, it eats nothing ; nor again in the autumn, before it retires : through the height of the summer it feeds voraciously, devouring all the food that comes in its way. I was much taken with its sagacity in discerning those that do it kind offices ; for, as soon as the good old lady comes in sight who has waited on it for more than thirty years, it hobbles towards its benefactress with awkward alacrity, but remains inattentive to strangers. Thus not only " the ox knoweth his owner, and the ass his master's crib,"[1] but the most abject reptile and torpid of beings distinguishes the hand that feeds it, and is touched with the feelings of gratitude.

P.S.—In about three days after I left Sussex, the tortoise retired into the ground under the hepatica.[2]

(1) Isaiah i. 3.
(2) Some time since, a man arrived in Oxford, bringing with him tortoises for sale. They passed their existence in a basket, where they were packed close, like so many bricks, standing on their tails, and their heads looking out of the basket. When I purchased one of them, the man emptied out his whole basketful upon the table, and then turned out the contents of four large pockets, until a large table was entirely covered with them.

The tortoise which I purchased was a very small one, and was tolerably lively, walking about the room, and always settling on the hearthrug. It had a great genius for climbing, and would sometimes spend nearly an hour in endeavouring to scale

LETTER XIV.

NATURAL AFFECTION OF ANIMALS.

SELBORNE, *March* 26, 1773.

DEAR SIR,—The more I reflect on the στοργὴ, or natural affection of animals, the more I am astonished at its effects. Nor is the violence of its affection more wonderful than the shortness of its duration. Thus every hen is in her turn the virago of the yard, in proportion to the helplessness of her brood; and will fly in the face of a dog or a sow in defence of those chickens, which in a few weeks she will drive before her with relentless cruelty.[1]

the fender, probably attracted by the heat. Unfit as the form of the creature may seem for such a purpose, it did contrive to scramble upon a footstool which was placed by the fender. Its method of attaining this elevation was as follows :—First it reared up against the footstool in the angle formed by it and the fender, and after several ineffectual attempts, succeeded in hitching the claws of one of its hind feet into the open work of the fender. On this it raised itself, and held on to the top of the stool by its fore feet, while it gained another step on the fender, and so managed to raise itself to such a height, that it only had to fall flat on the top of the footstool. When once there, it could hardly be induced to leave the elevation which it had gained with such difficulty.

Its food consisted of bread and milk, which it ate several times a-day, drinking the milk by scooping up some of it in its lower jaw, and then, by throwing its head back, the milk ran down its throat. Tortoises are generally long-lived, but this animal died within a few months after it came into my possession, in all probability because, for some days, its food was placed in a brass vessel.

Several days before its death it was very restless, and went about the room mewing like a young kitten, and made such a noise, that it had to be ejected during working hours. I could not for some time believe that the mewing could proceed from the tortoise, as the resemblance to that of a kitten was most exact.—J. G. W.

(1) I have known a common hedge-sparrow when sitting on the nest permit herself to be approached quite closely; nor would she leave her nest until the hand was placed close to her. I several times almost pushed her off her nest, examined the eggs, and replaced them, and in a few minutes saw the mother return to her post. I regret to say that some cruel boy destroyed both nest and young, when the latter were but a few days old.

A very affecting example of the attachment of the sheep to its young is told by the Ettrick Shepherd.—"I requested of my master to spare me a lamb for a ewe which he knew, and which was standing over a dead lamb in the end of the Hope about four miles from the house. He would not let me do it, but bid me let her stand over her lamb for a day or two, and perhaps a twin would be forthcoming. I did so; and faithfully did she stand to her charge. I visited her every morning and evening for the first eight days, and never found her above two or three yards

This affection sublimes the passions, quickens the invention, and sharpens the sagacity of the brute creation. Thus an hen, just become a mother, is no longer that placid bird she used to be; but, with feathers standing on end, wings hovering, and clucking note, she runs a out like one possessed. Dams will throw themselves in the way of the greatest danger in order to avert it from their progeny. Thus a partridge will tumble along before a sportsman, in order to draw away the dogs from her helpless covey. In the time of nidification, the most feeble birds will assault the most rapacious. All the *hirundines* of a village are up in arms at the sight of an hawk, whom they will persecute till he leaves that district. A very exact observer has often remarked, that a pair of ravens, nesting in the rock of Gibraltar, would suffer no vulture or eagle to rest near their station, but would drive them from the hill with an amazing fury: even the blue thrush, at the season of breeding, would dart out from the clefts of the rocks to chase away the kestrel or the sparrow-hawk. If you stand near the nest of a bird that has young, she will not be induced to betray them by an inadvertent fondness, but will wait about at a distance, with meat in her mouth, for an hour together.

Should I farther corroborate what I have advanced above by some anecdotes which I probably may have mentioned before in conversation, yet you will, I trust, pardon the repetition for the sake of the illustration.

The fly-catcher of the *Zoology* (the *stoparola* of Ray) builds every year in the vines that grow on the walls of my house.[1] A pair of these little birds had one year inadvertently placed

---

rom the lamb; and often as I went my rounds, she eyed me long ere I came near her, and kept stamping with her foot, and whistling through her nose, to frighten the dog way. He got a regular chase twice a-day, but however excited and fierce an ewe nay be, she never offers any resistance to mankind, being perfectly and meekly assive to them. The weather grew fine and warm, and the dead lamb soon decayed: ut still this affectionate and resolute creature kept hanging over the poor remains with an attachment that seemed to be nourished by hopelessness. It often drew ears from my eyes to see her hanging with such fondness over a few bones, mixed rith a small portion of wool. For the first fortnight she never quitted the spot; and or another week she visited it every morning and evening, uttering a few kindly and eart-piercing bleats, till at length every remnant of the offspring vanished, mixing rith the soil, or wafted away by the winds."—J. G. W.

(1) *Muscicapa grisola*, Linn.—J. G. W.

their nest on a naked bough, perhaps in a shady time, not being aware of the inconvenience that followed ; but an hot sunny season coming on before the brood was half fledged, the reflection of the wall became insupportable, and must inevitably have destroyed the tender young, had not affection suggested an expedient, and prompted the parent birds to hover over the nest all the hotter hours, while, with wings expanded, and mouths gaping for breath, they screened off the heat from their suffering offspring.

A farther instance I once saw of notable sagacity in a willow-wren. which had built in a bank in my fields. This bird a friend and myself had observed as she sat in her nest, but were particularly careful not to disturb her, though we saw she eyed us with some degree of jealousy. Some days after, as we passed that way, we were desirous of remarking how this brood went on; but no nest could be found, till I happened to take up a large bundle of long green moss, as it were carelessly thrown over the nest, in order to dodge the eye of any impertinent intruder.

A still more remarkable mixture of sagacity and instinct occurred to me one day, as my people were pulling off the lining of a hot-bed in order to add some fresh dung. From out of the side of this bed leaped an animal with great agility that made a most grotesque figure; nor was it without great difficulty that it could be taken, when it proved to be a large white-bellied field-mouse, with three or four young clinging to her teats by their mouths and feet. It was amazing that the desultory and rapid motions of this dam should not oblige her litter to quit their hold, especially when it appeared that they were so young as to be both naked and blind !

To these instances of tender attachment, many more of which might be daily discovered by those that are studious of nature, may be opposed that rage of affection, that monstrous perversion of the στοργή, which induces some females of the brute creation to devour their young, because their owners have handled them too freely, or removed them from place to place ! Swine, and sometimes the more gentle race of dogs and cats, are guilty of this horrid and preposterous murder. When I hear now and then of an abandoned mother that destroys her offspring, I am not so much amazed; since reason perverted, and the bad passions let loose, are capable of any enormity; but why the parental feelings of brutes, that usually flow in one most uniform tenor, should sometimes be so extravagantly diverted, I leave to abler philosophers than myself to determine.[1]

***

LETTER XV.

TEALS—WHITE OWLS—SWALLOWS—INSECTS ON BIRDS.

SELBORNE, *July* 8, 1773.

DEAR SIR,—Some young men went down lately to a pond on the verge of Wolmer Forest to hunt flappers, or young wild ducks, many of which were caught, and among the rest, some very minute yet well-fledged wild fowls alive, which, upon examination, I found to be teals. I did not know till then that teals ever bred in the south of England, and was much pleased with the discovery: this I look upon as a great stroke in natural history.

We have had, ever since I can remember, a pair of white owls that constantly breed under the eaves of this church. As I have paid good attention to the manner of life of these birds during their season of breeding, which lasts the summer through, the following remarks may not perhaps be unacceptable. About an hour before sunset (for then the mice begin to run) they sally forth in quest of prey, and hunt all round the hedges of meadows and small inclosures for them, which seem to be their only food. In this irregular country we can stand on an eminence, and see them beat the fields over like a setting-dog, and often drop down in the grass or corn. I have minuted these birds with my watch for an hour together, and have found that they return to their nests, the one or the other of them, about once in five minutes; reflecting, at the same time, on the adroitness that every

(1) This is not unfrequently the case with rabbits, at all events with tame rabbits. The dam will sometimes eat her young, if they have been disturbed during the first few days of their life.—J. G. W.

animal is possessed of, as far as regards the well-being of itself and offspring. But a piece of address which they show when they return loaded, should not, I think, be passed over in silence. As they take their prey with their claws, so they carry it in their claws to their nest; but as their feet are necessary in their ascent under the tiles, they constantly perch first on the roof of the chancel, and shift the mouse from their claws to their bill, that the feet may be at liberty to take hold of the plate on the wall, as they are rising under the eaves.

White owls seem not (but in this I am not positive) to hoot at all;[1] all that clamorous hooting appears to me to come from the wood kinds. The white owl does indeed snore and hiss in a tremendous manner; and these menaces well answer the intention of intimidating; for I have known a whole village up in arms on such an occasion, imagining the churchyard to be full of goblins and spectres. White owls also often scream horribly as they fly along: from this screaming, probably, arose the common people's imaginary species of screech-owl, which they superstitiously think attends the windows of dying persons. The plumage of the remiges of the wings of every species of owl that I have yet examined,

(1) *Strix flammea*. Jardine asserts that he has shot them while in the act. In the nest of the white or barn owl, both young and eggs are frequently found at the same time, the eggs being apparently hatched by the warmth derived from the young birds. The young remain in the nest for a much longer period than those of other birds. The owl has been seen to capture fish like the osprey. Like other birds of prey, it disgorges the bones, fur, &c. of its prey, in the form of pellets. Some which I have examined were principally composed of the bones and fur of mice, interspersed with the elytra or wing-cases of beetles, mostly Agonum, Abax Steropus, and here and there a Carabus.

When the owl is attacked, it throws itself on its back, and defends itself most vigorously with its claws and beak. When alarmed or angry, it expresses its displeasure by a smart snapping of the beak.

A barn owl in my possession lived for several weeks in a hamper, being fed upon mice, small birds, &c. After some time, as it appeared to require more air and exercise, I chained it to a post by a long light chain, fastened to a ring round its leg. This, however, proved the death of the poor owl, for it pulled so hard at the chain and was so restless, that it cut its leg to the bone, completely dividing the skin, and causing the leg to swell to such an extent, that the ring could only be got off by breaking it. The chain was immediately transferred to the other leg, being fastened to a band of soft leather instead of a ring, but the remedy came too late, for the owl's appetite failed, then became debased, and it finally died from the injury caused to its interior by a great pebble, which it had contrived to swallow.— J. G. W.

is remarkably soft and pliant. Perhaps it may be necessary that the wings of these birds should not make much resistance or rushing, that they may be able to steal through the air unheard upon a nimble and watchful quarry.

While I am talking of owls it may not be improper to mention what I was told by a gentleman of the county of Wilts. As they were grubbing a vast hollow pollard ash, that had been the mansion of owls for centuries, he discovered at the bottom a mass of matter that at first he could not account for. After some examination, he found that it was a congeries of the bones of mice, (and perhaps of birds and bats,) that had been heaping together for ages, being cast up in pellets out of the crops of many generations of inhabitants. For owls cast up the bones, fur, and feathers of what they devour, after the manner of hawks. He believes, he told me, that there were bushels of this kind of substance.

When brown owls hoot, their throats swell as big as an hen's egg. I have known an owl of this species live a full year without any water. Perhaps the case may be the same with all birds of prey. When owls fly, they stretch out their legs behind them, as a balance to their large heavy heads; for, as most nocturnal birds have large eyes and ears, they must have large heads to contain them. Large eyes, I presume, are necessary to collect every ray of light, and large concave ears to command the smallest degree of sound or noise.

The *hirundines* are a most inoffensive, harmless, entertaining, social, and useful tribe of birds; they touch no fruit in our gardens; delight, all except one species, in attaching themselves to our houses; amuse us with their migrations, songs, and marvellous agility; and clear our outlets from the annoyances of gnats and other troublesome insects. Some districts in the South Seas, near Guiaquil,[1] are desolated, it seems, by the infinite swarms of venomous mosquitoes, which fill the air, and render those coasts insupportable. It would be worth inquiring, whether any species of *hirundines* is found in these regions. Whoever contemplates the myriads of insects that sport in the sunbeams of a summer evening in

(1) See Ulloa's *Travels.*

this country, will soon be convinced to what a degree our atmosphere would be choked with them were it not for the friendly interposition of the swallow tribe.

Many species of birds have their peculiar lice : but the *hirundines* alone seem to be annoyed with *dipterous* insects, which infest every species, and are so large, in proportion to themselves, that they must be extremely irksome and injurious to them.   These are the *hippoboscæ hirundinis*,[1] with narrow subulated wings, abounding in every nest; and are hatched by the warmth of the bird's own body during incubation, and crawl about under its feathers.

A species of them is familiar to horsemen in the south of England, under the name of forest-fly, and, to some, of side-fly, from its running sideways, like a crab.[2]   It creeps under the tails and about the groins of horses, which, at their first coming out of the north, are rendered half-frantic by the tickling sensation; while our own breed little regards them.

The curious Reaumur discovered the large eggs, or rather *pupæ*, of these flies, as big as the flies themselves, which he hatched in his own bosom.   Any person that will take the trouble to examine the old nests of either species of swallows, may find in them the black shining cases, or skins, of the *pupæ* of these insects; but, for other particulars, too long for this place, we refer the reader to *L'Histoire d'Insectes* of that admirable entomologist, tom. iv. pl. 11.

LETTER XVI.

HOUSE-MARTIN.

SELBORNE, *Nov.* 20, 1773.

DEAR SIR,—In obedience to your injunctions, I sit down to give you some account of the house-martin, or martlet;[3] and if my monography of this little domestic and familiar bird should happen to meet with your approbation, I may probably

(1) *Craterina hirundinis*, Olf. This insect, together with the forest fly, is no dipterous, but belongs to the order *Homaloptera*, McLeay.—J. G. W.
(2) *Hippobosca equina*, Linn.—J. G. W.
(3) *Chelidon urbica.* See p. 95.—J. G. W.

soon extend my inquiries to the rest of the British *hirundines,*—the swallow, the swift, and the bank-martin.

A few house-martins begin to appear about the 16th of April; usually some few days later than the swallow. For some time after they appear, the *hirundines* in general pay no attention to the business of nidification, but play and sport about, either to recruit from the fatigue of their journey, if they do migrate at all, or else that their blood may recover its true tone and texture after it had been so long benumbed by the severities of winter. About the middle of May, if the weather be fine, the martin begins to think in earnest of providing a mansion for its family. The crust or shell of this nest seems to be formed of such dirt or loam as comes most readily to hand, and is tempered and wrought together with little bits of broken straws, to render it tough and tenacious. As this bird often builds against a perpendicular wall, without any projecting ledge under it, it requires its utmost efforts to get the first foundation firmly fixed, so that it may safely carry the superstructure. On this occasion the bird not only clings with its claws, but partly supports itself by strongly inclining its tail against the wall, making that a fulcrum; and thus steadied, it works and plasters the materials into the face of the brick or stone. But then, that this work may not, while it is soft and green, pull itself down by its own weight, the provident architect has prudence and forbearance enough not to advance her work too fast; but, by building only in the morning, and by dedicating the rest of the day to food and amusement, gives it sufficient time to dry and harden. About half an inch seems to be a sufficient layer for a day. Thus, careful workmen, when they build mud-walls (informed at first, perhaps, by these little birds), raise but a moderate layer at a time, and then desist, lest the work should become top-heavy, and so be ruined by its own weight. By this method, in about ten or twelve days, is formed an hemispheric nest, with a small aperture towards the top,—strong, compact, and warm, and perfectly fitted for all the purposes for which it was intended. But then, nothing is more common than for the house-sparrow, as soon as the shell is finished, to seize on it as its own, to eject the owner, and to line it after its own manner.

After so much labour is bestowed in erecting a mansion, as Nature seldom works in vain, martins will breed on, for several years together, in the same nest, where it happens to be well sheltered and secure from the injuries of weather. The shell, or crust, of the nest is a sort of rustic work full of knobs and protuberances on the outside: nor is the inside of those that I have examined smoothed with any exactness at all; but is rendered soft and warm, and fit for incubation, by a lining of small straws, grasses, and feathers ; and sometimes by a bed of moss interwoven with wool. In this nest, they tread, or engender, frequently during the time of building ; and the hen lays from three to five eggs.

At first, when the young are hatched, and are in a naked and helpless condition, the parent birds, with tender assiduity, carry out what comes away from their young. Were it not for this affectionate cleanliness, the nestlings would soon be burnt up and destroyed, in so deep and hollow a nest, by their own caustic excrement. In the quadruped creation, the same neat precaution is made use of ; particularly among dogs and cats, where the dams lick away what proceeds from their young. But, in birds, there seems to be a particular provision, that the dung of nestlings is enveloped in a tough kind of jelly, and, therefore, is the easier conveyed off, without soiling and daubing. Yet, as Nature is cleanly in all her ways, the young perform this office for themselves in a little time by thrusting their tails out at the aperture of their nest. As the young of small birds presently arrive at their ἡλικία, or full growth, they soon become impatient of confinement, and sit all day with their heads out at the orifice, where the dams, by clinging to the nest, supply them with food from morning to night. For a time, the young are fed on the wing by their parents: but the feat is done by so quick and almost imperceptible a sleight, that a person must have attended very exactly to their motions, before he would be able to perceive it. As soon as the young are able to shift for themselves, the dams immediately turn their thoughts to the business of a second brood : while the first flight, shaken off and rejected by their nurses, congregate in great flocks, and are the birds that are seen clustering and hovering, on sunny mornings and evenings, round towers and steeples, and on the roofs of churches and houses. These congregations usually begin to take place about the first week in August ; and, therefore, we may conclude that, by that time, the first flight is pretty well over. The young of this species do not quit their abodes altogether ; but the more forward birds get abroad some days before the rest. These, approaching the eaves of buildings, and playing about before them, make people think that several old ones attend one nest. They are often capricious in fixing on a nesting-place, beginning many edifices, and leaving them unfinished ; but, when once a nest is completed in a sheltered place, it serves for several seasons. Those which breed in a ready-finished house get the start, in hatching, of those that build new, by ten days or a fortnight. These industrious artificers are at their labours in the long days before four in the morning : when they fix their materials, they plaster them on with their chins, moving their heads with a quick vibratory motion. They dip and wash as they fly sometimes, in very hot weather, but not so frequently as swallows. It has been observed, that martins usually build to a north-east or north-west aspect, that the heat of the sun may not crack and destroy their nests : but instances are also remembered where they bred for many years in vast abundance in an hot stifled inn-yard, against a wall facing to the south.

Birds in general are wise in their choice of situation ; but, in this neighbourhood, every summer, is seen a strong proof to the contrary, at an house without eaves, in an exposed district, where some martins build, year by year, in the corners of the windows. But as the corners of these windows (which face to the south-east and south-west) are too shallow, the nests are washed down every hard rain ; and yet these birds drudge on to no purpose, from summer to summer, without changing their aspect or house. It is a piteous sight to see them labouring when half their nest is washed away, and bringing dirt "*generis lapsi sarcire ruinas.*" Thus is instinct a most wonderfully unequal faculty ; in some instances so much above reason ; in other respects, so far below it ! Martins love to frequent towns, especially if there are great lakes and rivers at hand ; nay, they even affect the close air of London. And I have not only seen them nesting in the Borough, but even in the Strand and Fleet-street; but then, it was obvious, from the dinginess of their aspect, that their feathers partook of the filth of that sooty atmosphere. Martins are, by far, the least agile of the four species; their wings and tails are short, and therefore they are not capable of such surprising turns, and quick and glancing evolutions, as the swallow. Accordingly, they make use of a placid, easy motion, in a middle region of the air, seldom mounting to any great height, and never sweeping along together over the surface of the ground or water. They do not wander far for food, but affect sheltered districts over some lake, or under some hanging wood, or in some hollow vale, especially in windy weather. They breed the latest of all the swallow kind: in 1772, they had nestlings on to October the twenty-first, and are never without unfledged young as late as Michaelmas.

As the summer declines, the congregating flocks increase in numbers daily by the constant accession of the second broods: till at last they swarm in myriads upon myriads round the villages on the Thames, darkening the face of the sky as they frequent the aits of that river, where they roost. They retire, the bulk of them I mean, in vast flocks together, about the beginning of October; but have appeared, of late years, in a considerable flight, in this neighbourhood, for one day or two, as late as November the third and sixth, after they were supposed to have been gone for more than a fort-night. They, therefore, withdraw with us the latest of any species. Unless these birds are very short lived indeed, or unless they do not return to the district where they are bred, they must undergo vast devastation somehow and somewhere; for the birds that return yearly bear no manner of proportion to the birds that retire.

House-martins are distinguished from their congeners by having their legs covered with soft downy feathers down to their toes. They are no songsters, but twitter, in a pretty inward, soft manner, in their nests. During the time of breeding, they are often greatly molested with fleas.

LETTER XVII.

SUSSEX DOWNS—SHEEP OF SUSSEX DOWNS—WHEATEAR—RING-OUSEL—
TORTOISE—ROOKERY.

RINGMER, near LEWES, *Dec.* 9, 1773.

DEAR SIR,—I received your last favour just as I was setting out for this place; and am pleased to find that my monography met with your approbation. My remarks are the result of many years' observation; and are, I trust, true on the whole; though I do not pretend to say that they are perfectly void of mistake, or that a more nice observer might not make many additions, since subjects of this kind are inexhaustible.

If you think my letter worthy the notice of your respectable Society, you are at liberty to lay it before them; and they will consider it, I hope, as it was intended, as a humble attempt to promote a more minute inquiry into natural history,—into the life and conversation of animals. Perhaps, hereafter, I may be induced to take the house-swallow under consideration; and from that proceed to the rest of the British *hirundines*.

Though I have now travelled the Sussex Downs upwards of thirty years, yet I still investigate that chain of majestic mountains with fresh admiration year by year; and I think I see new beauties every time I traverse it. The range, which runs from Chichester eastward as far as East Bourn, is about sixty miles in length, and is called the South Downs, properly speaking, only round Lewes. As you pass along, you command a noble view of the wold, or weald, on one hand, and the broad downs and sea on the other. Mr. Ray used to visit a family just at the foot of these hills, and was so ravished with the prospect from Plympton-plain, near Lewes, that he mentions those capes in his *Wisdom of God in the Works of Creation*, with the utmost satisfaction, and thinks them equal to anything he had seen in the finest parts of Europe.

For my own part, I think there is somewhat peculiarly sweet and amusing in the shapely figured aspect of chalk

(1) Mr. Courthope, of Danny.

hills, in preference to those of stone, which are rugged, broken, abrupt, and shapeless.

Perhaps I may be singular in my opinion, and not so happy as to convey to you the same idea, but I never contemplate these mountains without thinking I perceive somewhat analogous to growth in their gentle swellings and smooth fungus-like protuberances, their fluted sides, and regular hollows and slopes, that carry at once the air of vegetative dilatation and expansion; or, was there ever a time when these immense masses of calcareous matter were thrown into fermentation by some adventitious moisture,—were raised and leavened into such shapes by some plastic power, and so made to swell and heave their broad backs into the sky, so much above the less animated clay of the wild below?

By what I can guess from the admeasurements of the hills that have been taken round my house, I should suppose that these hills surmount the wild, at an average, at about the rate of five hundred feet.

RAM

One thing is very remarkable as to the sheep: from the westward, till you get to the river Adur, all the flocks have horns, and smooth white faces, and white legs; and a hornless sheep is rarely to be seen. But as soon as you pass that river eastward, and mount Beeding-hill, all the flocks at once become hornless, or, as they call them, poll-sheep ; and have, moreover, black faces, with a white tuft of wool on their foreheads, and speckled and spotted legs : so that you would think that the flocks of Laban were pasturing on one side of the stream, and the variegated breed of his son-in-law, Jacob, were cantoned on the other. And this diversity holds good respectively on each side, from the valley of Bramber and Beeding to the eastward, and westward all the whole length of the downs. If you talk with the shepherds on this subject, they tell you that the case has been so from time immemorial ; and smile at your simplicity if you ask them, whether the situation of these two different breeds might not be reversed? (However, an intelligent friend of mine near Chichester is determined to try the experiment ; and has, this autumn, at the hazard of being laughed at, introduced a parcel of black-faced hornless rams among his horned western ewes.) The black-faced poll-sheep have the shortest legs and the finest wool.[1]

(1) *Ovis aries.* The sheep form an essential part of the shepherd's barometer, as before rain they are always observed to eat very greedily.

Mild as the sheep may look, there is a mine of mischief in them, and especially in the rams, who not unfrequently assault the passer by with the greatest ferocity, and from the thickness of its skull, the onset even of a hornless ram is rather a formidable matter. An instance of this nature occurred not long since, which not only shows the fury and courage of the ram, but also a method of conquering the animal. The story is as follows.—Two gentlemen were fishing in one of the Welsh streams, and in the course of their piscatorial perambulations, entered a field where there were some sheep. Among them was an enormous ram with a pair of splendid horns. The ram, after examining the intruders for some time, made up his mind that they were to be expelled from the field, and accordingly charged full at one of them, and when repulsed made a charge at the other stranger. So furious was he, that the two fishermen, arming themselves with the butt end of their rods, separated about ten paces, so that when the ram charged at one, the other belaboured him. This went on for some time, until the ram was covered with blood, and staggering under the repeated blows that his fore legs had received ; yet he undauntedly continued his assaults, until his two antagonists could hardly wield their clubs from weariness, and were actually forced to beat a retreat.

Next day, they saw another gentleman going to fish in the same spot, and went to warn him of the ram. When they neared the spot, they heard cries for help, and found the gentleman kneeling on the ground with his rod extended, much as a soldier places himself, when ready to resist cavalry. There was the ram, apparently the worse for his yesterday's encounter, charging with the most determined ferocity. All three then united in attacking the infuriated ram, but with little success, as his fleece defended him from side blows, and his head was invulnerable as that of a brass battering-ram. Presently a farm-servant came up, who told the three combatants to desist from their attacks, and he would soon manage the ram. So

As I had hardly ever before travelled these downs at so late a season of the year, I was determined to keep as sharp a look-out as possible so near the southern coast, with respect to the summer short-winged birds of passage. We make great inquiries concerning the withdrawing of the swallow kind, without examining enough into the causes why this tribe is never to be seen in winter; for, *entre nous*, the disappearing of the latter is more marvellous than that of the former, and much more unaccountable. The *hirundines*, if they please, are certainly capable of migration; and yet, no doubt, are often found in a torpid state; but redstarts, nightingales, whitethroats, black-caps, &c. &c., are very ill provided for long flights; have never been once found, as I ever heard of, in a torpid state; and yet can never be supposed, in such troops, from year to year, to dodge and elude the eyes of the curious and inquisitive, which, from day to day, discern the other small birds that are known to abide our winters. But, notwithstanding all my care, I saw nothing like a summer bird of passage; and, what is more strange, not one wheatear, though they abound so in the autumn as to be a considerable perquisite to the shepherds that take them; and though many are seen to my knowledge all the winter through, in many parts of the south of England. The most intelligent shepherd tells me, that some few of these birds appear on the downs in March, and then withdraw to breed, probably, in warrens and stone quarries : now and then a nest is ploughed up in a fallow on the downs, under a fur row; but it is thought a rarity. At the time of wheat-harvest they begin to be taken in great numbers; are sent for sale in vast quantities to Brighthelmstone and Tunbridge, and appear

saying, he proceeded to cut a slight flexible stick, and walked up to the infuriated animal, while the three fishermen looked on, in no small curiosity to see how single antagonist would vanquish a foe that had already driven two enemies out of the field, and could not be overcome by three. The man, however, did not seem troubled by any misgivings on the subject, but put himself in the way of the ram who instantly charged at his new antagonist. The man stepped aside to let him pass, and as he rushed by, twisted his stick between the horns of the ram, who came to a stop, and seemed completely subdued by this apparently inadequate weapon. His conqueror then led him quietly out of the field and locked him up.

In all probability the stick formed with the horns so powerful a leverage on the animal's head, that it really could not move without serious injury to its head.—J. G. W.

at the tables of all the gentry that entertain with any degree of elegance. About Michaelmas they retire, and are seen no more till March. Though these birds are, when in season, in great plenty on the South Downs round Lewes, yet at East Bourn, which is the eastern extremity of those downs, they abound much more. One thing is very remarkable, that, though in the height of the season so many hundreds of dozens are taken, yet they are never seen to flock; and it is a rare thing to see more than three or four at a time : so that there must be a perpetual flitting and constant progressive succession. It does not appear that any wheatears are taken to the westward of Houghton-bridge, which stands on the river Arun.

I did not fail to look particularly after my new migration of ring-ousels; and to take notice whether they continued on the downs to this season of the year; as I had formerly remarked them in the month of October, all the way from Chichester to Lewes, wherever there were any shrubs and covert; but not one bird of this sort came within my observation. I only saw a few larks and whinchats, some rooks, and several kites and buzzards.

About midsummer, a flight of crossbills comes to the pine-groves about this house, but never makes any long stay.

The old tortoise, that I have mentioned in a former letter, still continues in this garden; and retired underground about the 20th of November, and came out again for one day on the 30th : it lies now buried in a wet swampy border under a wall facing to the south, and is enveloped at present in mud and mire !

Here is a large rookery round this house, the inhabitants of which seem to get their livelihood very easily; for they spend the greatest part of the day on their nest-trees when the weather is mild. These rooks retire every evening, all the winter, from this rookery, where they only call by the way, as they are going to roost in deep woods; at the dawn of day they always revisit their nest-trees, and are preceded a few minutes by a flight of daws that act as it were as their harbingers.

LETTER XVIII.

SWALLOWS.

SELBORNE, *Jan.* 2, 1769.

DEAR SIR,—The house-swallow,[1] or chimney-swallow, is, undoubtedly, the first comer of all the British *hirundines;* and appears in general on or about the 13th of April, as I have remarked from many years' observation. Not but now and then a straggler is seen much earlier; and, in particular, when I was a boy, I observed a swallow for a whole day together on a sunny warm Shrove-Tuesday; which day could not fall out later than the middle of March, and often happened early in February.

It is worth remarking, that these birds are seen first about lakes and mill-ponds; and it is also very particular, that, if these early visitors happen to find frost and snow, as was the case of the two dreadful springs of 1770 and 1771, they immediately withdraw for a time; a circumstance this, much more in favour of hiding than migration; since it is much more probable that a bird should retire to its hybernaculum just at hand, than return for a week or two only to warmer latitudes.

The swallow, though called the chimney-swallow, by no means builds altogether in chimneys, but often within barns and out-houses, against the rafters; and so she did in Virgil's time,—

———————————— " Ante
Garrula quàm tignis nidos suspendat hirundo."

" Before the noisy swallow's nest depends
From the strong beam that through the roof extends."

In Sweden, she builds in barns, and is called *ladu swala* (the barn-swallow) Besides, in the warmer parts of Europe there are no chimneys to houses, except they are English-built. In these countries she constructs her nest in porches, and gateways, and galleries, and open halls.

Here and there a bird may affect some odd, peculiar place; as we have known a swallow build down the shaft of an old

(1) *Hirundo rustica.*—J. G. W.

well, through which chalk had been formerly drawn up, for the purpose of manure ; but, in general, with us this *hirundo* breeds in chimneys, and loves to haunt those stacks where there is a constant fire—no doubt for the sake of warmth. Not that it can subsist in the immediate shaft where there is a fire ; but prefers one adjoining to that of the kitchen, and disregards the perpetual smoke of that funnel, as I have often observed with some degree of wonder.

Five or six, or more feet, down the chimney, does this little bird begin to form her nest, about the middle of May, which consists, like that of the house-martin, of a crust or shell composed of dirt or mud, mixed with short pieces of straw, to render it tough and permanent ; with this difference, that whereas the shell of the martin is nearly hemispheric, that of the swallow is open at the top, and like half a deep dish: his nest is lined with fine grasses and feathers, which are often collected as they float in the air.

Wonderful is the address which this adroit bird shows all day long, in ascending and descending with security through so narrow a pass. When hovering over the mouth of the funnel, the vibrations of her wings acting on the confined air, occasion a rumbling like thunder. It is not improbable that the dam submits to this inconvenient situation so low in the shaft in order to secure her broods from rapacious birds, and particularly from owls, which frequently fall down chimneys, perhaps in attempting to get at these nestlings.

The swallow lays from four to six white eggs, dotted with red specks ; and brings out her first brood about the last week in June, or the first week in July. The progressive method by which the young are introduced into life, is very amusing: first, they emerge from the shaft with difficulty enough, and often fall down into the rooms below: for a day or so they are fed on the chimney-top, and then are conducted to the dead leafless bough of some tree, where, sitting in a row, they are attended with great assiduity, and may then be called perchers. In a day or two more, they become fliers, but are still unable to take their own food ; therefore, they play about near the place where the dams are hawking for flies ; and when a mouthful is collected, at a certain signal given, the dam and the nestling advance, rising towards each

other, and meeting at an angle, the young one all the while
uttering such a little quick note of gratitude and complacency,
that a person must have paid very little regard to the wonders
of Nature that has not often remarked this feat.

The dam betakes herself immediately to the business of a
second brood as soon as she is disengaged from her first
which at once associates with the first broods of house-
martins, and with them congregates, clustering on sunny
roofs, towers, and trees. This *hirundo* brings out her second
brood towards the middle and end of August.

All the summer long is the swallow a most instructive
pattern of unwearied industry and affection; for, from morning to night, while there is a family to be supported, she
spends the whole day in skimming close to the ground,
and exerting the most sudden turns and quick evolutions.
Avenues, and long walks, under hedges and pasture-fields,
and mown meadows where cattle graze, are her delight, especially if there are trees interspersed, because in such spot
insects most abound. When a fly is taken, a smart snap
from her bill is heard, resembling the noise at the shutting
of a watch-case; but the motion of the mandibles is too quick
for the eye.

The swallow, probably the male bird, is the excubitor to
house-martins and other little birds, announcing the approach
of birds of prey; for as soon as an hawk appears, with a shrill
alarming note he calls all the swallows and martins about
him, who pursue in a body, and buffet and strike their enemy
till they have driven him from the village, darting down from
above on his back, and rising in a perpendicular line in perfect security. This bird also will sound the alarm and strike
at cats when they climb on the roofs of houses, or otherwise
approach the nests. Each species of *hirundo* drinks as it flies
along, sipping the surface of the water; but the swallow alone,
in general, washes on the wing, by dropping into a pool for
many times together. In very hot weather, house-martins
and bank-martins dip and wash a little.

The swallow is a delicate songster, and, in soft sunny
weather, sings both perching and flying; on trees in a kind
of concert, and on chimney-tops; is also a bold flier, ranging
to distant downs and commons even in windy weather, which

the other species seem much to dislike; nay, even frequenting
exposed sea-port towns, and making little excursions over
the salt water. Horsemen on wide downs are often closely
attended by a little party of swallows for miles together,
which play before and behind them, sweeping around, and
collecting all the skulking insects that are roused by the
trampling of the horses' feet. When the wind blows hard,
without this expedient, they are often forced to settle to pick
up their lurking prey.

This species feeds much on little *coleoptera*, as well as on
gnats and flies, and often settles on dug ground, or paths,
for gravels to grind and digest its food. Before they depart,
for some weeks, to a bird they forsake houses and chimneys,
and roost in trees, and usually withdraw about the beginning
of October, though some few stragglers may appear on, at
times, till the first week in November.

Some few pairs haunt the new and open streets of London
next the fields, but do not enter, like the house-martin, the
close and crowded parts of the city.

Both male and female are distinguished from their congeners by the length and forkedness of their tails. They are
undoubtedly the most nimble of all the species; and when
the male pursues the female in amorous chase, they then go
beyond their usual speed, and exert a rapidity almost too
quick for the eye to follow.

After this circumstantial detail of the life and discerning στοργὴ of the swallow, I shall add, for your further
amusement, an anecdote or two, not much in favour of her
sagacity.

A certain swallow built, for two years together, on the
handles of a pair of garden-shears, that were stuck up against
the walls in an out-house, and therefore must have her nest
spoiled whenever that implement was wanted. And, what
is stranger still, another bird of the same species built its
nest on the wings and body of an owl, that happened by
accident to hang dead and dry from the rafter of a barn.
This owl, with the nest on its wings, and with eggs in the
nest, was brought as a curiosity worthy the most elegant
private museum in Great Britain. The owner, struck with
the oddity of the sight, furnished the bringer with a large

shell, or conch, desiring him to fix it just where the owl hung. The person did as he was ordered; and the following year, a pair, probably the same pair, built their nest in the conch, and laid their eggs.

The owl and the conch make a strange, grotesque appearance, and are not the least curious specimens in that wonderful collection of art and nature.[1]

Thus is instinct in animals, taken the least out of its way, an undistinguishing, limited faculty, and blind to every circumstance that does not immediately respect self-preservation, or lead at once to the propagation or support of their species.

LETTER XIX.

VIRGIL'S HIRUNDO—WET WEATHER.

SELBORNE, *Feb.* 14, 1774.

DEAR SIR,—I received your favour of the eighth, and am pleased to find that you read my little history of the swallow with your usual candour ; nor was I the less pleased to find that you made objections where you saw reason.

As to the quotations, it is difficult to say precisely which species of *hirundo* Virgil might intend, in the lines in question, since the ancients did not attend to specific differences like modern naturalists; yet somewhat may be gathered enough to incline me to suppose, that, in the two passages quoted, the poet had his eye on the swallow.

In the first place, the epithet *garrula* suits the swallow well, who is a great songster, and not the martin, which is rather a mute bird, and when it sings, is so inward as scarce to be heard. Besides, if *tignum* in that place signifies a rafter rather than a beam, as it seems to me to do, then I think it must be the swallow that is alluded to, and not the martin, since the former does frequently build within the roof, against the rafters, while the latter always, as far as I have been able to observe, builds without the roof, against eaves and cornices.

As to the simile, too much stress must not be laid on it

---

(1) Sir Ashton Lever's Museum.

yet the epithet *nigra* speaks plainly in favour of the swallow, whose back and wings are very black ; while the rump of the martin is milk-white, its back and wings blue, and all its under part white as snow. Nor can the clumsy motions (comparatively clumsy) of the martin well represent the sudden and artful evolutions, and quick turns, which Juturna gave to her brother's chariot, so as to elude the eager pursuit of the enraged Æneas. The verb *sonat* also seems to imply a bird that is somewhat loquacious.[1]

We have had a very wet autumn and winter, so as to raise the springs to a pitch beyond anything since 1764, which was a remarkable year for floods and high waters. The land-springs, which we call levants, break out much on the downs of Sussex, Hampshire, and Wiltshire. The country people say, when the levants rise, corn will always be dear ; meaning, that when the earth is so glutted with water as to send forth springs on the downs and uplands, that the corn vales must be drowned : and so it has proved for these ten or eleven years past : for land-springs have never obtained more since the memory of man than during that period, nor has there been known a greater scarcity of all sorts of grain, considering the great improvements of modern husbandry. Such a run of wet seasons, a century or two ago, would, I am persuaded, have occasioned a famine. Therefore, pamphlets, and newspaper letters that talk of combinations, tend to inflame and mislead, since we must not expect plenty till Providence sends us more favourable seasons.

The wheat of last year, all round this district, and in the county of Rutland, and elsewhere, yields remarkably bad ; and our wheat on the ground, by the continual late sudden vicissitudes from fierce frost to pouring rains, looks poorly, and the turnips rot very fast.

(1) " *Nigra* velut magnas domini cum divitis ædes
    Pervolat, et pennis alta atria lustrat hirundo,
    Pabula parva legens, nidisque loquacibus escas :
    Et nunc porticibus vacuis, nunc humida circum
    Stagna *sonat*."

LETTER XX.

SAND-MARTIN.

SELBORNE, *February* 26, 1776.

DEAR SIR,—The sand-martin, or bank-martin, is by much the least of any of the British *hirundines*, and, as far as we have ever seen, the smallest known *hirundo ;* though Brisson asserts that there is one much smaller, and that is the *hirundo esculenta*.

SAND-MARTIN.

(1) *Cotile riparia*. The excavations of the Sand-martin are made with a close[d] beak. The holes dug by this bird are often perfectly circular, the radius of the circl[e] being the length of the bird from the beak to the legs. It does not, however, a[s] might be supposed, perch itself in the centre and work round, but clings with it[s] feet to the circumference, and pecks away at the centre.

The hole generally is cut straight for about two feet, and then makes a turn, at th[e] extremity of which are placed the eggs, beautifully delicate semitransparent littl[e] things. Many banks are perforated like honey-comb. The sandstone rocks abou[t] Ashbourn in Derbyshire were completely riddled with holes, and during the summe[r] months the air near the rocks was so filled with the martins that we used to brin[g] them down by first throwing among them a little stone, at which several made [a] dash, mistaking it for a fly, and were brought to the ground by a second and large[r] stone thrown after the first.

One morning while I was watching a large flock of sand-martins dashing abou[t] into and over just such a lane as is described by White, on page 23, a sparrow-haw[k] came over the bushes that fringed the top of the rocks, and found himself amon[g] the entire flock of martins. They immediately commenced an attack, scolding an[d] screaming at him on all sides, the boldest even striking at him with their wings Suddenly the noise ceased, their triumphant screams changed into harsh notes o[f] fear, and the place was deserted, not a bird being in sight except the ha[w]k, wh[o] with a struggling and screaming martin in his claws, slowly sailed over the bushe[s] and disappeared together with his prey.—J. G. W.

But it is much to be regretted that it is scarce possible for any observer to be so full and exact as he could wish in reciting the circumstances attending the life and conversation of this little bird, since it is *fera naturâ*, at least in this part of the kingdom, disclaiming all domestic attachments, and haunting wild heaths and commons where there are large lakes; while the other species, especially the swallow and house-martin, are remarkably gentle and domesticated, and never seem to think themselves safe but under the protection of man.

Here are in this parish, in the sand-pits and banks of the lake of Wolmer Forest, several colonies of these birds; and yet they are never seen in the village, nor do they at all frequent the cottages that are scattered about in that wild district. The only instance I ever remember where this species haunts any building, is at the town of Bishop's Waltham, in this county, where many sand-martins nestle and breed in the scaffold holes of the back wall of William of Wykeham's stables; but then this wall stands in a very sequestered and retired enclosure, and faces upon a large and beautiful lake. And, indeed, this species seems so to delight in large waters, that no instance occurs of their abounding but near vast pools or rivers; and, in particular, it has been remarked that they swarm in the banks of the Thames, in some places below London Bridge.

It is curious to observe with what different degrees of architectonic skill Providence has endowed birds of the same genus, and so nearly correspondent in their general mode of life; for, while the swallow and the house-martin discover the greatest address in raising and securely fixing crusts or shells of loam, as cunabula for their young, the bank-martin terebrates a round and regular hole in the sand or earth, which is serpentine, horizontal, and about two feet deep. At the inner end of this burrow does this bird deposit, in a good degree of safety, her rude nest, consisting of fine grasses and feathers, usually goose feathers, very inartificially laid together.

Perseverance will accomplish anything: though at first one would be disinclined to believe that this weak bird, with her soft and tender bill and claws, should ever be able to

bore the stubborn sand-bank, without entirely disabling herself; yet with these feeble instruments have I seen a pair of
them make great despatch, and could remark how much they
had scooped that day, by the fresh sand which ran down the
bank, and was of a different colour from that which lay loose
and bleached in the sun.

In what space of time these little artists are able to mine
and finish these cavities I have never been able to discover,
for reasons given above; but it would be a matter worthy of
observation, where it falls in the way of any naturalist, to
make his remarks. This I have often taken notice of, that
several holes of different depths are left unfinished at the
end of summer. To imagine that these beginnings were
intentionally made in order to be in the greater forwardness
for next spring, is allowing, perhaps, too much foresight and
skill to a simple bird. May not the cause of these *latebræ*
being left unfinished arise from their meeting in those places
with strata too harsh, hard, and solid for their purpose, which
they relinquish, and go to a fresh spot that works more
freely? or may they not in other places fall in with a soil as
much too loose and mouldering, liable to founder, and
threatening to overwhelm them and their labours?

One thing is remarkable, that, after some years, the old
holes are forsaken, and new ones bored; perhaps because the
old habitations grow foul and fetid from long use, or because
they may so abound with fleas as to become untenantable.
This species of swallow, moreover, is strangely annoyed with
fleas; and we have seen fleas, bed-fleas, (*pulex irritans*,)
swarming at the mouths of these holes, like bees on the
stools of their hives.

The following circumstance should by no means be
omitted,—that these birds do not make use of their caverns
by way of hybernacula, as might be expected; since banks
so perforated have been dug out with care in the winter,
when nothing was found but empty nests.

The sand-martin arrives much about the same time with
the swallow, and lays, as she does, from four to six white
eggs. But as this species is *cryptogame*, carrying on the
business of nidification, incubation, and the support of its
young in the dark, it would not be so easy to ascertain the

time of breeding, were it not for the coming forth of the
broods, which appear much about the time, or rather somewhat earlier than those of the swallow. The nestlings are
supported in common, like those of their congeners, with
gnats and other small insects, and sometimes they are fed
with *libellulæ* (dragon-flies) almost as long as themselves.

DRAGON-FLY.

In the last week in June we have seen a row of these
sitting on a rail, near a great pool, as perchers, and so
young and helpless, as easily to be taken by hand ; but
whether the dams ever feed them on the wing, as swallows
and house-martins do, we have never yet been able to
determine ; nor do we know whether they pursue and attack
birds of prey.

When they happen to breed near hedges and enclosures,
they are dispossessed of their breeding-holes, by the house-
sparrow, which is, on the same account, a fell adversary to
house-martins.

These *hirundines* are no songsters, but rather mute,
making only a little harsh noise when a person approaches
their nests. They seem not to be of a sociable turn, never,
with us, congregating with their congeners in the autumn.

Undoubtedly they breed a second time, like the house-martin and swallow; and withdraw about Michaelmas.

Though in some particular districts they may happen to abound, yet on the whole, in the south of England at least, is this much the rarest species; for there are few towns or large villages but what abound with house-martins; few churches, towers, or steeples but what are haunted by some swifts; scarce a hamlet or single cottage-chimney that has not its swallow; while the bank-martins, scattered here and there, live a sequestered life among some abrupt sand-hills, and in the banks of some few rivers.

These birds have a peculiar manner of flying, flitting about with odd jerks and vacillations, not unlike the motions of a butterfly.[1] Doubtless the flight of all *hirundines* is influenced by, and adapted to, the peculiar sort of insects which furnish their food. Hence it would be worth inquiry to examine what particular genus of insects affords the principal food of each respective species of swallow.

Notwithstanding what has been advanced above, some few sand-martins, I see, haunt the skirts of London, frequenting the dirty pools in St. George's Fields, and about Whitechapel. The question is where these build, since there are no banks or bold shores in that neighbourhood? Perhaps they nestle in the scaffold-holes of some old or new deserted building. They dip and wash as they fly sometimes, like the house-martin and swallow.

Sand-martins differ from their congeners in the diminutiveness of their size, and in their colour, which is what is usually called a mouse-colour. Near Valencia, in Spain, they are taken, says Willoughby, and sold in the markets for the table, and are called by the country people, probably from their desultory, jerking manner of flight, *Papillon de Montagna.*

(1) I have seen these little birds actually mistaken for white butterflies as they flitted about in the bright sunlight.—J. G. W.

LETTER XXI.

THE SWIFT.

SELBORNE, *Sept.* 28, 1774.

DEAR SIR,—As the swift, or black-martin is the largest of the British *hirundines*, so it is undoubtedly the latest comer: for I remember but one instance of its appearing before the last week in April; and in some of our late frosty harsh springs, it has not been seen till the beginning of May. This species usually arrives in pairs.

The swift, like the sand-martins, is very defective in architecture, making no crust, or shell, for its nest, but forming

SWIFT. [1]

(1) *Cypselus apus.* The swift is popularly termed "Jack Screamer," in allusion to its cry when on the wing, which is constant, shrill and piercing.

The chief point to be remarked in this bird, is the singular form of the foot, which is short, powerful, and has all the claws directed forward, probably to assist it in climbing into its nest. The nest, if so it can be called, is made under eaves of thatched roofs and other similar places, so that the bird has to clamber up the wall for some little distance before it can reach the nest. I must confess, however, that this reason does not quite satisfy me, as sparrows, redstarts, and other birds, build in similar situations, and yet their claws are directed both backwards and forwards.—J. G. W.

it of dry grasses and feathers, very rudely and inartificially put together. With all my attention to these birds, I have never been able once to discover one in the act of collecting or carrying in materials: so that I have suspected (since their nests are exactly the same) that they sometimes usurp upon the house-sparrows, and expel them, as sparrows do the house and sand-martin—well remembering that I have seen them squabbling together at the entrance of their holes, and the sparrows up in arms, and much disconcerted at these intruders; and yet I am assured by a nice observer in such matters, that they do collect feathers for their nests in Andalusia, and that he has shot them with such materials in their mouths.

Swifts, like sand-martins, carry on the business of nidification quite in the dark, in crannies of castles, and towers, and steeples, and upon the tops of the walls of churches, under the roof, and therefore cannot be so narrowly watched as those species that build more openly; but, from what I could ever observe, they begin nesting about the middle of May; and I have remarked, from eggs taken, that they have sat hard by the 9th of June. In general, they haunt tall buildings, churches, and steeples, and breed only in such; yet, in this village, some pairs frequent the lowest and meanest cottages, and educate their young under those thatched roofs. We remember but one instance where they breed out of buildings, and that is in the sides of a deep chalk pit near the town of Odiham, in this county, where we have seen many pairs entering the crevices, and skimming and squeaking round the precipices.

As I have regarded these amusive birds with no small attention, if I should advance something new and peculiar with respect to them, and different from all other birds, I might perhaps be credited, especially as my assertion is the result of many years' exact observation.

As the swift eats, drinks, and collects materials for its nest on the wing, it appears to live more in the air than any other bird, and to perform all functions there save those of sleeping and incubation.

This *hirundo* differs widely from its congeners in laying invariably but two eggs at a time, which are milk-white, long, and peaked at the small end; whereas the other species lay at each brood from four to six. It is a most alert bird, rising very early, and retiring to roost very late, and is on the wing in the height of summer at least sixteen hours. In the longest days it does not withdraw to rest till a quarter before nine in the evening, being the latest of all day birds. Just before they retire, whole groups of them assemble high in the air, and squeak and shoot about with wonderful rapidity. But this bird is never so much alive as in sultry thundery weather, when it expresses great alacrity, and calls forth all its powers. In hot mornings several getting together into little parties dash round the steeples and churches, squeaking as they go in a very clamorous manner: these, by nice observers, are supposed to be males serenading their sitting hens, and not without reason, since they seldom squeak till they come close to the walls or eaves, and since those within utter at the same time a little inward note of complacency.

When the hen has sat hard all day, she rushes forth just as it is almost dark, and stretches and relieves her weary limbs, and snatches a scanty meal for a few minutes, and then returns to her duty of incubation. Swifts, when wantonly and cruelly shot while they have young, discover a little lump of insects in their mouths, which they pouch and hold under their tongue. In general, they feed in a much higher district than other species; a proof that gnats and other insects do also abound to a considerable height in the air: they also range to vast distances; since locomotion is no labour to them, who are endowed with such wonderful powers of wing. Their powers seem to be in proportion to their levers; and their wings are longer in proportion than those of almost any other bird. When they mute, or ease themselves in flight, they raise their wings, and make them meet over their backs.

At some certain times, in the summer, I had remarked that swifts were hawking very low, for hours together, over pools and streams; and could not help inquiring into the object of their pursuit, that induced them to descend so much below their usual range. After some trouble I found that they were taking *phryganeæ, ephemeræ, libellulæ* (cadew-flies, may-flies, and dragon-flies), that were just emerged from their aurelia state. I then no longer wondered that they should be so willing to stoop for a prey that afforded them such plentiful and succulent nourishment.

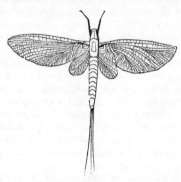

MAY-FLY.

They bring out their young about the middle or latter end of July; but as these never become perchers, nor, that ever I could discern, are fed on the wing by their dams, the coming forth of the young is not so notorious as in the other species.

On the 30th of last June I untiled the eaves of a house where many pairs build, and found in each nest only two squab, naked pulli. On the 8th of July I repeated the same inquiry, and found they had made very little progress towards a fledged state, but were still naked and helpless; from whence we may conclude, that birds whose way of life keeps them perpetually on the wing, would not be able to quit their nest till the end of the month. Swallows and martins that have numerous families, are continually feeding them every two or three minutes; while swifts, that have but two young to maintain, are much at their leisure, and do not attend on their nests for hours together.

Sometimes they pursue and strike at hawks that come in their way, but not with that vehemence and fury that swallows express on the same occasion. They are out all day long on wet days, feeding about, and disregarding still rain

(1) *Ephemera vulgata.*—J. G. W.

from whence two things may be gathered,—first, that many insects abide high in the air, even in rain ; and next, that the feathers of these birds must be well preened to resist so much wet. Windy, and particularly windy weather with heavy showers, they dislike, and on such days withdraw, and are scarcely ever seen.

There is a circumstance respecting the colour of swifts, which seems not to be unworthy our attention. When they arrive in the spring they are all over of a glossy dark soot colour, except their chins, which are white ; but, by being all day long in the sun and air, they become quite weather-beaten and bleached before they depart, and yet they return glossy again in the spring. Now, if they pursue the sun into lower latitudes, as some suppose, in order to enjoy a perpetual summer, why do they not return bleached ? Do they not rather, perhaps, retire to rest for a season, and at that juncture moult and change their feathers, since all other birds are known to moult soon after the season of breeding ?

Swifts are very anomalous in many particulars, dissenting from all their congeners, not only in the number of their young, but in breeding but once in summer ; whereas all the other British *hirundines* breed invariably twice. It is past all doubt that swifts can breed but once, since they withdraw in a short time after the flight of their young, and some time before their congeners bring out their second broods. We may here remark that, as swifts breed but once in a summer, and only two at a time, and the other *hirundines* twice, the latter, who lay from four to six eggs, increase, at an average, five times as fast as the former.

But in nothing are swifts more singular than in their early retreat. They retire, as to the main body of them, by the 10th of August, and sometimes a few days sooner ; and every straggler invariably withdraws by the 20th : while their congeners, all of them, stay till the beginning of October, many of them all through that month, and some occasionally to the beginning of November. This early retreat is mysterious and wonderful, since that time is often the sweetest season in the year. But what is more extra-ordinary, they begin to retire still earlier in the more southerly parts of Andalusia, where they can be nowise influenced by any defect of heat, or, as one might suppose, defect of food. Are they regulated in their motions with us by a failure of food, or by a propensity to moulting, or by a disposition to rest, after so rapid a life, or by what? This is one of those incidents in natural history that not only baffles our researches, but almost eludes our guesses!

These *hirundines* never perch on trees or roofs, and so never congregate with their congeners. They are fearless while haunting their nesting places, and are not to be scared with a gun, and are often beaten down with poles and cudgels as they stoop to go under the eaves. Swifts are much infested with those pests to the genus, called *hippoboscæ hirundinis*, and often wriggle and scratch themselves, in their flight, to get rid of that clinging annoyance.

Swifts are no songsters, and have only one harsh screaming note; yet there are ears to which it is not displeasing, from an agreeable association of ideas, since that note never occurs but in the most lovely summer weather.

They never settle on the ground but through accident, and when down can hardly rise, on account of the shortness of their legs and the length of their wings: neither can they walk, but only crawl; but they have a strong grasp with their feet, by which they cling to walls. Their bodies being flat, they can enter a very narrow crevice; and where they cannot pass on their bellies, they will turn up edgewise.

The particular formation of the foot discriminates the swift from all the British *hirundines*, and, indeed, from all other known birds, the *hirundo melba*, or great white-bellied swift of Gibraltar excepted; for it is so disposed as to carry " *omnes quatuor digitos anticos*," all its four toes forward: besides, the least toe, which should be the back toe, consists of one bone alone, and the other three only of two a-piece,— a construction most rare and particular, but nicely adapted to the purposes in which their feet are employed. This, and some peculiarities attending the nostrils and under mandible, have induced a discerning naturalist[1] to suppose that this species might constitute a genus *per se*.[2]

---

(1) John Antony Scopoli, of Carniola, M.D.
(2) See note on p. 221.—J. G. W.

In London, a party of swifts frequents the Tower, playing and feeding over the river just below the bridge; others haunt some of the churches of the Borough next the fields, but do not venture, like the house-martin, into the close, crowded part of the town.

The Swedes have bestowed a very pertinent name on this swallow, calling it *ring-swala*, from the perpetual rings, or circles, that it takes round the scene of its nidification.

Swifts feed on *coleoptera*, or small beetles with hard cases over their wings, as well as on the softer insects; but it does not appear how they can procure gravel to grind their food, as swallows do, since they never settle on the ground. Young ones, overrun with *hippoboscæ*, are sometimes found, under their nests, fallen to the ground, the number of vermin rendering their abode insupportable any longer. They frequent in this village several abject cottages; yet a succession still haunts the same unlikely roofs—a good proof this that the same birds return to the same spots. As they must stoop very low to get up under these humble eaves, cats lie in wait and sometimes catch them on the wing.

On the 5th of July, 1775, I again untiled part of a roof over the nest of a swift. The dam sat in the nest; but so strongly was she affected by natural στοργὴ for her brood which she supposed to be in danger, that regardless of her own safety, she would not stir, but lay sullenly by them, permitting herself to be taken in hand. The squab young we brought down and placed on the grass-plot, where they tumbled about, and were as helpless as a new-born child. While we contemplated their naked bodies, their unwieldy, disproportioned abdomina, and their heads too heavy for their necks to support, we could not but wonder when we reflected that these shiftless beings, in a little more than a fortnight, would be able to dash through the air almost with the inconceivable swiftness of a meteor, and, perhaps, in their emigration, must traverse vast continents and oceans as distant as the equator. So soon does Nature advance small birds to their ἡλικία, or state of perfection; while the progressive growth of men and large quadrupeds is slow and tedious!

LETTER XXII.

SELBORNE, *Sept.* 1774.

DEAR SIR,—By means of a straight cottage chimney, I had
an opportunity this summer of remarking, at my leisure, how
swallows ascend and descend through the shaft; but my
pleasure in contemplating the address with which this feat
was performed, to a considerable depth in the chimney, was
somewhat interrupted by apprehensions lest my eyes might
undergo the same fate with those of Tobit.[1]

Perhaps it may be some amusement to you to hear at what
times the different species of *hirundines* arrived this spring
in three very distant counties of this kingdom. With us,
the swallow was seen first on April the 4th; the swift on
April the 24th; the bank-martin on April the 12th; and
the house-martin not till April the 30th. At South Zele,
Devonshire, swallows did not arrive till April the 25th;
swifts, in plenty, on May the 1st; and house-martins not till
the middle of May. At Blackburn, in Lancashire, swifts
were seen April the 28th; swallows, April the 29th; house-
martins, May the 1st. Do these different dates, in such distant districts, prove anything for, or against migration?

A farmer near Weyhill fallows his land with two teams of
asses, one of which works till noon, and the other in the
afternoon. When these animals have done their work, they
are penned all night, like sheep, on the fallow. In the
winter, they are confined and foddered in the yard, and make
plenty of dung.

Linnæus says, that hawks "*paciscuntur inducias cum
avibus, quamdiu cuculus cuculat;*" but it appears to me that,
during that period, many little birds are taken and destroyed
by birds of prey, as may be seen by their feathers left in
lanes and under hedges.

The missel-thrush[2] is, while breeding, fierce and pugnacious,

---

(1) Tobit ii. 10.
(2) *Turdus viscivorus.* The nest of this bird is often most inartificially made. I
have seen them nearly the size of an ordinary hat, stuck in the most conspicuous

driving such birds as approach its nest, with great fury, to a distance. The Welsh call it *pen y llwyn*, the head or master of the coppice. He suffers no magpie, jay, or blackbird to enter the garden where he haunts, and is, for the time, a good guard to the new-sown legumens. In general, he is very successful in the defence of his family; but once I observed in my garden that several magpies came determined to storm the nest of a missel-thrush. The dams defended their mansion with great vigour, and fought resolutely; but numbers at last prevailed: they tore the nest to pieces, and swallowed the young alive.

MAGPIE.

parts of a tree. Whether the missel-thrush gains experience by losing its eggs and nest, I know not, but sometimes a nest is discovered which has been most carefully concealed. In some parts of England, this bird is called a "Jay-pie." This synonyme came under my notice while collecting the Wiltshire eggs. The rustics told me that the jay-pie built in the neighbourhood. And I immediately offered a certain price for jay's eggs, but was sadly discomfited by having to pay for two hatfulls of missel-thrush eggs. I may here observe, that the information of countrymen is scarcely ever to be depended on, as they have no notions of accurate observation, and moreover, have such a strange nomenclature. In many parts of England, for example, the large caterpillars of the Sphingidæ are called "Locusts."—J. G. W.

(1) *Pica caudata*. This is one of our most beautiful birds. Its dark glossy green plumage, diversified with bold white marks, is far more handsome than is generally supposed. The mischievous propensities of this bird are well known. There was a tame magpie residing in Wiltshire, who was an epitome of all the mischief that ever

In the season of nidification, the wildest birds are comparatively tame. Thus the ring-dove breeds in my fields, though they are continually frequented; and the missel-thrush, though most shy and wild in the autumn and winter, builds in my garden, close to a walk where people are passing all day long.

Wall-fruits abound with me this year; but my grapes, that used to be forward and good, are at present backward beyond all precedent. And this is not the worst of the story; for the same ungenial weather, the same black, cold solstice, has injured the more necessary fruits of the earth, and discoloured and blighted our wheat. The crop of hops promises to be very large.

Frequent returns of deafness incommode me sadly, and half disqualify me for a naturalist; for, when those fits are upon me, I lose all the pleasing notices and little intimations arising from rural sounds; and May is to me as silent and mute, with respect to the notes of birds, &c., as August. My eyesight is, thank God, quick and good; but with respect to the other sense, I am at times disabled,

" And Wisdom at one entrance quite shut out."

---

lurked in a bird's brain. This bird found a malicious enjoyment in pecking the unprotected ankles of little boys, not yet arrived at manly habiliments; and was such a terror to the female servants, that they were forced to pass his lurking-place armed with a broom. One of the servants having neglected this precaution, was actually found sitting down on the stones to protect her ankles, the magpie triumphantly pacing round her, until aid was brought, and the bird driven away. But to little boys and girls the magpie showed no mercy, springing out of its hiding-place, and chasing them completely along the garden walk.

It had also a great *penchant* for tearing and biting to pieces any papers that came in its way, probably because it had perceived that people valued them. One Sunday morning after the family had returned from church, the Rector found his study strewed with pamphlets, torn newspapers, &c., so that, until the delinquent was discovered, he really thought that thieves had been in the house. A magpie never seems to be happy unless it possesses a hiding-place; nor did this one form an exception to the general rule, as it had pecked a hole in the thatch of a barn, wherein to dispose its ill-gotten goods, and displayed great uneasiness if anybody approached it. Another magpie gained entrance into the chapel of Wadham College, Oxford, and remained quiet enough until the service had begun, when it gravely walked up the centre, bowing and saying, Pretty Mag! Pretty Mag! much to the discomposure of the junior members.—J. G. W.

LETTER XXIII.

GOSSAMER.

SELBORNE, *June* 8, 1775.

DEAR SIR,—On September the 21st, 1741, being then on a visit, and intent on field diversions, I rose before day-break: when I came into the enclosures, I found the stubbles and clover grounds matted all over with a thick coat of cobweb, in the meshes of which a copious and heavy dew hung so plentifully, that the whole face of the country seemed, as it were, covered with two or three setting-nets, drawn one over another. When the dogs attempted to hunt, their eyes were so blinded and hoodwinked that they could not proceed, but were obliged to lie down and scrape the incumbrances from their faces with their fore feet; so that, finding my sport interrupted, I returned home, musing in my mind on the oddness of the occurrence.

As the morning advanced, the sun became bright and warm, and the day turned out one of those most lovely ones which no season but the autumn produces,—cloudless, calm, serene, and worthy of the south of France itself.

About nine, an appearance very unusual began to demand our attention,—a shower of cobwebs falling from very elevated regions, and continuing, without any interruption, till the close of the day.

These webs are not single filmy threads, floating in the air in all directions, but perfect flakes or rags; some near an inch broad, and five or six long, which fell with a degree of velocity that showed they were considerably heavier than the atmosphere.

On every side, as the observer turned his eyes, he might behold a continual succession of fresh flakes falling into his sight, and twinkling like stars, as they turned their sides towards the sun.

How far this wonderful shower extended, would be difficult to say; but we know that it reached Bradley, Selborne, and Alresford, three places which lie in a sort of triangle, the shortest of whose sides is about eight miles in extent.

At the second of those places, there was a gentleman (for

whose veracity and intelligent turn we have the greatest
veneration) who observed it the moment he got abroad; but
concluded that, as soon as he came upon the hill above his
house, where he took his morning rides, he should be higher
than this meteor,' which he imagined might have been blown,
like thistle-down, from the common above; but, to his great
astonishment, when he rode to the most elevated part of the
down, 300 feet above his fields, he found the webs, in
appearance, still as much above him as before; still descending
into sight in a constant succession, and twinkling in the sun,
so as to draw the attention of the most incurious.

Neither before nor after, was any such fall observed;
but on this day the flakes hung in the trees and hedges so
thick that a diligent person sent out might have gathered
baskets full.

The remark that I shall make on these cobweb-like appearances, called gossamer, is, that strange and superstitious as
the notions about them were formerly, nobody in these
days doubts but that they are the real production of small
spiders, which swarm in the fields in fine weather in autumn,
and have a power of shooting out webs from their tails, so as
to render themselves buoyant, and lighter than air. But
why these apterous insects,[1] should that day take such a
wonderful aerial excursion, and why their webs should at once
become so gross and material as to be considerably more
weighty than air, and to descend with precipitation, is a
matter beyond my skill. If I might be allowed to hazard a
supposition, I should imagine that those filmy threads, when
first shot, might be entangled in the rising dew, and so drawn
up, spiders and all, by a brisk evaporation, into the region
where clouds are formed; and if the spiders have a power of
coiling and thickening their webs in the air, as Dr. Lister
says they have [see his *Letters* to Mr. Ray], then, when
they were become heavier than the air, they must fall.

Every day in fine weather, in autumn chiefly, do I see
those spiders shooting out their webs and mounting aloft

(1) Spiders are not insects, but belong to the order *Arachnida*. It is a *sine qu-*
non, that an insect, exclusive of other differences, must possess six legs, neithe
more nor less. Now a spider possessing eight legs, is of course excluded from th
insects.—J. G. W.

they will go off from your finger, if you will take them into your hand. Last summer one alighted on my book as I was reading in the parlour; and, running up to the top of the page, and shooting out a web, took his departure from thence. But what I most wondered at was, that it went off with considerable velocity in a place where no air was stirring; and I am sure that I did not assist it with my breath. So that these little crawlers seem to have, while mounting, some locomotive power without the use of wings, and to move in the air faster than the air itself.[1]

LETTER XXIV.

SOCIALITY OF BRUTES.

SELBORNE, *Aug.* 15, 1775.

DEAR SIR,—There is a wonderful spirit of sociality in the brute creation, independent of sexual attachment: the congregation of gregarious birds in the winter is a remarkable instance.

Many horses, though quiet with company, will not stay one minute in a field by themselves: the strongest fences cannot restrain them. My neighbour's horse will not only not stay by himself abroad, but he will not bear to be left alone in a strange stable without discovering the utmost impatience, and endeavouring to break the rack and manger with his fore feet. He has been known to leap out at a stable-window, through which dung was thrown, after company; and yet, in other respects, is remarkably quiet. Oxen and cows will not fatten by themselves; but will neglect the finest pasture that is not recommended by society. It would be needless to instance in sheep, which constantly flock together.

But this propensity seems not to be confined to animals of the same species; for we know a doe, still alive, that was brought up from a little fawn with a dairy of cows; with

(1) Many experiments have been made on the power of spiders to shoot out their threads without the aid of wind, but owing to the very great difficulty of entirely preventing all movement of the air, the experiments have not proved very satisfactory. Some say that the phenomenon may be accounted for by electrical agency.— J. G. W.

them it goes a-field, and with them it returns to the yard.
The dogs of the house take no notice of this deer, being used
to her; but, if strange dogs come by, a chase ensues; while
the master smiles to see his favourite securely leading her
pursuers over hedge, or gate, or stile, till she returns to the
cows, who, with fierce lowings, and menacing horns, drive the
assailants quite out of the pasture.

Even great disparity of kind and size does not always
prevent social advances and mutual fellowship. For a very
intelligent and observant person has assured me that in the

HORSE

former part of his life, keeping but one horse, he happened
also on a time to have but one solitary hen.[1] These two

(1) There have been several examples of the friendships formed by horses with very
dissimilar animals. The celebrated racer, the Godolphin Arabian, had contracted a
great friendship for a cat, who used to sit on his back, or nestle as closely to him as
possible. The cat's love for the horse was very strong, for when her friend died,
she soon pined away and followed him.

There was another Arabian horse, whose ferocity and untameable disposition were
so great, that only one groom would ever venture to approach him. A sheep, however,
formed an exception to his antipathies, and was accustomed to live in the san   table
with him.—J. G. W.

incongruous animals spent much of their time together, in a lonely orchard, where they saw no creature but each other. By degrees, an apparent regard began to take place between these two sequestered individuals. The fowl would approach the quadruped with notes of complacency, rubbing herself gently against his legs; while the horse would look down with satisfaction, and move with the greatest caution and circumspection, lest he should trample on his diminutive companion. Thus, by mutual good offices, each seemed to console the vacant hours of the other : so that Milton, when he puts the following sentiment in the mouth of Adam, seems to be somewhat mistaken :—

> " Much less can bird with beast, or fish with fowl,
> So well converse, nor with the ox the ape."

LETTER XXV.

GIPSIES.

SELBORNE, *Oct.* 2, 1775.

DEAR SIR,—We have two gangs, or hordes of gipsies, which infest the south and west of England, and come round in their circuit two or three times in the year. One of these tribes calls itself by the noble name of Stanley, of which I have nothing particular to say; but the other is distinguished by an appellative somewhat remarkable. As far as their harsh gibberish can be understood, they seem to say that the name of their clan is Curleople. Now the termination of this word is apparently Grecian; and, as Mezeray and the gravest historians all agree that these vagrants did certainly migrate from Egypt and the East, two or three centuries ago, and so spread by degrees over Europe, may not this family name, a little corrupted, be the very name they brought with them from the Levant? It would be matter of some curiosity could one meet with an intelligent person among them, to inquire whether, in their jargon, they still retained any Greek words: the Greek radicals will appear in hand, foot, head, water, earth, &c. It is possible that, amidst their cant and corrupted dialect, many mutilated remains of their native language might still be discovered.

With regard to these peculiar people, the gipsies, one thing is very remarkable, and especially as they came from a warmer climate, and that is, that while other beggars lodge in barns, stables, and cow-houses, these sturdy savages seem to pride themselves in braving the severities of winter, and in living *sub dio* the whole year round. Last September was as wet a month as ever was known ; and yet, during those deluges, did a young gipsy girl lie in the midst of one of our hop-gardens, on the cold ground, with nothing over her but a piece of a blanket, extended on a few hazel-rods bent hoop fashion, and stuck into the earth at each end, in circumstances too trying for a cow in the same condition; yet within this garden there was a large hop-kiln, into the chambers of which she might have retired, had she thought shelter an object worthy her attention.

Europe itself, it seems, cannot set bounds to the rovings of these vagabonds; for Mr. Bell, in his return from Peking, met a gang of these people on the confines of Tartary, who were endeavouring to penetrate those deserts, and try their fortune in China.[1]

Gipsies are called in French, *Bohemians;* in Italian and modern Greek, *Zingani.*

LETTER XXVI.

RUSH CANDLES—GREAT GOLDEN MAIDEN-HAIR BESOMS.

"Hic —— tædæ pingues, hic plurimus ignis
Semper, et assiduâ postes fuligine nigri."

SELBORNE, *Nov.* 1, 1775.

I SHALL make no apology for troubling you with the detail of a very simple piece of domestic economy, being satisfied that you think nothing beneath your attention that tends to utility. The matter alluded to is the use of rushes instead of candles, which I am well aware prevails in many districts besides this; but as I know there are countries also where it

(1) See BELL's *Travels in China.*

"During one of these deluges, did a young Gipsey Girl lie in the midst of one of our hop-gardens, on the cold ground."—P. 236.

QUEEN ANNE VIEWING THE RED DEER IN WOLMER FOREST.

does not obtain, and as I have considered the subject with some degree of exactness, I shall proceed in my humble story, and leave you to judge of the expediency.

The proper species of rush for this purpose seems to be the *juncus conglomeratus,* or common soft rush, which is to be found in most moist pastures, by the sides of streams, and under hedges. These rushes are in the best condition in the height of summer; but may be gathered, so as to serve the purpose well, quite on to autumn. It would be needless to add, that the largest and longest are best. Decayed labourers, women, and children, make it their business to procure and prepare them. As soon as they are cut, they must be flung into water and kept there, for otherwise they will dry and shrink, and the peel will not run. At first, a person would find it no easy matter to divest a rush of its peel, or rind, so as to leave one regular, narrow, even rib from top to bottom, that may support the pith; but this, like other feats, soon becomes familiar, even to children; and we have seen an old woman, stone blind, performing this business with great despatch, and seldom failing to strip them with the nicest regularity. When these *junci* are thus far prepared, they must lie out on the grass to be bleached, and take the dew for some nights, and afterwards be dried in the sun.

Some address is required in dipping these rushes in the scalding fat or grease; but this knack also is to be attained by practice. The careful wife of an industrious Hampshire labourer obtains all her fat for nothing, for she saves the scummings of her bacon-pot for this use; and if the grease abounds with salt, she causes the salt to precipitate to the bottom, by setting the scummings over a warm oven. Where hogs are not much in use, and especially by the sea-side, the coarser animal oils will come very cheap. A pound of common grease may be procured for fourpence; and about six pounds of grease will dip a pound of rushes; and one pound of rushes may be bought for one shilling; so that a pound of rushes, medicated and ready for use, will cost three shillings. If men that keep bees will mix a little wax with the grease, it will give it a consistency, and render it more cleanly, and make the rushes burn longer: mutton-suet would have the same effect.

A good rush, which measured in length two feet four inches and a half, being minuted, burnt only three minutes short of an hour; and a rush of still greater length has been known to burn one hour and a quarter.

These rushes give a good clear light. Watch-lights (coated with tallow), it is true, shed a dismal one—"darkness visible;" but then the wicks of those have two ribs of the rind, or peel, to support the pith, while the wick of the dipped rush has but one. The two ribs are intended to impede the progress of the flame, and make the candle last.

In a pound of dry rushes, avoirdupois, which I caused to be weighed and numbered, we found upwards of one thousand six hundred individuals. Now, suppose each of these burns one with another only half an hour, then a poor man will purchase eight hundred hours of light, a time exceeding thirty-three entire days, for three shillings. According to this account, each rush, before dipping, cost one thirty-third of a farthing, and one eleventh afterwards. Thus a poor family will enjoy five and a half hours of comfortable light for a farthing. An experienced old housekeeper assures me, that one pound and a half of rushes completely supplies his family the year round, since working people burn no candle in the long days, because they rise and go to bed by daylight.

Little farmers use rushes much in the short days, both morning and evening, in the dairy and kitchen; but the very poor, who are always the worst economists, and therefore must continue very poor, buy a halfpenny candle every evening, which in their blowing, open rooms, does not burn much more than two hours. Thus have they only two hours light for their money, instead of eleven.

While on the subject of rural economy, it may not be improper to mention a pretty implement of housewifery that we have seen nowhere else; that is, little neat besoms which our foresters make from the stalks of the *polytricum commune* or great golden maiden-hair, which they call silk-wood, and find plenty in the bogs. When this moss is well combed and dressed, and divested of its outer skin, it becomes of a beautiful bright chestnut colour; and being soft and pliant, is very proper for the dusting of beds, curtains, carpets, hangings, &c.

If these besoms were known to the brush-makers in town, it is probable they might come much in use for the purpose above mentioned.[1]

LETTER XXVII.

PROPENSITY OF AN IDIOT BOY.

SELBORNE, *Dec.* 12, 1775.

DEAR SIR,—We had in this village, more than twenty years ago, an idiot boy, whom I well remember, who, from a child, showed a strong propensity to bees; they were his food, his amusement, his sole object. And as people of this cast have seldom more than one point in view, so this lad exerted all his few faculties on this one pursuit. In the winter he dozed away his time, within his father's house, by the fire-side, in a kind of torpid state, seldom departing from the chimney corner; but in the summer he was all alert, and in quest of his game in the fields, and on sunny banks. Honey-bees, humble-bees, and wasps, were his prey wherever he found them: he had no apprehensions from their stings, but would seize them with naked hands,[2] and at once disarm them of their weapons, and suck their bodies for the sake of their honey-bags. Sometimes he would fill his bosom, between his shirt and his skin, with a number of these captives: and sometimes would confine them in bottles. He was a very *merops apiaster*, or bee-bird, and very injurious to men that kept bees; for he would slide into their bee-gardens, and sitting down before the stools, would rap with his finger on the hives, and so take the bees as they came out. He has been known to overturn hives for the sake of honey, of which he was passionately fond. Where metheglin was making, he would linger round the tubs and vessels, begging a draught of what he called bee-wine. As he ran about, he used to make a humming noise with his lips, resembling the buzzing of bees. This lad was lean and sallow, and of a cadaverous complexion; and, except in his favourite pursuit, in which he was wonderfully adroit, discovered no manner of understanding. Had

---

(1) A besom of this sort is to be seen in Sir Ashton Lever's Museum.

(2) This agrees with Mr. Cotton's remark, that if a bee is seized suddenly and handled fearlessly, it is so much astonished that it forgets to use its sting.—J. G. W

his capacity been better, and directed to the same object, he had perhaps abated much of our wonder at the feats of a more modern exhibitor of bees; and we may justly say of him now,

> "Thou,
> Had thy presiding star propitious shone,
> Shouldst Wildman be."

When a tall youth, he was removed from hence to a distant village, where he died, as I understand, before he arrived at manhood.

LETTER XXVIII.

SUPERSTITIONS OF SELBORNE.

SELBORNE, *Jan.* 8, 1776.

DEAR SIR,—It is the hardest thing in the world to shake off superstitious prejudices: they are sucked in, as it were, with our mother's milk; and growing up with us at a time when they take the fastest hold, and make the most lasting impressions, become so interwoven into our very constitutions, that the strongest good sense is required to disengage ourselves from them. No wonder, therefore, that the lower people retain them their whole lives through, since their minds are not invigorated by a liberal education, and therefore not enabled to make any efforts adequate to the occasion.

Such a preamble seems to be necessary before we enter on the superstitions of this district, lest we should be suspected of exaggeration in a recital of practices too gross for this enlightened age.

But the people of Tring, in Hertfordshire, would do well to remember, that no longer ago than the year 1751, and within twenty miles of the capital, they seized on two superannuated wretches, crazed with age, and overwhelmed with infirmities, on a suspicion of witchcraft; and, by trying experiments, drowned them in a horse-pond.

In a farm-yard, near the middle of this village, stands at this day a row of pollard-ashes, which, by the seams and long cicatrices down their sides, manifestly show that in former times they have been cleft asunder. These trees, when young and flexible, were severed and held open by wedges, while ruptured children stripped naked were pushed through the apertures, under a persuasion that by such a process the poor babes would be cured of their infirmity. As soon as the operation was over, the tree in the suffering part was plastered with loam, and carefully swathed up. If the parts coalesced and soldered together, as usually fell out where the feat was performed with any adroitness at all, the party was cured; but where the cleft continued to gape, the operation, it was supposed, would prove ineffectual. Having occasion to enlarge my garden not long since, I cut down two or three such trees, one of which did not grow together.

We have several persons now living in the village, who, in their childhood, were supposed to be healed by this superstitious ceremony, derived down, perhaps, from our Saxon ancestors, who practised it before their conversion to Christianity.[1]

At the south corner of the Plestor, or area near the church, there stood, about twenty years ago, a very old, grotesque,

(1) Even in the present day, these superstitious customs have not been entirely banished. Only last year (1852) an account appeared in the papers, of a superstitious ceremony performed with great solemnity in a country village. A donkey was standing in the midst of a crowd of country gossips, and a man was gravely passing a little boy a certain number of times round the animal's body. When this part of the ceremony had been completed, a piece of bread was offered to the donkey, and as it bit the bread, the father cut off the part that was not yet bitten, and then nipping off some of the donkey's hairs, he strewed them upon the bread, and gave he delicate morsel to the child to eat. This ceremony was considered to be an infallible cure for epilepsy.

As to omens of salt, winding sheets, strangers and letters in the candle, coffins in the fire, they are too well known to be mentioned; but there is a singular superstition which is, I believe, not so generally known. I was once engaged in the pursuit of twirling a chair round, when an ancient nurse, who had been looking on with great horror, begged me to desist, as twirling a chair was the work of the devil. Before the chair was replaced, I was directed to turn it in the opposite direction, an operation which I performed with becoming solemnity. If the old lady is still alive, I fear she must have lost her senses altogether in these table-turning days, when no less than four evil spirits get into a table in the course of half an hour, and give their names in the politest manner imaginable.

Bee superstition still reigns with undiminished sway. I knew a case where a lady who kept bees, died. Her servant immediately tapped at each hive, and informed them of the death of their mistress, and that they would have to work under another owner. A piece of black cloth was then fastened upon each hive, and the stock was then considered safe, as, if bees are not made acquainted with any change in the family, they always die! At a wedding, a white favour must be affixed to each hive, and the circumstances narrated. The attention of the bees is requested by rapping the hive with the door-key, no other instrument being sufficiently respectful.—
. G. W.

hollow pollard-ash, which for ages had been looked on with no
small veneration as a shrew-ash.   Now, a shrew-ash is an ash
whose twigs or branches, when gently applied to the limbs of
cattle, will immediately relieve the pains which a beast suffers
from the running of a shrew-mouse over the part affected;

SHREW-MOUSE.

for it is supposed that a shrew-mouse is of so baneful and
deleterious a nature, that wherever it creeps over a beast, be
it horse, cow, or sheep, the suffering animal is afflicted with
cruel anguish, and threatened with the loss of the use of the
limb.   Against this accident, to which they were continually
liable, our provident forefathers always kept a shrew-ash at
hand, which, when once medicated, would maintain its virtue
for ever.   A shrew-ash was made thus:[2]—Into the body of
the tree a deep hole was bored with an auger, and a poor
devoted shrew-mouse was thrust in alive, and plugged in, no
doubt, with several quaint incantations, long since forgotten.
As the ceremonies necessary for such a consecration are no
longer understood, all succession is at an end, and no such
tree is known to exist in the manor or hundred.

As to that on the Plestor,

> "The late vicar stubb'd and burnt it,"

when he was way-warden, regardless of the remonstrances of
the by-standers, who interceded in vain for its preservation,
urging its power and efficacy, and alleging that it had been

> "Religione patrum multos servata per annos:"
> "With reverential awe preserved for years."

---

(1) *Sorex araneus.* There are two other species of shrew inhabiting Eng.and
The Water-shrew, *sorex fodiens*, for which see note on p. 104, and the Oared Shrew
Sorex remifer. The bite of the shrew was thought to be poisonous, and only to be
cured by cutting the shrew in halves, and laying them on the bite.—J. G. W.

(2) For a similar practice, see PLOT's *Staffordshire.*

LETTER XXIX.

DRIPPING OF TREES—PONDS ON HILLS.

SELBORNE, *Feb.* 7, 1776.

DEAR SIR,—In heavy fogs, on elevated situations especially, trees are perfect alembics; and no one that has not attended to such matters can imagine how much water one tree will distil in a night's time, by condensing the vapour, which trickles down the twigs and boughs, so as to make the ground below quite in a float. In Newton-lane, in October, 1775, on a misty day, a particular oak in leaf dropped so fast that the cart-way stood in puddles, and the ruts ran with water, though the ground in general was dusty.

In some of our smaller islands in the West Indies, if I mistake not, there are no springs or rivers; but the people are supplied with that necessary element, water, merely by the dripping of some large tall trees, which, standing in the bosom of a mountain, keep their heads constantly enveloped with fogs and clouds, from which they dispense their kindly, never ceasing moisture; and so render those districts habitable by condensation alone.

Trees in leaf have such a vast proportion more of surface than those that are naked, that, in theory, their condensations should greatly exceed those that are stripped of their leaves; but, as the former imbibe also a great quantity of moisture, it is difficult to say which drip most: but this I know, that deciduous trees, that are entwined with much ivy, seem to distil the greatest quantity. Ivy leaves are smooth, and thick, and cold, and therefore condense very fast; and besides, evergreens imbibe very little. These facts may furnish the intelligent with hints concerning what sorts of trees they should plant round small ponds that they would wish to be perennial; and show them how advantageous some trees are in preference to others.

Trees perspire profusely, condense largely, and check evaporation so much, that woods are always moist; no wonder, therefore, that they contribute much to pools and streams.

That trees are great promoters of lakes and rivers appears from a well-known fact in North America; for, since the

woods and forests have been grubbed and cleared, all bodies
of water are much diminished; so that some streams that
were very considerable a century ago, will not now drive a
common mill.[1] Besides, most woodlands, forests, and chases
with us, abound with pools and morasses, no doubt for the
reason given above.

To a thinking mind, few phenomena are more strange than
the state of little ponds on the summits of chalk hills, many
of which are never dry in the most trying droughts of
summer;—on chalk hills, I say, because in many rocky and
gravelly soils springs usually break out pretty high on the
sides of elevated grounds and mountains; but no person
acquainted with chalky districts will allow that they ever saw
springs in such a soil but in valleys and bottoms, since the
waters of so pervious a stratum as chalk all lie on one dead
level, as well-diggers have assured me again and again.

Now, we have many such little round ponds in this district; and one in particular on our sheep-down, three hundred
feet above my house, which, though never above three feet
deep in the middle, and not more than thirty feet in diameter, and containing perhaps not more than two or three
hundred hogsheads of water, yet never is known to fail, though
it affords drink for three hundred or four hundred sheep, and
for at least twenty head of large cattle besides. This pond,
it is true, is overhung with two moderate beeches, that, doubtless, at times afford it much supply; but then we have others
as small, that, without the aid of trees, and in spite of evaporation from sun and wind, and perpetual consumption by
cattle, yet constantly maintain a moderate share of water
without overflowing in the wettest seasons, as they would do
if supplied by springs. By my journal of May, 1775, it appears that "the small and even considerable ponds in the
vales are now dried up, while the small ponds on the very top
of hills are but little affected." Can this difference be ac
counted for from evaporation alone, which certainly is more
prevalent in bottoms? or rather have not those elevated pool
some unnoticed recruits, which in the night-time counter
balance the waste of the day, without which the cattle alon

must soon exhaust them? And here it will be necessary to enter more minutely into the cause. Dr. Hales, in his *Vegetable Statics*, advances, from experiment, that "the moister the earth is, the more dew falls on it in a night; and more than a double quantity of dew falls on an equal surface of moist earth." Hence we see that water, by its coolness, is enabled to assimilate to itself a large quantity of moisture nightly by condensation; and that the air, when loaded with fogs and vapours, and even with copious dews, can alone advance a considerable and never failing resource. Persons that are much abroad, and travel early and late, such as shepherds, fishermen, &c., can tell what prodigious fogs prevail in the night on elevated downs, even in the hottest parts of summer; and how much the surfaces of things are drenched by those swimming vapours, though to the senses all the while little moisture seems to fall.

LETTER XXX.

THE CUCKOO.

SELBORNE, *April* 3, 1776.

DEAR SIR,—Monsieur Herissant, a French anatomist, seems persuaded that he has discovered the reason why cuckoos do not hatch their own eggs; the impediment, he supposes, arises from the internal structure of their parts, which incapacitates them for incubation. According to this gentleman, the crop, or craw of a cuckoo, does not lie before the sternum at the bottom of the neck, as in the *gallinæ, columbæ,* &c., but immediately behind it, on and over the bowels, so as to make a large protuberance in the belly.[1]

Induced by this assertion, we procured a cuckoo; and, cutting open the breast-bone, and exposing the intestines to sight, found the crop lying as mentioned above. This stomach was large and round, and stuffed hard, like a pincushion, with food, which, upon nice examination, we found to consist of various insects, such as small scarabs, spiders, and dragon-flies—the last of which we have seen cuckoos catching on the wing, as they were just emerging out of the aurelia state. Among this farrago also were to be seen

(1) *Histoire de l'Académie Royale,* 1752.

maggots, and many seeds which belonged either to goose-berries, currants, cranberries, or some such fruit; so that these birds apparently subsist on insects and fruits, nor was there the least appearance of bones, feathers, or fur, to support the idle notion of their being birds of prey.[1]

The sternum in this bird seemed to us to be remarkably short, between which and the anus lay the crop, or craw, and immediately behind that the bowels, against the back-bone.

It must be allowed, as this anatomist observes, that the crop, placed just below the bowels, must, especially when full, be in a very uneasy situation during the business of incubation; yet the test will be, to examine whether birds that are actually known to sit for certain, are not formed in a similar manner. This inquiry I proposed to myself to make with a fern-owl, or goat-sucker, as soon as opportunity offered; because, if their information proves the same, the reason for incapacity in the cuckoo will be allowed to have been taken up somewhat hastily.

Not long after, a fern-owl was procured, which, from its habits and shape, we suspected might resemble the cuckoo in its internal construction. Nor were our suspicions ill grounded; for, upon the dissection, the crop, or craw, also lay behind the sternum, immediately on the viscera, between them and the skin of the belly. It was bulky, and stuffed hard with large *phalænæ*, moths of several sorts, and their eggs, which, no doubt, had been forced out of these insects by the action of swallowing.

Now, as it appears that this bird, which is so well known to practise incubation, is formed in a similar manner with cuckoos, Monsieur Herissant's conjecture that cuckoos are incapable of incubation from the disposition of their intestines, seems to fall to the ground; and we are still at a loss for the cause of that strange and singular peculiarity in the instance of the *cuculus canorus*.

We found the case to be the same with the ring-tail hawk, in respect to formation; and, as far as I can recollect, with the swift; and probably it is so with many more sorts of birds that are not granivorous.

(1) See note on p. 166.—J. G. W.

LETTER XXXI.

THE VIPER.

SELBORNE, *April 29*, 1776.

DEAR SIR,—On August the 4th, 1775, we surprised a large viper, which seemed very heavy and bloated, as it lay in the grass, basking in the sun. When we came to cut it up, we found that the abdomen was crowded with young, fifteen in number ; the shortest of which measured full seven inches, and were about the size of full-grown earth-worms. This little fry issued into the world with the true viper spirit about them, showing great alertness as soon as disengaged from the belly of the dam : they twisted and wriggled about, and set themselves up, and gaped very wide when touched with a stick, showing manifest tokens of menace and defiance, though as yet they had no manner of fangs that we could find, even with the help of our glasses.

To a thinking mind, nothing is more wonderful than that early instinct which impresses young animals with the notion of the situation of their natural weapons, and of using them properly in their own defence, even before those weapons subsist or are formed. Thus a young cock will spar at his adversary before his spurs are grown, and a calf or lamb will push with their heads before their horns are sprouted. In the same manner did these young adders attempt to bite before their fangs were in being.[1] The dam, however, was furnished with very formidable ones, which we lifted up, (for they fold down when not used,) and cut them off with the point of our scissors.

There was little room to suppose that this brood had ever been in the open air before, and that they were taken in for refuge, at the mouth of the dam, when she perceived that danger was approaching ; because then, probably, we should have found them somewhere in the neck, and not in the abdomen.

(1) I have seen a little tiny kitten, hardly able to see, pounce upon a dead mouse, and growl in the most savage way if approached.—J. G. W.

LETTER XXXII.

THE HOG.

THE natural term of a hog's life is little known, and the reason is plain—because it is neither profitable nor convenient to keep that turbulent animal to the full extent of its time; however, my neighbour, a man of substance, who had no occasion to study every little advantage to a nicety, kept a half-bred Bantam sow, who was as thick as she was long, and whose belly swept on the ground, till she was advanced to her seventeenth year; at which period she showed some tokens of age by the decay of her teeth, and the decline of her fertility.

For about ten years, this prolific mother produced two litters in the year, of about ten at a time, and once above twenty at a litter; but, as there were near double the number of pigs to that of teats, many died. From long experience in the world, this female was grown very sagacious and artful. She used to open all the intervening gates, and march by herself to a distant farm, and would return by the same means. At the age of about fifteen, her litters began to be reduced to four or five; and such a litter she exhibited when in her fatting-pen. She proved, when fat, good bacon, juicy and tender; the rind, or sward, was remarkably thin. At a moderate computation, she was allowed to have been the fruitful parent of three hundred pigs—a prodigious instance of fecundity in so large a quadruped! She was killed in spring, 1775.

LETTER XXXIII.

AFFECTION IN A CAT.

SELBORNE, *May* 9, 1776.

" ———Admôrunt ubera tigres."

DEAR SIR,—We have remarked in a former letter how much incongruous animals, in a lonely state, may be attached to each other from a spirit of sociality; in this, it may not be amiss to recount a different motive, which has been known to create as strange a fondness.

My friend had a little helpless leveret brought to him, which the servants fed with milk in a spoon, and about the same time his cat kittened, and the young were dispatched and buried. The hare was soon lost, and supposed to be gone the way of most foundlings, to be killed by some dog or cat. However, in about a fortnight, as the master was sitting in his garden, in the dusk of the evening, he observed his cat, with tail erect, trotting towards him, and calling with little short inward notes of complacency, such as they use towards their kittens, and something gambolling after, which proved to be the leveret that the cat had supported with her milk, and continued to support with great affection.[1]

Thus was a graminivorous animal nurtured by a carnivorous and predaceous one!

Why so cruel and sanguinary a beast as a cat, of the ferocious genus of *felis*, the *murium leo*, "the lion of mice," as Linnæus calls it, should be affected with any tenderness towards an animal which is its natural prey, is not so easy to determine.

This strange affection probably was occasioned by that desiderium, those tender maternal feelings, which the loss of her kittens had awakened in her breast; and by the complacency and ease she derived to herself from procuring her teats to be drawn, which were too much distended with milk; till, from habit, she became as much delighted with this foundling, as if it had been her real offspring.

This incident is no bad solution of that strange circumstance which grave historians, as well as the poets, assert, of exposed children being sometimes nurtured by female wild beasts that probably had lost their young. For it is not one whit more marvellous that Romulus and Remus, in their infant state, should be nursed by a she-wolf, than that a poor little sucking leveret should be fostered and cherished by a bloody grimalkin.

(1) A fox bitch has been known to steal a puppy and rear it as her own offspring. —J. G. W.

LETTER XXXIV.

WORMS.

Selborne, *May* 20, 1777.

Dear Sir,—Lands that are subject to frequent inundations are always poor; and, probably, the reason may be, because the worms are drowned.[1] The most insignificant insects and reptiles are of much more consequence, and have much more influence in the economy of Nature, than the incurious are aware of; and are mighty in their effect, from their minuteness, which renders them less an object of attention; and from their numbers and fecundity. Earth-worms, though in appearance a small and despicable link in the chain of Nature, yet, if lost, would make a lamentable chasm.[2] For to say

(1) This is not the case at Oxford, for there is a large field which is flooded annually, and sometimes twice in a year, remaining under water for a long while, yet the grass of this field is always thick and well grown, and indeed in some places it is so rank that it cannot be used as hay.—J. G. W.

(2) *Lumbricus terrestris.* The earth-worm belongs to that class of creatures called Annelides, that is, ringed animals. The body of the earth-worm is surrounded by about one hundred and twenty rings, placed very close to each other. The rings are joined by four bands of muscle, which pass along the entire length of the worm; and the space within the rings is filled up with numerous circular bands of soft muscular structure. Every one must be acquainted with the power possessed by worms, of lengthening or shortening their bodies at pleasure. When they wish to shorten their bodies, they merely contract the four long muscular bands, but if they wish to elongate themselves, they contract the little circular muscles between the rings, which movement, by lessening the diameter of the worm, of course increases its length. The long muscular bands are placed two along the back, and the other two along the under surface of the worm. If, then, the creature wishes to arch its body, the feat is easily accomplished by shortening the lower pair of bands, while the others are left free to expand. This movement of course brings the body into an arch.

The onward progress of the earth-worm is made like that of the snake, viz. by advancing part of the body, and holding by that, while the remainder of the body is drawn up. The under surface of the snake is furnished with large horny scales, the edges of which hold firmly to the ground, while the hinder parts are being drawn forward. In the worm the place of these scales is supplied by very minute hairs, which are placed not only at the under surface, but entirely surround the body at each ring. When the worm pushes its body forwards, the stiff short hairs hold the advanced part of its body firm, while the hinder parts are fixing themselves.

The respiration of the earth-worm is accomplished through a series of very minute pores placed on its sides. It is rather a singula fact, that the blood of the earth worms is red, like that of the vertebrated animals.

The earth-worm never leaves its hole entirely, always retaining hold by a few of its caudal rings, by means of which it is enabled to withdraw itself into its hiding place on the slightest alarm.—J. G. W.

nothing of half the birds, and some quadrupeds, which are almost entirely supported by them, worms seem to be the great promoters of vegetation, which would proceed but lamely without them, by boring, perforating, and loosening the soil, and rendering it pervious to rains and the fibres of plants, by drawing straws and stalks of leaves into it; and, most of all, by throwing up such infinite numbers of lumps of earth, called worm-casts, which, being their excrement, is a fine manure for grain and grass. Worms probably provide new soils for hills and slopes where the rain washes the earth away; and they affect slopes, probably, to avoid being flooded. Gardeners and farmers express their detestation of worms; the former, because they render their walks unsightly, and make them much work; and the latter, because, as they think, worms eat their green corn. But these men would find, that the earth without worms would soon become cold, hard-bound, and void of fermentation; and, consequently, sterile: and, besides, in favour of worms, it should be hinted, that green corn, plants and flowers, are not so much injured by them as by many species of *coleoptera* (scarabs), and *tipulæ* (long-legs), in their larva or grub-state; and by unnoticed myriads of small shell-less snails, called slugs, which silently and imperceptibly make amazing havoc in the field and garden.[1]

These hints we think proper to throw out, in order to set the inquisitive and discerning to work.

A good monography of worms would afford much entertainment, and information, at the same time; and would open a large and new field in natural history. Worms work most in the spring, but by no means lie torpid in the dead months; are out every mild night in the winter, as any person may be convinced that will take the pains to examine his grass plots with a candle; are hermaphrodites, and consequently very prolific.

(1) Farmer Young, of Norton-farm, says, that this spring (1777) about four acres of his wheat in one field was entirely destroyed by slugs, which swarmed on the blades of corn, and devoured it as fast as it sprang.

LETTER XXXV.

TORPIDITY OF SWALLOWS.

SELBORNE, *Nov. 22*, 1777.

DEAR SIR,—You cannot but remember that the 26th and 27th of last March were very hot days; so sultry, that every body complained, and were restless under those sensations to which they had not been reconciled by gradual approaches.

This sudden summer-like heat was attended by many summer coincidences; for, on those two days, the thermometer rose to sixty-six in the shade; many species of insects revived and came forth; some bees swarmed in this neighbourhood; the old tortoise, near Lewes, awakened, and came forth out of its dormitory; and, what is most to my present purpose, many house-swallows appeared, and were very alert in many places, and particularly at Cobham, in Surrey.

But as that short warm period was succeeded as well as preceded by harsh, severe weather, with frequent frosts and ice, and cutting winds, the insects withdrew, the tortoise retired again into the ground, and the swallows were seen no more until the 10th of April, when the rigour of the spring abating, a softer season began to prevail.

Again, it appears by my journals for many years past, that house-martins retire, to a bird, about the beginning of October; so that a person not very observant of such matters would conclude that they had taken their last farewell; but then it may be seen in my diaries, also, that considerable flocks have discovered themselves again in the first week of November, and often on the fourth day of that month, only for one day; and that not as if they were in actual migration, but playing about at their leisure, and feeding calmly, as if no enterprise of moment at all agitated their spirits. And this was the case in the beginning of this very month; for, on the 4th of November, more than twenty house-martins, which, in appearance, had all departed about the 7th of October, were seen again, for that one morning only, sporting between my fields and the Hanger, and feasting on insects which swarmed in that sheltered district. The preceding day was wet and blustering, but the fourth was dark, and mild, and soft, the wind at south-west, and the thermometer at $58\frac{1}{2}$, a pitch not common at that season of the year. Moreover, it may not be amiss to add in this place, that whenever the thermometer is above 50, the bat comes flitting out in every autumnal and winter month.

From all these circumstances laid together, it is obvious that torpid insects, reptiles, and quadrupeds, are awakened from their profoundest slumbers by a little untimely warmth, and, therefore, that nothing so much promotes this death-like stupor as a defect of heat. And, further, it is reasonable to suppose, that two whole species, or at least many individuals of these two species of British *hirundines*, do never leave this island at all, but partake of the same benumbed state; for we cannot suppose that, after a month's absence, house-martins can return from southern regions to appear for one morning in November, or that house-swallows should leave the districts of Africa to enjoy, in March, the transient summer of a couple of days.

LETTER XXXVI.

LEPROSY—GARDENS.

SELBORNE, *Jan.* 8, 1778.

DEAR SIR,—There was in this little village several years ago, a miserable pauper who from his birth was afflicted with a leprosy, as far as we are aware, of a singular kind, since it affected only the palms of his hands and the soles of his feet. This scaly eruption usually broke out twice in the year, at the spring and fall; and by peeling away left the skin so thin and tender that neither his hands nor his feet were able to perform their functions; so that the poor object was half his time on crutches, incapable of employ, and languishing in a tiresome state of indolence and inactivity. His habit was lean, lank, and cadaverous. In this sad plight he dragged on a miserable existence, a burden to himself and his parish, which was obliged to support him, till he was relieved by death, at more than thirty years of age.

The good women, who love to account for every defect in

children by the doctrine of longing, said his mother felt a violent propensity for oysters, which she was unable to gratify, and that the black rough scurf on his hands and feet were the shells of that fish. We knew his parents, neither of whom were lepers; his father, in particular, lived to be far advanced in years.

In all ages, the leprosy has made dreadful havoc among mankind. The Israelites seem to have been greatly afflicted with it from the most remote times, as appears from the peculiar and repeated injunctions given them in the Levitical law.[1] Nor was the rancour of this foul disorder much abated in the last period of their commonwealth, as may be seen in many passages of the New Testament.

Some centuries ago this horrible distemper prevailed all over Europe; and our forefathers were by no means exempt, as appears by the large provision made for objects labouring under this calamity. There was an hospital for female lepers in the diocese of Lincoln, a noble one near Durham, three in London and Southwark, and perhaps many more in or near our great towns and cities. Moreover, some crowned heads, and other wealthy and charitable personages, bequeathed large legacies to such poor people as languished under this hopeless infirmity.

It must, therefore, in these days be to a humane and thinking person a matter of equal wonder and satisfaction when he contemplates how nearly this pest is eradicated, and observes that a leper is now a rare sight. He will, moreover, when engaged in such a train of thought, naturally inquire for the reason. This happy change, perhaps, may have originated and been continued from the much smaller quantity of salted meat and fish now eaten in these kingdoms —from the use of linen next the skin—from the plenty of better bread—and from the profusion of fruits, roots, legumes, and greens, so common in every family. Three or four centuries ago, before there were any enclosures, sown grasses, field turnips, or field carrots, or hay, all the cattle that had grown fat in summer, and were not killed for winter use, were turned out soon after Michaelmas to shift as they could through the dead months: so that no fresh

(1) See Leviticus, chap. xiii. and xiv.

meat could be had in winter or spring. Hence the marvellous
account of the vast stores of salted flesh found in the larder
of the eldest Spenser,[1] in the days of Edward the Second,
even so late in the spring as the 3d of May. It was from
magazines like these that the turbulent barons supported in
idleness their riotous swarms of retainers, ready for any
disorder or mischief. But agriculture has now arrived at
such a pitch of perfection, that our best and fattest meats are
killed in the winter ; and no man needs eat salted flesh,
unless he prefer it, that has money to buy fresh.

One cause of this distemper might be, no doubt, the
quantity of wretched fresh and salt fish consumed by the
commonalty at all seasons, as well as in Lent, which our poor
now would hardly be persuaded to touch.

The use of linen changes, shifts or shirts, in the room of
sordid or filthy woollen, long worn next the skin, is a matter
of neatness comparatively modern, but must prove a great
means of preventing cutaneous ails. At this very time,
woollen instead of linen prevails among the poorer Welsh, who
are subject to foul eruptions.

The plenty of good wheaten bread that now is found
among all ranks of people in the south, instead of that miserable sort which used in old days to be made of barley or
beans, may contribute not a little to the sweetening their
blood, and correcting their juices ; for the inhabitants of
mountainous districts to this day are still liable to the itch
and other cutaneous disorders, from a wretchedness and
poverty of diet.

As to the produce of a garden, every middle-aged person
of observation may perceive, within his own memory, both
in town and country, how vastly the consumption of
vegetables is increased. Green stalls in cities now support
multitudes in a comfortable state, while gardeners get for-
tunes. Every decent labourer also has his garden, which is
half his support, as well as his delight ; and common farmers
provide plenty of beans, peas, and greens, for their hinds to
eat with their bacon ; and those few that do not are despised
for their sordid parsimony, and looked upon as regardless of
the welfare of their dependents. Potatoes have prevailed

(1) Viz. six hundred bacons, eighty carcasses of beef, and six hundred muttons.

in this little district, by means of premiums, within these
twenty years only, and are much esteemed here now by the
poor, who would scarce have ventured to taste them in the
last reign.

Our Saxon ancestors certainly had some sort of cabbage,
because they call the month of February sprout-cale ; but
long after their days the cultivation of gardens was little
attended to.   The religious, being men of leisure, and keeping
up a constant correspondence with Italy, were the first
people among us who had gardens and fruit-trees in any
perfection, within the walls of their abbeys[1] and priories.
The barons neglected every pursuit that did not lead to war,
or tend to the pleasure of the chase.

It was not till gentlemen took up the study of horticulture
themselves that the knowledge of gardening made such hasty
advances.   Lord Cobham, Lord Ila, and Mr. Waller, of
Beaconsfield, were some of the first people of rank that pro-
moted the elegant science of ornamenting, without despising
the superintendence of the kitchen quarters and fruit walls.

A remark made by the excellent Mr. Ray, in his *Tour of
Europe*, at once surprises us, and corroborates what has been
advanced above ; for we find him observing, so late as his
days, that " the Italians use several herbs for sallets, which
are not yet, or have not been but lately used in England, viz.
selleri (celery), which is nothing else but the sweet smallage,
the young shoots whereof, with a little of the head of the
root cut off, they eat raw with oil and pepper."   And further,
he adds, "curled endive blanched is much used beyond seas,
and for a raw sallet, seemed to excel lettuce itself."   Now
this journey was undertaken no longer ago than in the
year 1663.

(1) " In monasteries, the lamp of knowledge continued to burn, however dimly
In them, men of business were formed for the state.   The art of writing was culti
vated by the monks; they were the only proficients in mechanics, *gardening*, and
architecture."—See DALRYMPLE'S *Annals of Scotland*.

LETTER XXXVII.

ECHOES.

SELBORNE, *Feb.* 12, 1778.

" Fortè puer, comitum seductus ab agmine fido,
    Dixerat, ecquis adest ? et, adest, responderat echo.
Hic stupet ; utque aciem partes divisit in omnes ;
    Voce, veni, clamat magnâ.   Vocat illa vocantem."

DEAR SIR,—In a district so diversified as this, so full of hollow vales and hanging woods, it is no wonder that echoes should abound.   Many we have discovered, that return the cry of a pack of dogs, the notes of a hunting-horn, a tunable ring of bells, or the melody of birds, very agreeably ; but we were still at a loss for a polysyllabical articulate echo, till a young gentleman, who had parted from his company in a summer evening walk, and was calling after them, stumbled upon a very curious one in a spot where it might least be expected.   At first he was much surprised, and could not be persuaded but that he was mocked by some boys ; but repeating his trials in several languages, and finding his respondent to be a very adroit polyglot, he then discerned the deception.

This echo, in an evening before rural noises cease, would repeat ten syllables most articulately and distinctly, especially if quick dactyls were chosen.   The last syllables of

" Tityre, tu patulæ recubans ——"

were as audibly and intelligibly returned as the first ; and there is no doubt, could trial have been made, but that at midnight, when the air is very elastic, and a dead stillness prevails, one or two syllables more might have been obtained ; but the distance rendered so late an experiment very inconvenient.

Quick dactyls, we observed, succeeded best ; for when we

came to try its powers in slow, heavy, embarrassed spondees
of the same number of syllables,

"Monstrum horrendum, informe, ingens ——"

we could perceive a return of but four or five.[1]

All echoes have some one place to which they are returned
stronger and more distinct than to any other; and that is
always the place that lies at right angles with the object of
repercussion, and is not too near, nor too far off.   Buildings
or naked rocks, re-echo much more articulately than hanging
woods or vales; because, in the latter, the voice is as it were
entangled, and embarrassed in the covert, and weakened in
the rebound.

The true object of this echo, as we found by various experiments, is the stone-built, tiled hop-kiln in Gally Lane,
which measures in front 40 feet, and from the ground to the
eaves 12 feet.   The true *centrum phonicum*, or just distance,
is one particular spot in the King's Field, in the path to
Norehill, on the very brink of the steep balk above the hollow
cart-way.   In this case, there is no choice of distance: but
the path, by mere contingency, happens to be the lucky, the
identical spot, because the ground rises or falls so immediately,
if the speaker either retires or advances, that his mouth
would at once be above or below the object.

We measured this polysyllabical echo with great exactness,
and found the distance to fall very short of Dr. Plot's rule
for distant articulation; for the Doctor, in his *History of
Oxfordshire*, allows 120 feet for the return of each syllable
distinctly; hence this echo, which gives ten distinct syllables,

(1) A few years ago there was an echo in Wiltshire, produced by a barn and a wall
whose imitative powers were quite equal to that mentioned by White.   There was,
however, only one very small space where the full powers of the echo were displayed.   We discovered it by accident, as we were returning at night over the Wiltshire
downs, and beguiling our journey by sundry catches.   As we arrived at a rise in the
road, the last chords of the catch were clearly returned, and after some examination
of the locality, we discovered the reason.
There is a very high, almost conical hill in Derbyshire, called Thorpe Cloud.   It
forms the entrance to Dovedale, and one side of it sweeps down close to the river
Dove.   I once fired a pistol at the summit for the sake of the experiment.   The
sound of the pistol was quite faint, and for a short time there was silence, but in
a few seconds the report came rolling back from all the rocky prominences of the
beautiful dale, like a peal of thunder, and continued uninterrupted for several
minutes. The pistol, when fired in the vale itself, did not produce nearly so many
echoes, as when it was fired on the summit of Thorpe Cloud.—J. G. W.

ought to measure 400 yards, or 120 feet to each syllable; whereas our distance is only 258 yards, or near 75 feet to each syllable. Thus our measure falls short of the Doctor's as five to eight; but then it must be acknowledged, that this candid philosopher was convinced afterwards, that some latitude must be admitted of in the distance of echoes according to time and place.

When experiments of this sort are making, it should always be remembered, that weather and the time of day have a vast influence on an echo; for a dull, heavy, moist air deadens and clogs the sound; and hot sunshine renders the air thin and weak, and deprives it of all its springiness; and a ruffling wind quite defeats the whole. In a still, clear dewy evening, the air is most elastic; and perhaps the later the hour the more so.

Echo has always been so amusing to the imagination, that the poets have personified her; and in their hands she has been the occasion of many a beautiful fiction. Nor need the gravest man be ashamed to appear taken with such a phenomenon, since it may become the subject of philosophical or mathematical inquiries.

One should have imagined that echoes, if not entertaining, must at least have been harmless and inoffensive: yet Virgil advances a strange notion, that they are injurious to bees. After enumerating some probable and reasonable annoyances, such as prudent owners would wish far removed from their bee-gardens, he adds,

——————— " Aut ubi concava pulsu
Saxa sonant, vocisque offensa resultat imago."

"Or where the hollow rocks emit a sound,
And echoed voices from the cliffs rebound."

This wild and fanciful assertion will hardly be admitted by the philosophers of these days, especially as they all now seem agreed that insects are not furnished with any organs of hearing at all.[1] But if it should be urged, that, though they cannot hear, yet perhaps they may feel the repercussion of

(1) This is still a disputed point. Yet it is somewhat singular, that many insects should be furnished with such ample means of making a noise, if others are not to hear them. Moreover, bee-masters of the present day always prefer to place the hives at a distance from all loud noises.—J. G. W.

sounds, I grant it is possible they may. Yet that these impressions are distasteful or hurtful I deny, because bees, in good summers, thrive well in my outlet, where the echoes are very strong; for this village is another Anathoth, a place of responses, or echoes. Besides, it does not appear from experiment that bees are in any way capable of being affected by sounds: for I have often tried my own with a large speaking trumpet held close to their hives, and with such an exertion of voice as would have hailed a ship at the distance of a mile, and still these insects pursued their various employments undisturbed, and without showing the least sensibility or resentment.

Some time since its discovery, this echo is become totally silent, the object or hop-kiln remains: nor is there any mystery in this defect, for the field between is planted as a hop-garden, and the voice of the speaker is totally absorbed and lost among the poles and entangled foliage of the hops. And when the poles are removed in autumn, the disappointment is the same; because a tall quick-set hedge, nurtured up for the purpose of shelter to the hop-ground, entirely interrupts the impulse and repercussion of the voice: so that till those obstructions are removed, no more of its garrulity can be expected.

Should any gentleman of fortune think an echo in his park or outlet a pleasant incident, he might build one at little or no expense. For, whenever he had occasion for a new barn, stable, dog-kennel, or the like structure, it would be only needful to erect this building on the gentle declivity of a hill, with a like rising opposite to it, at a few hundred yards distance; and perhaps success might be the easier insured could some canal, lake, or stream intervene. From a seat at the phonic centre, he and his friends might amuse themselves sometimes of an evening with the prattle of this loquacious nymph; of whose complacency and decent reserve, more may be said than can with truth of every individual of her sex since she is

> "Quæ nec *reticere* loquenti,
> Nec *prior* ipsa loqui, didicit resonabilis echo."

> "The vocal echo ne'er withholds reply,
> But ne'er intrudes."

P.S. The classic reader will, I trust, pardon the following lovely quotation, so finely describing echoes, and so poetically accounting for their causes from popular superstition.

" Quæ benè quom videas, rationem reddere possis
Tute tibi atque aliis, quo pacto per loca sola
Saxa pareis formas verborum ex ordine reddant,
Palanteis comites quom monteis inter opacos
Quærimus, et magnâ dispersos voce ciemus.
Sex etiam, aut septem loca vidi reddere voces
Unam quom jaceres : ita colles collibus ipsis
Verba repulsantes iterabant dicta referre.
Hæc loca capripedes Satyros, Nymphasque tenere
Finitimi fingunt, et Faunos esse loquuntur ;
Quorum noctivago strepitu, ludoque jocanti
Adfirmant volgo taciturna silentia rumpi,
Chordarumque sonos fieri, dulceisque querelas,
Tibia quas fundit digitis pulsata canentum ;
Et genus agricolûm latè sentiscere, quom Pan
Pinea semiferi capitis velamina quassans,
Unco sæpe labro calamos percurrit hianteis,
Fistula silvestrem ne cesset fundere musam."

LUCRETIUS, lib. iv. l. 576.

LETTER XXXVIII.

SWIFTS.

SELBORNE, *May* 13, 1778.

DEAR SIR,—Among the many singularities attending those amusing birds, the swifts, I am now confirmed in the opinion that we have every year the same number of pairs invariably ; at least, the result of my inquiry has been exactly the same for a long time past. The swallows and martins are so numerous, and so widely distributed over the village, that it is hardly possible to recount them ; while the swifts, though they do not all build in the church, yet so frequently haunt it, and play and rendezvous round it, that they are easily enumerated. The number that I constantly find are eight pairs, about half of which reside in the church, and the rest in some of the lowest and meanest thatched cottages. Now, as these eight pairs—allowance being made for accidents— breed yearly eight pairs more, what becomes annually of this increase ? and what determines, every spring, which pairs shall visit us, and re-occupy their ancient haunts ?

Ever since I have attended to the subject of ornithology, I have always supposed that the sudden reverse of affection, that ἀντιστόργη, which immediately succeeds in the feathered kind to the most passionate fondness, is the occasion of an equal dispersion of birds over the face of the earth. Without this provision, one favourite district would be crowded with inhabitants, while others would be destitute and forsaken. But the parent birds seem to maintain a jealous superiority, and to oblige the young to seek for new abodes; and the rivalry of the males in many kinds prevents their crowding the one on the other. Whether the swallows and house-martins return in the same exact number annually is not easy to say, for reasons given above; but it is apparent, as I have remarked before in my *Monographies*, that the numbers returning bear no manner of proportion to the numbers retiring.

LETTER XXXIX.

BOTANY.

SELBORNE, *June* 2, 1778.

DEAR SIR,—The standing objection to botany has always been, that it is a pursuit that amuses the fancy and exercises the memory, without improving the mind or advancing any real knowledge; and, where the science is carried no farther than a mere systematic classification, the charge is but too true. But the botanist that is desirous of wiping off this aspersion, should be by no means content with a list of names; he should study plants philosophically, should investigate the laws of vegetation, should examine the powers and virtues of efficacious herbs, should promote their cultivation, and graft the gardener, the planter, and the husbandman on the phytologist. Not that system is by any means to be thrown aside—without system the field of Nature would be a pathless wilderness—but system should be subservient to, not the main object of, pursuit.

Vegetation is highly worthy of our attention, and in itself is of the utmost consequence to mankind, and productive of many of the greatest comforts and elegancies of life. To plants we owe timber, bread, beer, honey, wine, oil, linen, cotton, &c.—what not only strengthens our hearts, and exhilarates our spirits, but what secures us from inclemencies of weather, and adorns our persons. Man, in his true state of nature, seems to be subsisted by spontaneous vegetation; in middle climes, where grasses prevail, he mixes some animal food with the produce of the field and garden: and it is towards the polar extremes only, that, like his kindred bears and wolves, he gorges himself with flesh alone, and is driven to what hunger has never been known to compel the very beasts – to prey upon his own species.[1]

The productions of vegetation have had a vast influence on the commerce of nations, and have been the great promoters of navigation, as may be seen in the articles of sugar, tea, tobacco, opium, ginseng, betel, pepper, &c. As every climate has its peculiar produce, our natural wants bring a mutual intercourse: so that by means of trade, each distant part is supplied with the growth of every latitude. But, without the knowledge of plants and their culture, we must have been content with our hips and haws, without enjoying the delicate fruits of India, and the salutiferous drugs of Peru.

Instead of examining the minute distinctions of every various species of each obscure genus, the botanist should endeavour to make himself acquainted with those that are useful. You shall see a man readily ascertain every herb of the field, yet hardly know wheat from barley, or at least one sort of wheat or barley from another.

---

(1) See the late voyages to the South Seas.—WHITE. This is hardly a correct opinion, as in the Friendly Islands and others, where cannibalism is most prevalent, the vegetation is particularly luxuriant, and appears perfectly capable of supporting the inhabitants without their having recourse to animal food at all. Cannibalism, indeed, is on many occasions either a religious, a warlike, or a penal custom. In the first case, human sacrifices are offered to the deities and then eaten; in the second case, the bodies of slaughtered enemies are devoured, partly in derision, partly in the hope of transfusing into the persons of the victims some portion of the warlike spirit that had animated the slain; and in the third case, the criminal is actually eaten alive, as a punishment for certain offences.

There is even a place where the aged people voluntarily surrender themselves to be eaten, considerately choosing the time of year, when the limes are most plentiful, for the juice of the lime is used as sauce.

But it is very seldom that man is induced to prey upon his own species by hunger, and the few instances that are known, are but exceptions.—J. G. W.

But of all sorts of vegetation the grasses seem to be most neglected ; neither the farmer nor the grazier seem to distinguish the annual from the perennial, the hardy from the tender, nor the succulent and nutritive from the dry and juiceless.

The study of grasses would be of great consequence to a northerly and grazing kingdom. The botanist that could improve the sward of the district where he lived, would be an useful member of society : to raise a thick turf on a naked soil, would be worth volumes of systematic knowledge ; and he would be the best commonwealth's man that could occasion the growth of "two blades of grass where one alone was seen before."

LETTER XL.

BOTANY OF SELBORNE—VERNAL AND AUTUMNAL CROCUS.

SELBORNE, *July* 3, 1778.

DEAR SIR,—In a district so diversified with such a variety of hill and dale, aspects and soils, it is no wonder that great choice of plants should be found. Chalks, clays, sands, sheep-walks and downs, bogs, heaths, woodlands, and champaign fields, cannot but furnish an ample *flora*. The deep rocky lanes abound with *filices*, and the pastures and moist woods with *fungi*. If in any branch of botany we may seem to be wanting, it must be in the large aquatic plants, which are not to be expected on a spot far removed from rivers, and lying up amidst the hill-country at the spring-heads. To enumerate all the plants that have been discovered within our limits, would be a needless work ; but a short list of the more rare, and the spots where they are to be found, may neither be unacceptable nor unentertaining.

Helleborus fœtidus, stinking hellebore, bear's-foot, or setter-wort—all over the Highwood and Coneycroft-hanger ; this continues a great branching plant the winter through, blossoming about January, and is very ornamental in shady walks and shrubberies. The good women give the leaves powdered to children troubled with worms; but it is a violent remedy, and ought to be administered with caution.

Helleborus viridis, green hellebore—in the deep stony lane, on the left hand just before the turning to Norton farm, and at the top of Middle Dorton, under the edge; this plant dies down to the ground early in autumn, and springs again about February, flowering almost as soon as it appears above ground.

Vaccinium oxycoccus, creeping bilberries, or cranberries—in the bogs of Bin's pond;

Vaccinium myrtillus, whortle, or bilberries—on the dry hillocks of Wolmer Forest;

Drosera rotundifolia, round-leaved sundew—in the bogs of Bin's pond;

Drosera longifolia, long-leaved sundew—in the bogs of Bin's pond;

Comarum palustre, purple comarum, or marsh cinque-foil —in the bogs of Bin's pond;

Hypericum androsæmum, Tutsan, St. John's wort—in the stony, hollow lanes;

Vinca minor, less periwinkle—in Selborne-hanger and Shrub-wood;

Monatropa hypopithys, yellow monotropa, or bird's nest— in Selborne-hanger under the shady beeches, to whose roots it seems to be parasitical—at the north-west end of the Hanger;

Chlora perfoliata, *Blackstonia perfoliata*, *Hudsoni*, perfoliated yellow-wort—on the banks in the King's Field;

Paris quadrifolia, herb Paris, true love, or one-berry—in the Church-litten coppice;

Chrysosplenium oppositifolium, opposite golden saxifrage— in the dark and rocky hollow lanes;

Gentiana amarella, autumnal gentian, or fellwort—on the Zig-zag and Hanger;

Lathræa squammaria, tooth-wort—in the Church-litten coppice, under some hazels near the foot-bridge, in Trimming's garden hedge, and on the dry wall opposite Grange-yard;

Dipsacus pilosus, small teasel—in the Short and Long Lith;

Lathyrus sylvestris, narrow-leaved, or wild lathyrus—in the bushes at the foot of the Short Lith, near the path;

Ophrys spiralis, ladies' traces—in the Long Lith, and towards the south corner of the common;

Ophrys nidus avis, bird's-nest ophrys—in the Long Lith, under the shady beeches among the dead leaves, in Great Dorton among the bushes, and on the Hanger plentifully;

Serapias latifolia, helleborine—in the High-wood under the shady beeches;

Daphne laureola, spurge-laurel—in Selborne-hanger and the High-wood;

Daphne mezereum, the mezereon—in Selborne-hanger, among the shrubs at the south-east end, above the cottages;

Lycoperdon tuber, truffles—in the Hanger and High-wood;

Sambucus ebulus, dwarf-elder, wal-wort, or dane-wort—among the rubbish and ruined foundations of the Priory.

Of all the propensities of plants, none seem more strange than their different periods of blossoming. Some produce their flowers in the winter, or very first dawnings of spring; many when the spring is established; some at midsummer, and some not till autumn. When we see the *helleborus fœtidus* and *helleborus niger* blowing at Christmas, the *helleborus hyemalis* in January, and the *helleborus viridis* as soon as ever it emerges out of the ground, we do not wonder, because they are kindred plants that we expect should keep pace the one with the other; but other congenerous vegetables differ so widely in their time of flowering, that we cannot but admire. I shall only instance at present in the *crocus sativus,* the vernal and the autumnal crocus, which have such an affinity, that the best botanists only make them varieties of the same genus, of which there is only one species, not being able to discern any difference in the corolla, or in the internal structure. Yet the vernal crocus expands its flowers by the beginning of March at farthest, and often in very rigorous weather; and cannot be retarded but by some violence offered; while the autumnal (the saffron) defies the influence of the spring and summer, and will not blow till most plants begin to fade and run to seed. This circumstance is one of the wonders of the creation, little noticed because a common occurrence; yet ought not to be overlooked on account of its being familiar, since it would be as difficul to be explained as the most stupendous phenomenon in nature.

> "Say, what impels, amidst surrounding snow
> Congeal'd, the crocus' flamy bud to glow?
> Say, what retards, amidst the summer's blaze,
> Th' autumnal bulb, till pale, declining days?
> The GOD of SEASONS; whose pervading power
> Controls the sun, or sheds the fleecy shower:
> He bids each flower his quickening word obey,
> Or to each lingering bloom enjoins delay."

LETTER XLI.

THE FLIGHT OF BIRDS.

SELBORNE, *Aug.* 7, 1778.

"Omnibus animalibus reliquis certus et uniusmodi, et in suo cuique genere incessus est: aves solæ vario meatu feruntur, et in terrâ, et in äere."—PLIN. *Hist Nat.* lib. x. cap. 38.

"All other animals have a certain, definite, and peculiar gait; birds alone move in a varied manner both on the ground and in the air."

DEAR SIR,—A good ornithologist should be able to distinguish birds by their air, as well as by their colours and shape, on the ground as well as on the wing, and in the bush as well as in the hand. For, though it must not be said that every species of birds has a manner peculiar to itself, yet there is somewhat in most genera at least that at first sight discriminates them, and enables a judicious observer to pronounce upon them with some certainty. Put a bird in motion,

> "Et vera incessu patuit."
> "And it is truly declared by its gait."

Thus kites and buzzards sail round in circles,[1] with wings expanded and motionless; and it is from their gliding manner that the former are still called, in the north of England, gleads, from the Saxon verb *glidan,* to glide. The kestrel, or windhover, has a peculiar mode of hanging in the air in one place, his wings all the while being briskly agitated.[2] Hen-

(1) The hawk tribe are now becoming so scarce, that, with the exception of the kestrel and the sparrow-hawk, the British ornithologist who, like White, is confined mostly to one spot, has very few opportunities of seeing them on the wing at all. The kite is now very seldom seen, and the buzzard still more rarely.—J. G. W.

(2) The observer will not fail to notice the resemblance of the flight in this bird

KITE [1]

harriers fly low over heaths or fields of corn, and beat the
ground regularly like a pointer or setting dog. Owls move in
a buoyant manner, as if lighter than the air; they seem to
want ballast.[2] There is a peculiarity belonging to ravens that
must draw the attention even of the most incurious—they
spend all their leisure time in striking and cuffing each other
on the wing in a kind of playful skirmish; and when they
move from one place to another, frequently turn on their
backs with a loud croak, and seem to be falling on the ground.
When this odd gesture betides them, they are scratching
themselves with one foot, and thus lose the centre of gravity.

to that of the various hovering insects, such as the humming-bird moth, or the
Syrphus Volucella and *Bombylius* flies. The flight of insects is quite as distinctive as
that of birds, and the practical entomologist can discover an insect when on the
wing, merely by its flight.—J. G. W.

(1) *Millva regalis.*—J. G. W.

(2) This is very true. Owls always appear as if they were *blown* about by the
wind, like a bunch of feathers, without any effort of their own. Their noiseless
flight strengthens this idea.—J. G. W.

HEN HARRIER.[1]

Rooks sometimes dive and tumble in a frolicsome manner;[2] crows and daws swagger in their walk; woodpeckers fly *volatu*

WOODPECKER.[3]

(1) *Circus cyaneus.*—J. G. W.

(2) This manœuvre is sometimes called the shooting of rooks, and is supposed to indicate the approach of rain.—J. G. W.

(3) *Picus major.* The Great Spotted Woodpecker.—J. G. W

undoso, opening and closing their wings at every stroke, and so are always rising and falling in curves. All of this genus use their tails, which incline downwards, as a support while they run up trees. Parrots, like all other hooked-clawed

MACAW. [1]

birds, walk awkwardly, and make use of their bill as a third foot, climbing and descending with ridiculous caution. All the *gallinæ* parade and walk gracefully, and run nimbly; but fly with difficulty, with an impetuous whirring, and in a straight line. Magpies and jays flutter with powerless wings, and make no despatch; herons seem encumbered with too much sail for their light bodies; but these vast hollow wings are necessary in carrying burdens, such as large fishes, and the like; pigeons, and particularly the sort called smiters, have a way of clashing their wings, the one against the other, over their backs, with a loud snap; another variety, called tumblers, turn themselves over in the air. Some birds have movements peculiar to the season of love; thus ring-doves,

(1) *Macrocercus Ararauna.* The Blue and Yellow Macaw.—J. G. W.

though strong and rapid at other times, yet, in the spring, hang about on the wing in a toying and playful manner; thus the cock-snipe, while breeding, forgetting his former flight,

GREENFINCH [1]

fans the air like a windhover; and the greenfinch, in particular, exhibits such languishing and faltering gestures as to

KINGFISHER. [2]

appear like a wounded and dying bird; the kingfisher darts along like an arrow; fern-owls, or goat-suckers, glance in the dusk over the tops of trees like a meteor; starlings, as it were,

(1) *Fringilla chloris.*—J. G. W.   (2) *Alcedo hispida.*—J. G. W.

swim along; while missel-thrushes use a wild and desultory
flight; swallows sweep over the surface of the ground and
water, and distinguish themselves by rapid turns and quick
evolutions; swifts dash round in circles; and the bank-martin
moves with frequent vacillations, like a butterfly. Most of
the small birds fly by jerks, rising and falling as they advance. Most small birds hop; but wagtails and larks walk,
moving their legs alternately. Skylarks rise and fall perpendicularly as they sing; woodlarks hang poised in the air;
and titlarks rise and fall in large curves, singing in their
descent. The whitethroat uses odd jerks and gesticulations
over the tops of hedges and bushes. All the duck kind
waddle; divers and auks walk as if fettered, and stand erect

AUK. [1]

on their tails; these are the *compedes* of Linnæus. Geese and
cranes, and most wild fowls, move in figured flights, often
changing their position. The secondary remiges of Tringæ,
wild ducks, and some others, are very long, and give their

(1) *Alca impennis.* The Great Auk.—J. G. W.

wings, when in motion, an hooked appearance. Dabchicks,[1] moor-hens, and coots, fly erect, with their legs hanging down.

MOOR-HEN [2]

COOT. [3]

(1) *Podiceps minor*, sometimes called the Little Grebe. The foot of the grebe is rather a remarkable structure. Most aquatic birds have webbed feet, but in the grebes each toe is separately flattened. In the stomach of the grebes is usually to be found a hard and compact mass of their own feathers, apparently analogous to the masses of hairs so often found in the stomachs of cows. The grebes are all very expert divers, and disappear so instantaneously that it is impossible to shoot them with a flint gun.—J. G. W.

(2) *Gallinula chloropus*. The moor-hen, or water-hen, as it is sometimes called, is also an excellent diver. It swims with a very graceful nodding of the head at every stroke, and on land it can run with considerable rapidity, constantly flicking up its tail, so as to show the white feathers beneath. When the water-hen leaves her nest, she has a common practice of scraping dried grass and reeds over her eggs, so as to completely conceal them.—J. G. W.

(3) *Fulica atra*. The nest of the Coot is usually made in the shallow water of a

and hardly make any despatch; the reason is plain, their wings are placed too forward out of the true centre of gravity; as the legs of auks and divers are situated too backward.

LETTER XLII.

THE LANGUAGE OF BIRDS AND FOWLS.

SELBORNE, Sept. 9, 1778.

DEAR SIR,—From the motion of birds, the transition is natural enough to their notes and language, of which I shall say something. Not that I would pretend to understand their language like a vizier, who, by the recital of a conversation which passed between two owls, reclaimed a sultan, before delighting in conquest and devastation; but I would be thought only to mean, that many of the winged tribes have various sounds and voices adapted to express their various passions, wants, and feelings, such as anger, fear, love, hatred, hunger, and the like. All species are not equally eloquent; some are copious and fluent, as it were in their utterance, while others are confined to a few important sounds; no bird, like the fish[2] kind, is quite mute, though some are rather silent. The language of birds is very ancient, and like other ancient modes of speech, very elliptical: little is said, but much is meant and understood.

The notes of the eagle kind are shrill and piercing; and about the season of nidification much diversified, as I have been often assured by a curious observer of Nature, who long resided at Gibraltar, where eagles abound. The note of our hawks much resemble those of the king of birds. Owls have very expressive notes; they hoot in a fine vocal

marsh, or close to the banks of a pond, and is composed of a huge mass of reeds and grass. I once took five eggs out of a coot's nest in Swindon reservoir. The nest looked more like an island than the work of a bird, and was built in three feet water, its foundation resting on a little sunken hillock.—J. G. W.

(1) See Spectator, vol. vii. No. 512.

(2) The tench are said to utter a kind of croaking sound, somewhat resembling that of a frog.—J. G. W.

sound, much resembling the *vox humana,* and reducible by a pitch-pipe to a musical key. This note seems to express complacency and rivalry among the mates; they use also a quick call and a horrible scream; and can snore and hiss when they mean to menace. Ravens, besides their loud croak, can exert a deep and solemn note, that makes the woods to echo; the amorous sound of a crow is strange and ridiculous; rooks, in the breeding season, attempt sometimes, in the gaiety of their hearts, to sing, but with no great success; the parrot kind have many modulations of voice, as appears by their aptitude to learn human sounds; doves coo in an amorous and mournful manner, and are emblems of despairing lovers; the woodpecker sets up a sort of loud and hearty laugh; the fern-owl, or goat-sucker, from the dusk till day-break, serenades his mate with the clattering of castanets. All the tuneful *passeres* express their complacency by sweet modulations, and a variety of melody. The swallow, as has been observed in a former letter, by a shrill alarm bespeaks the attention of the other *hirundines,* and bids them be aware that the hawk is at hand. Aquatic and gregarious birds, especially the nocturnal, that shift their quarters in the dark, are very noisy and loquacious; as cranes, wild-geese, wild-ducks, and the like: their perpetual clamour prevents them from dispersing and losing their companions.[1]

In so extensive a subject, sketches and outlines are as much as can be expected: for it would be endless to instance in all the infinite variety of the feathered nation. We shall, therefore, confine the remainder of this letter to the few domestic fowls of our yards, which are most known, and, therefore, best understood. At first,—the peacock, with his gorgeous train, demands our attention; but, like most of the gaudy birds, his notes are grating and shocking to the ear: the yelling of cats, and the braying of an ass, are not more disgustful. The voice of the goose is trumpet-like, and clanking; and once saved the Capitol at Rome, as grave

---

(1) For the same reason, sheep when they are quietly in their field are nearly silent, only an occasional baa being heard; but when they begin to move, every mother-sheep cries out for her young one, the lamb answers it, and others appear to join from sympathy.—J. G. W.

historians assert : the hiss also of the gander is formidable, and full of menace, and " protective of his young." Among ducks the sexual distinction of voice is remarkable ; for, while the quack of the female is loud and sonorous, the voice of the drake is inward, and harsh, and feeble, and scarce discernible. The cock-turkey struts and gobbles to his

TURKEY. [1]

mistress in a most uncouth manner ; he hath also a pert and petulant note when he attacks his adversary. When a hen-turkey leads forth her young brood, she keeps a watchful eye ; and if a bird of prey appear, though ever so high in the air, the careful mother announces the enemy with a little inward moan, and watches him with a steady and attentive look ; but, if he approach, her note becomes earnest and alarming, and her outcries are redoubled.

No inhabitants of a yard seem possessed of such a variety

(1) *Meleagris gallopavo.*— J. G. W.

of expression, and so copious a language, as common poultry. Take a chicken of four or five days old, and hold it up to a window where there are flies, and it will immediately seize its prey with little twitterings of complacency; but if you tender it a wasp or a bee, at once its note becomes harsh, and expressive of disapprobation and a sense of danger. When a pullet is ready to lay, she intimates the event by a joyous and easy soft note. Of all the occurrences of their life, that of laying seems to be the most important; for, no sooner has a hen disburdened herself, than she rushes forth with a clamorous kind of joy, which the cock and the rest of his mistresses immediately adopt. The tumult is not confined to the family concerned, but catches from yard to yard, and spreads to every homestead within hearing, till at last the whole village is in an uproar. As soon as a hen becomes a mother, her new relation demands a new language; she then runs clucking and screaming about, and seems agitated as if possessed. The father of the flock has also a considerable vocabulary; if he finds food, he calls a favourite concubine to partake; and if a bird of prey passes over, with a warning voice he bids his family beware. The gallant chanticleer has, at command, his amorous phrases, and his terms of defiance. But the sound by which he is best known is his crowing: by this he has been distinguished in all ages as the countryman's clock or larum—as the watchman that proclaims the divisions of the night. Thus the poet elegantly styles him

> "The crested cock, whose clarion sounds
> The silent hours."

A neighbouring gentleman, one summer, had lost most of his chickens by a sparrow-hawk, that came gliding down between a fagot pile and the end of his house to the place where the coops stood. The owner, inwardly vexed to see his flock thus diminishing, hung a setting net adroitly between the pile and the house, into which the caitiff dashed, and was entangled. Resentment suggested the law of retaliation; he therefore clipped the hawk's wings, cut off his talons, and, fixing a cork on his bill, threw him down among the brood-hens. Imagination cannot paint the scene that

ensued ; the expressions that fear, rage, and revenge inspired
were new, or at least such as had been unnoticed before.
The exasperated matrons upbraided—they execrated—they
insulted—they triumphed. In a word, they never desisted
from buffeting their adversary till they had torn him in a
hundred pieces.

LETTER XLIII.

HELIOTROPES.

SELBORNE.

".. ———— Monstrent
*   *   *   *
Quid tantùm Oceano properent se tingere soles
Hyberni; vel quæ tardis mora noctibus obstet."

"———— They show
*   *   *
Why winter-suns so rapidly descend,
And what delays the tardy nights extend."

GENTLEMEN who have outlets might contrive to make
ornament subservient to utility; a pleasing eye-trap might
also contribute to promote science ; an obelisk in a garden or
park might be both an embellishment and an heliotrope.

Any person that is curious, and enjoys the advantage of
a good horizon, might, with little trouble, make two heliotropes, the one for the winter, the other for the summer
solstice; and these two erections might be constructed with
very little expense; for two pieces of timber frame-work,
about ten or twelve feet high, and four feet broad at the base,
and close lined with plank, would answer the purpose.

The erection for the former should, if possible, be placed
within sight of some window in the common sitting parlour ;
because men, at that dead season of the year, are usually
within doors at the close of the day; while that of the latter
might be fixed for any given spot in the garden or outlet,
whence the owner might contemplate, in a fine summer's
evening, the utmost extent that the sun makes to the northward at the season of the longest days. Now nothing would
be necessary but to place these two objects with so much

exactness, that the westerly limb of the sun, at setting, might but just clear the winter heliotrope to the west of it, on the shortest day, and that the whole disc of the sun, at the longest day, might exactly, at setting, also clear the summer heliotrope to the north of it.

By this simple expedient, it would soon appear that there is no such thing, strictly speaking, as a solstice; for, from the shortest day, the owner would, every clear evening, see the disc advancing, at its setting, to the westward of the object; and, from the longest day, observe the sun retiring backwards every evening, at its setting, towards the object westward, till, in a few nights, it would set quite behind it, and so by degrees to the west of it; for when the sun comes near the summer solstice, the whole disc of it would at first set behind the object: after a time, the northern limb would first appear, and so every night gradually more, till at length the whole diameter would set northward of it for about three nights; but, on the middle night of the three, sensibly more remote than the former or following. When beginning its recess from the summer tropic, it would continue more and more to be hidden every night, till at length it would descend quite behind the object again; and so nightly more and more to the westward.

LETTER XLIV.

HILLS—FALL OF A CLIFF.

SELBORNE.

" ———— Mugire videbis
Sub pedibus terram, et descendere montibus ornos."

WHEN I was a boy, I used to read, with astonishment and implicit assent, accounts in Baker's *Chronicle* of walking hills and travelling mountains. John Philips, in his *Cyder*, alludes to the credit that was given to such stories, with a delicate but quaint vein of humour, peculiar to the author of the *Splendid Shilling:*—

> " I nor advise, nor reprehend, the choice
>   Of Marcley Hill; the apple nowhere finds
>   A kinder mould: yet 'tis unsafe to trust
>   Deceitful ground: who knows but that, once more,
>   This mount may journey, and, his present site
>   Forsaking, to thy neighbour's bounds transfer
>   Thy goodly plants, affording matter strange
>   For law debates!"

But, when I came to consider better, I began to suspect that, though our hills may never have journeyed far, yet that the ends of many of them have slipped and fallen away at distant periods, leaving the cliffs bare and abrupt. This seems to have been the case with Nore and Whetham Hills, and especially with the ridge between Harteley Park and Wardle-ham, where the ground has slid into vast swellings and furrows, and lies still in such romantic confusion as cannot be accounted for from any other cause. A strange event, that happened not long since, justifies our suspicions; which, though it befel not within the limits of this parish, yet as it was within the hundred of Selborne, and as the circumstances were singular, may fairly claim a place in a work of this nature.

The months of January and February, in the year 1774, were remarkable for great melting snows and vast gluts of rain; so that, by the end of the latter month, the land-springs, or levants, began to prevail, and to be near as high as in the memorable winter of 1764. The beginning of March also went on in the same tenor, when, in the night between the 8th and 9th of that month, a considerable part of the great woody hanger at Hawkley was torn from its place, and fell down, leaving a high free-stone cliff naked and bare, and resembling the steep side of a chalk pit. It appears that this huge fragment, being, perhaps, sapped and undermined by waters, foundered, and was ingulfed, going down in a perpendicular direction; for a gate, which stood in the field on the top of the hill, after sinking with its posts for thirty or forty feet, remained in so true and upright a position, as to open and shut with great exactness, just as in its first situation. Several oaks also are still standing, and in a state of vegetation, after taking the same desperate leap. That great part of this prodigious mass was absorbed in some gulf

below, is plain also, from the inclining ground at the bottom
of the hill, which is free and unencumbered, but would have
been buried in heaps of rubbish, had the fragment parted and
fallen forward. About a hundred yards from the foot of this
hanging coppice, stood a cottage by the side of a lane; and
two hundred yards lower, on the other side of the lane, was
a farm-house, in which lived a labourer and his family; and
just by, a stout new barn. The cottage was inhabited by an
old woman and her son, and his wife. These people, in the
evening, which was very dark and tempestuous, observed that
the brick floors of their kitchens began to heave and part, and
that the walls seemed to open, and the roofs to crack; but
they all agree that no tremor of the ground, indicating an
earthquake, was ever felt, only that the wind continued to
make a most tremendous roaring in the woods and hangers.
The miserable inhabitants, not daring to go to bed, remained
in the utmost solicitude and confusion, expecting every
moment to be buried under the ruins of their shattered
edifices. When daylight came, they were at leisure to contemplate the devastations of the night. They then found
that a deep rift, or chasm, had opened under their houses, and
torn them, as it were, in two, and that one end of the barn
had suffered in a similar manner: that a pond near the
cottage had undergone a strange reverse, becoming deep at
the shallow end, and so *vice versa*: that many large oaks were
removed out of their perpendicular, some thrown down, and
some fallen into the heads of neighbouring trees; and that
a gate was thrust forward, with its hedge, full six feet, so as to
require a new track to be made to it. From the foot of the
cliff, the general course of the ground, which is pasture,
inclines in a moderate descent for half a mile, and is interspersed with some hillocks, which were rifted in every
direction, as well towards the great woody hanger as from it.
In the first pasture the deep clefts began, and, running across
the lane and under the buildings, made such vast shelves that
the road was impassable for some time; and so over to an
arable field on the other side, which was strangely torn and
disordered. The second pasture field, being more soft and
springy, was protruded forward without many fissures in the
turf, which was raised in long ridges resembling graves, lying

at right angles to the motion. At the bottom of this enclosure, the soil and turf rose many feet against the bodies of some oaks that obstructed their further course, and terminated this awful commotion.

The perpendicular height of the precipice, in general, is twenty-three yards; the length of the lapse or slip, as seen from the fields below, one hundred and eighty-one: and a partial fall, concealed in the coppice, extends seventy yards more; so that the total length of this fragment that fell was two hundred and fifty-one yards. About fifty acres of land suffered from this violent convulsion; two houses were entirely destroyed; one end of a new barn was left in ruins, the walls being cracked through the very stones that composed them; a hanging coppice was changed to a naked rock; and some grass grounds and an arable field so broken and rifted by the chasms, as to be rendered, for a time, neither fit for the plough, nor safe for pasturage, till considerable labour and expense had been bestowed in levelling the surface, and filling in the gaping fissures.

LETTER XLV.

FIELD-CRICKETS.

SELBORNE.

" Resonant arbusta."
" The groves resound."

THERE is a steep abrupt pasture field, interspersed with furze, close to the back of this village, well known by the name of the Short Lithe, consisting of a rocky dry soil, and inclining to the afternoon sun. This spot abounds with the *gryllus campestris*, or field-cricket;[1] which, though frequent in these parts, is by no means a common insect in many other countries.

As their cheerful summer cry cannot but draw the attention of a naturalist, I have gone down to examine the economy of these *grylli*, and study their mode of life; but they are

(1) *Acheta campestris.*—FABRICIUS.

so shy and cautious, that it is no easy matter to get a sight of them; for, feeling a person's footsteps as he advances, they stop short in the midst of their song, and retire backward nimbly into their burrows, where they lurk till all suspicion of danger is over.

At first we attempted to dig them out with a spade, but without any great success; for either we could not get to the

FIELD CRICKET

bottom of the hole, which often terminated under a great stone; or else in breaking up the ground, we inadvertently squeezed the poor insect to death. Out of one so bruised, we took a multitude of eggs, which were long and narrow, of a yellow colour, and covered with a very tough skin. By this accident we learned to distinguish the male from the female; the former of which is shining black, with a golden stripe across his shoulders;[1] the latter is more dusky, more capacious about the abdomen, and carries a long sword-shaped weapon at her tail, which probably is the instrument with which she deposits her eggs in crannies and safe receptacles.[2]

(1)   " The vaulting grasshopper of glossy green."

(2) This instrument is termed the ovipositor, and is found in many insects, of which the most conspicuous are the grasshoppers or locusts, and the ichneumon

Where violent methods will not avail, more gentle means will often succeed; and so it proved in the present case: for, though a spade be too boisterous and rough an implement, a pliant stalk of grass, gently insinuated into the caverns, will probe their windings to the bottom, and quickly bring out the inhabitant; and thus the humane inquirer may gratify his curiosity without injuring the object of it. It is remarkable, that though these insects are furnished with long legs behind, and brawny thighs for leaping, like grasshoppers; yet when driven from their holes, they show no activity, but crawl along in a shiftless manner, so as easily to be taken: and again, though provided with a curious apparatus of wings, yet they never exert them when there seems to be the greatest occasion. The males only make that shrilling noise, perhaps out of rivalry and emulation, as is the case with many animals which exert some sprightly note during their breeding-time: it is raised by a brisk friction of one wing against the other. They are solitary beings, living singly male or female, each as it may happen; but there must be a time when they have some intercourse, and then the wings may be useful, perhaps during the hours of night. When the males meet they will fight fiercely, as I found by some which I put into the crevices of a dry stone wall, where I should have been glad to have made them settle: for though

---

flies. The ovipositor of the grasshopper is composed of six blades, two upper, and four lower, which not only serve to make a hole in the earth, but convey the egg between them safely to its position. The common crane-fly, or, as it is more familiarly termed, "Daddy-longlegs," also deposits its egg on the ground by means of a similar instrument. The ovipositor of the ichneumon flies is in some instances longer than the creature itself; and is used partly for exploring, and partly as a guide for the eggs. The ichneumons in particular require this instrument, as many of them lay their eggs in the bodies of hairy caterpillars. It is a curious sight to witness one of these little flies settle upon a caterpillar, and inflict numberless wounds in which it deposits its eggs, perfectly heedless of all the twistings and wrigglings with which the creature very naturally endeavours to shake off its assailant. Some caterpillars are so filled with the ichneumon larvæ, that they occupy the entire space between the spine and the vital organs. I have sometimes raised a heap as high as a thimble of these creatures taken out of a moderately sized caterpillar.

The ichneumons are very useful to the world at large, but very annoying to the entomologist, who when he has with great trouble procured and reared a caterpillar of a rare moth, just when he expects to see its change into the chrysalis state take place, only sees a multitude of yellow silken cases spun all round the caterpillar's empty skin.—J. G. W.

they seemed distressed by being taken out of their knowledge, yet the first that got possession of the chinks, would seize on any that were obtruded upon them, with a vast row of serrated fangs. With their strong jaws, toothed like the shears of a lobster's claws, they perforate and round their curious regular cells, having no fore-claws to dig, like the mole-cricket. When taken in hand, I could not but wonder that they never offered to defend themselves, though armed with such formidable weapons. Of such herbs as grow before the mouths of their burrows, they eat indiscriminately; and on a little platform, which they make just by, they drop their dung; and never in the day-time seem to stir more than two or three inches from home. Sitting in the entrance of their caverns, they chirp all night as well as day, from the middle of the month of May to the middle of July; and in hot weather, when they are most vigorous, they make the hills echo; and in the still hours of darkness, may be heard to a considerable distance. In the beginning of the season, their notes are more faint and inward; but become louder as the summer advances, and so die away again by degrees.

Sounds do not always give us pleasure according to their sweetness and melody; nor do harsh sounds always displease. We are more apt to be captivated or disgusted with the associations which they promote, than with the notes themselves. Thus the shrilling of the field-cricket, though sharp and stridulous, yet marvellously delights some hearers, filling their minds with a train of summer ideas of everything that is rural, verdurous, and joyous.

About the 10th of March, the crickets appear at the mouths of their cells, which they then open and bore, and shape very elegantly. All that ever I have seen at that season were in their pupa state, and had only the rudiments of wings lying under a skin, or coat, which must be cast before the insect can arrive at its perfect state:[1] from whence I should suppose that the old ones of last year do not always survive the winter. In August their holes begin to be obliterated, and the insects are seen no more till spring.

---

(1) We have observed that they cast these skins in April, which are then seen lying at the mouths of their holes.

Not many summers ago, I endeavoured to transplant a colony to the terrace in my garden, by boring deep holes in the sloping turf. The new inhabitants stayed some time, and fed and sung; but wandered away, by degrees, and were heard at a farther distance every morning; so that it appears that, on this emergency, they made use of their wings in attempting to return to the spot from which they were taken.

One of these crickets, when confined in a paper cage, and set in the sun, and supplied with plants moistened with water, will feed and thrive, and become so merry and loud as to be irksome in the same room where a person is sitting : if the plants are not wetted, it will die.

LETTER XLVI.

HOUSE-CRICKETS.

SELBORNE.

" Far from all resort of mirth,
Save the cricket on the hearth."
MILTON'S *Il Penseroso.*

DEAR SIR,—While many other insects must be sought after in fields, and woods, and waters, the *gryllus domesticus*, or house-cricket, resides altogether within our dwellings, intruding itself upon our notice whether we will or no. This species delights in new-built houses, being, like the spider, pleased with the moisture of the walls; and, besides, the softness of the mortar enables them to burrow and mine between the joints of the bricks or stones, and to open communications from one room to another. They are particularly fond of kitchens and bakers' ovens, on account of their perpetual warmth.

Tender insects that live abroad either enjoy only the short period of one summer, or else doze away the cold uncomfortable months in profound slumbers; but these, residing as it were in a torrid zone, are always alert and merry;[1] a good

---

(1) A few years since, I heard a great chirping of crickets under the grates in the kitchen of Merton College. I got a few newspapers, lighted them, and pushed them

Christmas fire is to them like the heats of the dog-days. Though they are frequently heard by day, yet is their natural time of motion only in the night. As soon as it grows dusk, the chirping increases, and they come running forth, and are

HOUSE CRICKET [1]

from the size of a flea to that of their full stature. As one should suppose, from the burning atmosphere which they inhabit, they are a thirsty race, and show a great propensity for liquids, being found frequently drowned in pans of water, milk, broth, or the like.[2] Whatever is moist they affect; and, therefore, often gnaw holes in wet woollen stockings and aprons that are hung to the fire; they are the housewife's barometer, foretelling her when it will rain; and are prognostics sometimes, she thinks, of ill or good luck; of the death of a near relation, or the approach of an absent lover.

under the grates, when out came the crickets, absolutely swarming on the floor, giving me a choice of all possible ages. I once kept a pair in my study under an inverted glass vase. One of them had lost a leg, but was so active with his one member, that he made his escape, and was within a day or two chirping merrily. He feeds on pieces of moistened bread, that I place near his home, which is behind a large water-vessel, stocked with fish, caddis worms, shells, and other aquatic creatures.—J. G. W.

(1) *Acheta domesticus.*

(2) Cockroaches are quite as thirsty beings. One day at dinner I found in my portion of pie two baked cockroaches, which had evidently scrambled in for the sake of the moisture.—J. G. W.

By being the constant companions of her solitary hours, they naturally become the objects of her superstition. These crickets are not only very thirsty, but very voracious; for they will eat the scummings of pots, and yeast, salt, and crumbs of bread, and any kitchen offal or sweepings. In the summer we have observed them to fly, when it became dusk, out of the windows, and over the neighbouring roofs. This feat of activity accounts for the sudden manner in which they often leave their haunts, as it does for the method by which they come to houses where they were not known before. It is remarkable that many sorts of insects seem never to use their wings but when they have a mind to shift their quarters and settle new colonies. When in the air, they move *volatu undoso*, in waves, or curves, like woodpeckers, opening and shutting their wings at every stroke, and so are always rising or sinking.

When they increase to a great degree, as they did once in the house where I am now writing, they become noisome pests, flying into the candles, and dashing into people's faces; but may be blasted and destroyed by gunpowder discharged into their crevices and crannies. In families, at such times, they are, like Pharaoh's plague of frogs, "in their bed-chambers, and upon their beds, and in their ovens, and in their kneading-troughs."[1] Their shrilling noise is occasioned by a brisk attrition of their wings. Cats catch hearth-crickets, and, playing with them as they do with mice, devour them. Crickets may be destroyed, like wasps, by phials half filled with beer, or any liquid, and set in their haunts; for, being always eager to drink, they will crowd in till the bottles are full.

---

(1) Exod. viii. 3.

LETTER XLVII.

MOLE-CRICKETS.

SELBORNE.

How diversified are the modes of life, not only of incongruous, but even of congenerous animals! and yet their specific distinctions are not more various than their propensities. Thus, while the field-cricket delights in sunny, dry banks, and the house-cricket rejoices amidst the glowing heat of the kitchen hearth or oven, the *gryllus gryllotalpa* (the mole-cricket) haunts most moist meadows, and frequents the sides of ponds and banks of streams, performing all its functions in a swampy, wet soil. With a pair of fore-feet, curiously adapted to the purpose, it burrows and works under ground like the mole, raising a ridge as it proceeds, but seldom throwing up hillocks.

As mole-crickets often infest gardens by the sides of canals, they are unwelcome guests to the gardener, raising up ridges in their subterraneous progress, and rendering the walks unsightly. If they take to the kitchen quarters, they occasion great damage among the plants and roots, by destroying whole beds of cabbages, young legumes, and flowers. When dug out, they seem very slow and helpless, and make no use of their wings by day; but at night they come abroad, and make long excursions, as I have been convinced by finding stragglers in a morning in improbable places. In fine weather, about the middle of April, and just at the close of day, they begin to solace themselves with a low, dull, jarring note, continued for a long time without interruption, and not unlike the chattering of the fern-owl, or goat-sucker, but more inward.

About the beginning of May they lay their eggs, as I was once an eye-witness; for a gardener, at a house where I was on a visit, happening to be mowing, on the 6th of that month, by the side of a canal, his scythe struck too deep, pared off a large piece of turf, and laid open to view a curious scene of domestic economy :—

"Ingentem lato dedit ore fenestram :
Apparet domus intus, et atria longa patescunt:
Apparent . . . . . . penetralia."

"A yawning breach of monstrous size he made ;
The inmost house is now to sight display'd ;
The admitted light with sudden lustre falls
On the long galleries and the splendid halls."

MOLE-CRICKET.[1]

There were many caverns and winding passages leading to
a kind of chamber, neatly smoothed and rounded, about
the size of a moderate snuff-box.   Within this secret nursery
were deposited near an hundred eggs, of a dirty yellow
colour, and enveloped in a tough skin ; but too lately excluded
to contain any rudiments of young, being full of a viscous
substance.   The eggs lay but shallow, and within the influence of the sun, just under a little heap of fresh moved
mould, like that which is raised by ants.

When mole-crickets fly, they move *cursu undoso*, rising
and falling in curves, like the other species mentioned
before.   In different parts of the kingdom people call them
fen-crickets, churr-worms, and eve-churrs, all very apposite
names.

(1) *Gryllotalpa vulgaris.* There are some places where the mole-cricket is found more
abundantly than others.   There is a village near Oxford, called Besselsleigh, where
they abound.   The soil is sandy, and interspersed with grass.   I paid a visit there lately,
and made inquiries of the country people after the mole-cricket, but no one had
ever heard of it, except one man, who averred that he had often seen them, and
could show me some.   After much talk on the subject, it turned out that he meant
the mole, imagining that the terminal "cricket" was an Oxford elongation of the
word.   After digging for a whole afternoon in the sand without the slightest
success, I returned home, and afterwards learned that I ought to have asked for
"croakers."—J. G. W.

Anatomists, who have examined the intestines of these insects, astonish me with their accounts; for they say that, from the structure, position, and number of their stomachs, or maws, there seems to be good reason to suppose that this and the two former species ruminate, or chew the cud, like many quadrupeds![1]

LETTER XLVIII.

HIMANTOPUS.

SELBORNE, *May* 7, 1779.

IT is now more than forty years that I have paid some attention to the ornithology of this district, without being able to exhaust the subject: new occurrences still arise as long as any inquiries are kept alive.

In the last week of last month, five of those most rare birds, too uncommon to have obtained an English name, but known to naturalists by the terms of *himantopus,* or *loripes,* and *charadrius himantopus,* were shot upon the verge of Frinsham Pond, a large lake belonging to the Bishop of

---

(1) This is an error. The digestive organs of this insect are certainly very singular, but I can see nothing in their formation to warrant such a supposition. Indeed, so far from the food being returned into the mouth, there is an organ, apparently placed for the purpose of preventing it. The upper part of the digestive organs, from which the anatomists spoken of drew this singular inference, are as follows: At a short distance from the mouth, the throat widens laterally into a crop, analogous to that of fowls. From the crop, the food passes into the gizzard, an organ also analogous to the gizzard of fowls, and furnished with no less than two hundred and seventy teeth. The coats of the gizzard are very muscular, and might by their pressure force some of the food back again into the throat, were it not furnished with a kind of valvular apparatus which effectually retains the food. Just below the gizzard, and actually surrounding and hiding it until they are separated by the hand of the anatomist, are two enormously large sacs or cæca, as they are called. These apparently serve to receive the solid food from the gizzard, while the more liquid passes on at once. There is however nothing analogous to the ruminating apparatus of the cow, and the anatomists who gave their opinion that the mole-cricket ruminated, must have based their assertion upon a very cursory examination of the digestive organs.

It is rather singular that the intestine of the mole-cricket is nearly invariably infested with worms about the eighth of an inch in length.—J. G. W.

(2) *Himantopus candidus:* the Black-winged Stilt. This very rare bird is distributed over many parts of the world, but does not appear to be common in any locality. It has several times been found in England, and its visits do not seem to be confined to any season of the year.—J. G. W.

Winchester, and lying between Wolmer Forest and the town
of Farnham, in the county of Surrey. The pond-keeper says
there were three brace in the flock; but that, after he had
satisfied his curiosity, he suffered the sixth to remain unmo-
lested. One of these specimens I procured, and found the
length of the legs to be so extraordinary, that, at first sight,
one might have supposed the shanks had been fastened on
to impose on the credulity of the beholder: they were legs
in *caricatura;* and had we seen such proportions on a Chinese or Japan screen, we should have made large allowances
for the fancy of the draughtsman. These birds are of the
plover family, and might, with propriety, be called the *stilt-
plovers.* Brisson, under that idea, gives them the apposite

STILT-PLOVER.

name of *l'échasse.* My specimen, when drawn, and stuffed
with pepper, weighed only four ounces and a quarter, though
the naked part of the thigh measured three inches and a half,
and the legs four inches and a half. Hence we may safely

assert, that these birds exhibit, weight for inches, incomparably the greatest length of legs of any known bird. The
flamingo, for instance, is one of the most long-legged birds,

FLAMINGO [1]

and yet it bears no manner of proportion to the himantopus ;
for a cock flamingo weighs, at an average, about four pounds
avoirdupoise ; and his legs and thighs measure usually about
twenty inches. But four pounds are fifteen times and a
fraction more than four ounces and a quarter ; and if four
ounces and a quarter have eight inches of legs, four pounds
must have one hundred and twenty inches and a fraction of
legs, viz. somewhat more than ten feet,—such a monstrous

(1) *Phœnicopteros rubra.*—J. G. W.

proportion as the world never saw! If you should try the
experiment in still larger birds, the disparity would sti ¹
increase. It must be matter of great curiosity to see the
stilt-plover move; to observe how it can wield such a length
of lever with such feeble muscles as the thighs seem to be
furnished with. At best, one should expect it to be but a
bad walker: but what adds to the wonder is, that it has no
back toe. Now, without that steady prop to support its
steps, it must be liable, in speculation, to perpetual vacillations,
and seldom able to preserve the true centre of gravity.

The old name of *himantopus* is taken from Pliny; and, by
an awkward metaphor, implies that the legs are as slender
and pliant as if cut out of a thong of leather. Neither
Willughby nor Ray, in all their curious researches, either at
home or abroad, ever saw this bird. Mr. Pennant never
met with it in all Great Britain, but observed it often in the
cabinets of the curious at Paris. Hasselquist says, that it
migrates to Egypt in the autumn; and a most accurate
observer of nature has assured me, that he has found it on
the banks of the streams in Andalusia.

Our writers record it to have been found only twice in
Great Britain. From all these relations it plainly appears,
that these long-legged plovers are birds of South Europe,
and rarely visit our island; and when they do, are wanderers
and stragglers, and impelled to make so distant and northern
an excursion, from motives or accidents, for which we are
not able to account. One thing may fairly be deduced, that
these birds come over to us from the Continent, since nobody
can suppose that a species not noticed once in an age, and of
such a remarkable make, can constantly breed unobserved in
this kingdom.

LETTER XLIX.

TORTOISE.

SELBORNE, *April* 21, 1780.

DEAR SIR,—The old Sussex tortoise, that I have mentioned to you so often, is become my property. I dug it out of its winter dormitory in March last, when it was enough awakened to express its resentments by hissing ; and, packing it in a box with earth, carried it eighty miles in post-chaises. The rattle and hurry of the journey so perfectly roused it, that when I turned it out on a border, it walked twice down to the bottom of my garden : however, in the evening, the weather being cold, it buried itself in the loose mould, and continues still concealed.

As it will be under my eye, I shall now have an opportunity of enlarging my observations on its mode of life and propensities ; and perceive already, that, towards the time of coming forth, it opens a breathing-place in the ground near its head, requiring, I conclude, a freer respiration as it becomes more alive. This creature not only goes under the earth from the middle of November to the middle of April, but sleeps great part of summer; for it goes to bed, in the longest days, at four in the afternoon, and often does not stir in the morning till late. Besides, it retires to rest for every shower, and does not move at all in wet days.

When one reflects on the state of this strange being, it is a matter of wonder to find that Providence should bestow such a profusion of days, such a seeming waste of longevity, on a reptile that appears to relish it so little as to squander more than two-thirds of its existence in a joyless stupor, and be lost to all sensation for months together in the profoundest of slumbers.

While I was writing this letter, a moist and warm afternoon, with the thermometer at 50, brought forth troops of shell-snails; and, at the same juncture, the tortoise heaved up the mould and put out its head; and the next morning came forth, as it were raised from the dead, and walked about till four in the afternoon. This was a curious coincidence—

a very amusing occurrence—to see such a similarity of feelings between two φερέοικοι,—for so the Greeks call both the shell-snail and the tortoise.

SNAIL. [1]

Summer birds are, this cold and backward spring, unusually late: I have seen but one swallow yet. This conformity with the weather convinces me more and more that they sleep in the winter.

MORE PARTICULARS RESPECTING THE OLD FAMILY TORTOISE.

BECAUSE we call this creature an abject reptile, we are too apt to undervalue his abilities, and to depreciate his powers of instinct. Yet he is, as Mr. Pope says of his lord,

"Much too wise to walk into a well;"

and has so much discernment as not to fall down an haha, but to stop and withdraw from the brink with the readiest precaution.

Though he loves warm weather, he avoids the hot sun; because his thick shell, when once heated, would, as the poet says of solid armour, "scald with safety." He therefore spends the more sultry hours under the umbrella of a large cabbage-leaf, or amidst the waving forests of an asparagus bed.

But as he avoids the heat in summer, so, in the decline of the year, he improves the faint autumnal beams by getting within the reflection of a fruit wall; and, though he never

(1) *Helix aspersa.*—J. G. W.

has read that planes inclining to the horizon receive a greater share of warmth, he inclines his shell, by tilting it against the wall, to collect and admit every feeble ray.

Pitiable seems the condition of this poor embarrassed reptile: to be cased in a suit of ponderous armour, which he cannot lay aside; to be imprisoned, as it were, within his own shell, must preclude, we should suppose, all activity and disposition for enterprise. Yet there is a season of the year (usually the beginning of June) when his exertions are remarkable. He then walks on tiptoe, and is stirring by five in the morning; and traversing the garden, examines every wicket and interstice in the fences, through which he will escape if possible; and often has eluded the care of the gardener, and wandered to some distant field. The motives that impel him to undertake these rambles seem to be of the amorous kind. His fancy then becomes intent on sexual attachments, which transport him beyond his usual gravity, and induce him to forget for a time his ordinary solemn deportment.

LETTER L.

HOUSE-MARTINS.

SELBORNE, *Sept.* 12, 1781.

I HAVE now read your *Miscellanies* through with much care and satisfaction; and am to return you my best thanks for the honourable mention made in them of me as a naturalist, which I wish I may deserve.

In some former letters, I expressed my suspicions that many of the house-martins do not depart in winter far from this village. I therefore determined to make some search about the south-east end of the hill, where I imagined they might slumber out the uncomfortable months of winter. But supposing that the examination would be made to the best advantage in the spring, and observing that no martins had appeared by the 11th of April last, on that day I employed some men to explore the shrubs and cavities of the suspected spot. The persons took pains, but without any success; however, a remarkable incident occurred in the midst of our pursuit,—while the labourers were at work, a house-martin, the first that had been seen this year, came down the village in the sight of several people, and went at once into a nest, where it stayed a short time, and then flew over the houses; for some days after, no martins were observed, not till the 16th of April, and then only a pair.   Martins in general were remarkably late this year.

LETTER LI.

SWIFTS.

SELBORNE, *Sept.* 9, 1781.

I HAVE just met with a circumstance respecting swifts, which furnishes an exception to the whole tenor of my observations ever since I have bestowed any attention on that species of *hirundines.*   Our swifts, in general, withdrew this year about the first day of August, all save one pair, which in two or three days was reduced to a single bird.   The perseverance of this individual made me suspect that the strongest of motives, that of an attachment to her young, could alone occasion so late a stay.   I watched, therefore, till the 24th of August, and then discovered that, under the eaves of the church, she attended upon two young, which were fledged, and now put out their white chins from a crevice.   These remained till the 27th, looking more alert every day, and seeming to long to be on the wing.   After this day, they were missing at once; nor could I ever observe them with their dam coursing round the church in the act of learning to fly, as the first broods evidently do.   On the 31st, I caused the eaves to be searched; but we found in the nest only two callow, dead, stinking swifts, on which a second nest had been formed.   This double nest was full of the black shining cases of the *hippoboscæ hirundinis.*

The following remarks on this unusual incident are obvious. The first is, that though it may be disagreeable to swifts to remain beyond the beginning of August, yet that they can subsist longer is undeniable.   The second is, that this uncommon event, as it was owing to the loss of the first brood, so it corroborates my former remark, that swifts breed regularly but once; since, was the contrary the case, the occurrence above could neither be new nor rare.

P.S. One swift was seen at Lyndon, in the county of Rutland, so late as the 3d of September.

LETTER LII.

INSECTS—THE COCCUS—APHIDES.

As I have sometimes known you make inquiries about several kinds of insects, I shall here send you an account of one sort which I little expected to have found in this kingdom. I had often observed that one particular part of a vine, growing on the walls of my house, was covered in the autumn with a black dust-like appearance, on which the flies fed eagerly; and the shoots and leaves thus affected did not thrive, nor did the fruit ripen. To this substance I applied my glasses; but could not discover that it had anything to do with animal life, as I at first expected: but upon a closer examination behind the larger boughs, we were surprised to find that they were coated over with husky shells, from whose sides proceeded a cotton-like substance, surrounding a multitude of eggs. This curious and uncommon production put me upon recollecting what I have heard and read concerning the *coccus vitis viniferæ* of Linnæus; which, in the south of Europe, infests many vines, and is a horrid and loathsome pest. As soon as I had turned to the accounts given of this insect, I saw at once that it swarmed on my vine; and did not appear to have been at all checked by the preceding winter, which had been uncommonly severe.

Not being then at all aware that it had anything to do with England, I was much inclined to think that it came from Gibraltar, among the many boxes and packages of plants and birds which I had formerly received from thence; and especially as the vine infested grew immediately under my study window, where I usually kept my specimens. True it is, that I had received nothing from thence for some years:

but as insects, we know, are conveyed from one country to another in a very unexpected manner, and have a wonderful power of maintaining their existence till they fall into a *nidus* proper for their support and increase, I cannot but suspect still that these *cocci* came to me originally from Andalusia. Yet, all the while, candour obliges me to confess, that Mr. Lightfoot has written me word that he once, and but once saw these insects on a vine at Weymouth, in Dorsetshire ; which it is here to be observed, is a seaport town, to which the *coccus* might be conveyed by shipping.

As many of my readers may possibly never have heard of this strange and unusual insect, I shall here transcribe a passage from a *Natural History of Gibraltar*. written by the Reverend John White, late Vicar of Blackburn, in Lancashire, but not yet published.

"In the year 1770, a vine, which grew on the east side of my house, and which had produced the finest crop of grapes for years past, was suddenly overspread on all the woody branches with large lumps of a white fibrous substance, resembling spiders' webs, or rather raw cotton. It was of a very clammy quality, sticking fast to everything that touched it, and capable of being spun into long threads. At first I suspected it to be the product of spiders, but could find none. Nothing was to be seen connected with it, but many brown oval husky shells, which by no means looked like insects, but rather resembled bits of dry bark of the vine. The tree had a plentiful crop of grapes set, when this pest appeared upon it; but the fruit was manifestly injured by this foul encumbrance. It remained all the summer, still increasing, and loaded the woody and bearing branches to a vast degree. I often pulled off great quantities by handfuls; but it was so slimy and tenacious that it could by no means be cleared. The grapes never filled to their natural perfection, but turned watery and vapid. Upon perusing the works afterwards of M. de Reaumur, I found this matter perfectly described and accounted for. Those husky shells which I had observed, were no other than the female *coccus*, from whose sides this cotton-like substance exudes, and serves as a covering and security for their eggs."

To this account I think proper to add, that though the

female *cocci* are stationary, and seldom remove from the place to which they stick, yet the male is a winged insect; and that the black dust which I saw was undoubtedly the excrement of the females, which is eaten by ants as well as flies. Though the utmost severity of our winter did not destroy these insects, yet the attention of the gardener, in a summer or two, has entirely relieved my vine from this filthy annoyance.

As we have remarked above, that insects are often conveyed from one country to another in a very unaccountable manner, I shall here mention an emigration of small *aphides* which was observed in the village of Selborne, no longer ago than August the 1st, 1785.

APHIS.[1]

At about three o'clock in the afternoon of that day, which was very hot, the people of this village were surprised by a shower of *aphides*, or smother-flies, which fell in these parts. Those that were walking in the street at that juncture found themselves covered with these insects, which settled also on the hedges and gardens, blackening all the vegetables where they alighted. My annuals were discoloured with them, and the stalks of a bed of onions were quite coated over for six days after. These armies were then, no doubt, in a state of emigration, and shifting their quarters: and might have come, as we know, from the great hop plantations of Kent or Sussex, the wind being all that day in the easterly quarter. They were observed, at the same time, in great clouds, about Farnham, and all along the vale from Farnham to Alton.[2]

(1) Magnified about five diameters.—J. G. W.

(2) For various methods by which several insects shift their quarters, see DER-HAM's *Physico-Theology*.

LETTER LIII.

GOLD AND SILVER FISHES.

DEAR SIR,—When I happen to visit a family where gold
and silver fishes are kept in a glass bowl, I am always
pleased with the occurrence, because it offers me an opportunity of observing the actions and propensities of those
beings, with whom we can be little acquainted in their
natural state.   Not long since, I spent a fortnight at the
house of a friend, where there was such a *vivary*, to which
I paid no small attention, taking every occasion to remark
what passed within its narrow limits.   It was here that I
first observed the manner in which fishes die.   As soon as
the creature sickens, the head sinks lower and lower, and it
stands, as it were, on its head; till, getting weaker, and
losing all poise, the tail turns over, and, at last, it floats on
the surface of the water, with its belly uppermost.   The
reason why fishes, when dead, swim in that manner, is very
obvious; because, when the body is no longer balanced by
the fins of the belly, the broad muscular back preponderates
by its own gravity, and turns the belly uppermost, as lighter,
from its being a cavity, and because it contains the swimming bladders, which contribute to render it buoyant.[1]   Some
that delight in gold and silver fishes, have adopted a notion
that they need no aliment.   True it is, that they will subsist
for a long time without any apparent food but what they can
collect from pure water frequently changed; yet they must
draw some support from animalcula, and other nourishment
supplied by the water; because, though they seem to eat
nothing, yet the consequences of eating often drop from them.
That they are best pleased with such *jejune* diet may easily
be confuted, since, if you toss them crumbs, they will seize

(1) When my own fish die, they always remain for some time on the surface of the
water. They then act as White describes, but very often float on their backs for
several hours before their death. I have had often considerable difficulty in cap-
turing a fish, that lay on its back apparently dead; as at the first touch of the hand
it gave a great flounce, and darted about with great rapidity, but soon subsided into
the position assumed by the bad shepherd, according to the Eton Grammar.—J. G. W

them with great readiness, not to say greediness: however, bread should be given sparingly, lest, turning sour, it corrupt the water. They will also feed on the water-plant called *lemna* (duck's meat), and also on small fry.

GOLD FISH [1]

When they want to move a little, they gently protrude themselves with their *pinnæ pectorales;* but it is with their strong muscular tails only that they, and all fishes, shoot along with such inconceivable rapidity. It has been said, that the eyes of fishes are immoveable; but these apparently turn them forward or backward, in their sockets, as their occasions require. They take little notice of a lighted candle, though applied close to their heads, but flounce, and seem much frightened, by a sudden stroke of the hand against the support whereon the bowl is hung; especially when they have been motionless, and are perhaps asleep. As fishes have no eyelids, it is not easy to discern when they are sleeping or not, because their eyes are always open.

Nothing can be more amusing than a glass bowl containing such fishes: the double refractions of the glass and water

(1) *Cyprinus auratus.* Gold-fish will eat bread and many other artificial aliments. If the fish are kept in a glass globe, they should be transferred to a basin while the water in the globe is being changed, and there fed, or the water will be discoloured. The lower of the two fishes represented is the gold-fish, the upper figure represents the barbel.—J. G. W.

represent them, when moving, in a shifting and changeable variety of dimensions, shades, and colours; while the two mediums, assisted by the concavo-convex shape of the vessel, magnify and distort them vastly; not to mention that the introduction of another element and its inhabitants into our parlours engages the fancy in a very agreeable manner.

Gold and silver fishes, though originally natives of China and Japan, yet are become so well reconciled to our climate, as to thrive and multiply very fast in our ponds and stews. Linnæus ranks this species of fish under the genus of *cyprinus*, or carp, and calls it *cyprinus auratus.*[1]

Some people exhibit this sort of fish in a very fanciful way; for they cause a glass bowl to be blown with a large hollow space within, that does not communicate with it. In this cavity they put a bird occasionally, so that you may see a goldfinch or a linnet hopping, as it were, in the midst of the water, and the fishes swimming in a circle round it. The simple exhibition of the fishes is agreeable and pleasant; but in so complicated a way becomes whimsical and unnatural, and liable to the objection due to him,

> " Qui variare cupit rem prodigialitèr unam."

> " Who loves to vary every single thing
> Prodigiously."

LETTER LIV.

HOUSE-MARTINS.

October 10, 1781.

DEAR SIR,—I think I have observed before, that much the most considerable part of the house-martins withdraw from hence about the first week in October; but that some, the latter broods, I am now convinced, linger on till the middle of that month; and that, at times, once perhaps in two or

---

(1) There are several ponds at Oxford stocked with these beautiful fish, which have increased not only in number, but in size. One pond-full came to an untimely end. The pond was partly supplied and warmed by the waste water from a steam engine. One day, some verdigris that had formed in some part of the engine was washed into the pond, and poisoned every one of the fish. Warmed water, however, is not necessary, for there are several ponds in Oxford which are frozen over every year, and yet support plenty of gold-fish, which may often be seen swimming about under the ice.—J. G. W.

three years, a flight, for one day only, has shown itself in the first week in November.

Having taken notice, in October, 1780, that the last flight was numerous, amounting perhaps to one hundred and fifty, and that the season was soft and still, I resolved to pay uncommon attention to these late birds, to find, if possible, where they roosted, and to determine the precise time of their retreat. The mode of life of the latter *hirundines* is very favourable to such a design, for they spend the whole day in the sheltered district between me and the Hanger, sailing about in a placid, easy manner, and feasting on those insects which love to haunt a spot so secure from ruffling winds. As my principal object was to discover the place of their roosting, I took care to wait on them before they retired to rest, and was much pleased to find that, for several evenings together, just at a quarter past five in the afternoon, they all scudded away in great haste towards the south-east, and darted down among the low shrubs above the cottages at the end of the hill. This spot, in many respects, seems to be well calculated for their winter residence, for, in many parts, it is as steep as the roof of any house, and, therefore, secure from the annoyances of water; and it is, moreover, clothed with beechen shrubs, which, being stunted and bitten by sheep, make the thickest covert imaginable, and are so entangled as to be impervious to the smallest spaniel; besides, it is the nature of underwood beech never to cast its leaf all the winter, so that, with the leaves on the ground and those on the twigs, no shelter can be more complete. I watched them on to the thirteenth and fourteenth of October, and found their evening retreat was exact and uniform; but after this they made no regular appearance. Now and then a straggler was seen; and, on the twenty-second of October, I observed two, in the morning, over the village, and with them my remarks for the season ended.

From all these circumstances put together, it is more than probable that this lingering flight, at so late a season of the year, never departed from the island. Had they indulged me that autumn with a November visit, as I much desired, I presume that, with proper assistants, I should have settled the matter past all doubt; but though the third of November was a sweet day, and, in appearance, exactly suited to my wishes, yet not a martin was to be seen, and so I was forced, reluctantly, to give up the pursuit.

I have only to add, that were the bushes, which cover some acres, and are not my own property, to be grubbed and carefully examined, probably those late broods, and perhaps the whole aggregate body of the house-martins of this district, might be found there, in different secret dormitories; and that, so far from withdrawing into warmer climes, it would appear that they never depart three hundred yards from the village.

LETTER LV.

INSTINCT—MUSIC.

THEY who write on natural history, cannot too frequently advert to instinct, that wonderful limited faculty, which, in some instances, raïses the brute creation, as it were, above reason, and in others, leaves them so far below it. Philosophers have defined instinct to be that secret influence by which every species is impelled naturally to pursue, at all times, the same way, or track, without any teaching or example; whereas reason, without instruction, would often vary, and do that by many methods which instinct effects by one alone. Now, this maxim must be taken in a qualified sense, for there are instances in which instinct does vary and conform to the circumstances of place and convenience.

It has been remarked, that every species of bird has a mode of nidification peculiar to itself, so that a schoolboy would at once pronounce on the sort of nest before him. This is the case among fields, and woods, and wilds; but, in the villages round London, where mosses, and gossamer and cotton from vegetables, are hardly to be found, the nest of the chaffinch has not that elegant finished appearance, nor is it so beautifully studded with lichens, as in a more rural district, and the wren is obliged to construct its house with straws and dry grasses, which do not give it that rotundity and compactness so remarkable in the edifices of that little architect. Again, the regular nest of the house-martin is hemispheric but where a raf⁺er, or a joist, or a cornice, may happen to stand in the way, the nest is so contrived as to conform to the obstruction, and becomes flat, or oval, or compressed.[1]

In the following instances, instinct is perfectly uniform and consistent. There are three creatures—the squirrel, the field-

SQUIRREL.[2]

mouse and the nut-hatch (*Sitta europœa*), which live much on hazel-nuts, and yet they open them each in a different way. The first, after rasping off the small end, splits the shell into two with his long fore-teeth, as a man does with his knife; the

(1) " Each creature has a wisdom for its good :
    The pigeons feed their tender offspring, crying,
    When they are callow, but withdraw their food
    When they are fledged, that need may teach them flying."—HERBERT.

This change of nidification is very common in insects. There are few insects who in captivity do not contrive to make some kind of a domicile from the means given them. This more especially applies to the larvæ of the Lepidoptera, when about to undergo the great change from a caterpillar into a chrysalis. If the caterpillar of the Puss-Moth, (*Cerura vinula*) be placed in a tin box, and thereby deprived of the wood necessary for forming its curious hut, it makes for itself a semi-transparent case, formed apparently out of a gelatine-like substance secreted by the insect itself, and which in its natural state would only serve to fasten together the little chips of wood with which it makes the wall of its house so firm, that a strong knife is required to pierce it.

I have now before me a still more curious instance of adaptation of the natural instinct to peculiar purposes. I had placed the caterpillar of the gold-tailed moth *Porthesia chrysorrhœa*) in a box, which also contained a remarkably fine dragon-fly. On opening the box some days afterwards, the caterpillar was found not only to have made a habitation for itself under such disadvantageous circumstances, but even to have taken advantage of the means placed within its reach to add to the warmth or security of its dwelling place. It had bitten off the wings of the dragon fly, cut them into small pieces, and woven them into the substance of its web, so that its habitation presented the most extraordinary appearance. Similar instances must have occurred to every practical entomologist.—J. G. W.

(2) *Sciurus Europæus.*—J. G. W.

second nibbles a hole with his teeth, so regular as if drilled
with a wimble, and yet so small that one would wonder how
the kernel can be extracted through it; while the last picks
an irregular ragged hole with its bill; but as this artist has
no paws to hold the nut firm while he pierces it, like an
adroit workman, he fixes it as it were in a vice, in some cleft
of a tree, or in some crevice, when, standing over it, he
perforates the stubborn shell. We have often placed nuts in
the chink of a gate-post, where nut-hatches have been known
to haunt, and have always found that those birds have readily
penetrated them. While at work they make a rapping noise
that may be heard at a considerable distance.

You that understand both the theory and practical part of
music, may best inform us why harmony or melody should so
strangely affect some men, as it were, by recollection, for days
after a concert is over. What I mean, the following passage
will most readily explain:—

"Præhabebat porrò vocibus humanis, instrumentisque
harmonicis, musicam illam avium: non quod aliâ quoque non
delectaretur; sed quod ex musicâ humanâ relinqueretur in
animo continens quædam, attentionemque et somnum con-
turbans agitatio: dum ascensus, exscensus, tenores, ac mu-
tationes illæ sonorum et consonantiarum, euntque, redeuntque
per phantasiam:—cum nihil tale relinqui possit ex modula-
tionibus avium, quæ, quod non sunt perinde a nobis imitabiles,
non possunt perinde internam facultatem commovere."
GASSENDUS, _in Vitâ Peireskii._—" He preferred, also, the music
of birds to vocal and instrumental harmony; not that he did
not take pleasure in any other, but because there was left in
the mind some constant agitation, disturbing the sleep and
the attention, whilst the several variations of sound and
concord go and return through the imagination, when no such
effect can be produced by the modulation of birds, because, as
they are not equally imitable by us, they cannot equally
excite the internal faculty."

This curious quotation strikes me much by so well repre-
senting my own case, and by describing what I have so often
felt, but never could so well express. When I hear fine music,
I am haunted with passages therefrom night and day; and
especially at first waking, which, by their importunity, give

more uneasiness than pleasure: elegant lessons still tease my imagination, and recur irresistibly to my recollection at seasons, and even when I am desirous of thinking of more serious matters.[1]

LETTER LVI.

PETTICHAPS—PEREGRINE FALCON.

A RARE, and I think a new, little bird frequents my garden, which, I have great reason to think, is the pettichaps: it is

PETTICHAPS.[2]

(1) Some persons, myself among the number, find that the constant rattling of a railway carriage is apt to resolve itself into musical sounds, usually of the most strange and wild description, utterly irrespective of the will, which cannot call them up or even guide them, but is irresistibly carried on with the stream of harmony. On one night journey, I well recollect one strange piece of music that stood out as it were from the rest. A full chorus was being executed, when suddenly, all the voices ceased except one soprano, which commenced a simple melody, such as I never heard before, and which I could not recollect afterwards. Soon the remainder of the choir joined, and while the soprano still kept to the same air, the chorus began to execute the most extraordinary variations upon it, the chords changing and winding about in the most wild and wonderful manner imaginable. So clear was the whole performance, that, had I been a musician, I could have written the entire piece of music.

As to the involuntary recollections of music described by White, there are few who have not been haunted by them. There was an instance of a gentleman, a good musician, and of a very nervous temperament, being actually driven into a brain fever by the constant recurrence of a musical passage, which never left him day or night.—J. G. W.

(2) *Sylvia hortensis.*—J. G. W.

common in some parts of the kingdom; and I have received formerly several dead specimens from Gibraltar. This bird much resembles the white-throat, but has a more white, or rather silvery, breast and belly; is restless and active like the willow-wrens, and hops from bough to bough, examining every part for food: it also runs up the stems of the crown imperials, and putting its head into the bells of those flowers sips the liquor which stands in the nectarium of each petal. Sometimes it feeds on the ground like the hedge-sparrow, by hopping about on the grass-plots and mown walks.

One of my neighbours, an intelligent and observing man, informs me, that in the beginning of May, and about ten minutes before eight o'clock in the evening, he discovered a great cluster of house-swallows, thirty, at least, he supposes, perching on a willow that hung over the verge of James Knight's upper pond. His attention was first drawn by the twittering of these birds, which sat motionless in a row on the bough, with their heads all one way, and by their weight pressing down the twig, so that it nearly touched the water. In this situation he watched them till he could see no longer. Repeated accounts of this sort, spring and fall, induce us greatly to suspect, that house-swallows have some strong attachment to water, independent of the matter of food; and though they may not retire into that element, yet they may conceal themselves in the banks of pools and rivers during the uncomfortable months of winter.

One of the keepers of Wolmer Forest sent me a peregrine falcon, which he shot on the verge of that district, as it was devouring a wood-pigeon. The *Falco peregrinus*, or haggard falcon, is a noble species of hawk, seldom seen in the southern counties. In winter 1767, one was killed in the neighbouring parish of Faringdon, and sent by me to Mr. Pennant into North Wales.[1] Since that time, I have met with none till now. The specimen mentioned above was in fine preservation, and not injured by the shot: it measured forty-two inches from wing to wing, and twenty-one from beak to tail, and weighed two pounds and a half standing weight. This species is very robust, and wonderfully formed for rapine: its breast was plump and muscular; its thighs long, thick and brawny; and its legs remarkably short and well-set: the feet

(1) See Letters X. and XI. to Thomas Pennant, Esq.

were armed with most formidable, sharp, long talons: the eyelids and cere of the bill were yellow; but the irides of the eyes dusky: the beak was thick and hooked, and of a dark

PEREGRINE FALCON.[1]

colour, and had a jagged process near the end of the upper mandible on each side: its tail, or train, was short in proportion to the bulk of its body; yet the wings, when closed, did not extend to the end of the train. From its large and fair proportions, it might be supposed to have been a female; but I was not permitted to cut open the specimen. For one of the birds of prey, which are usually lean, this was in high case; in its craw were many barleycorns, which probably came from the crop of the wood-pigeon, on which it was feeding when shot: for voracious birds do not eat grain; but, when devouring their quarry, with undistinguishing vehemence swallow bones and feathers, and all matters, indiscriminately. This falcon was probably driven from the mountains in North Wales or Scotland, where they are known to breed, by rigorous weather and deep snows that had lately fallen.

(1) *Falco peregrinus.* This beautiful bird is the falcon so generally used in the olden times for taking the heron. The language of falconry had many names for the bird, according to its age. Thus a wild falcon was called a "haggard," a young bird was named "eyas" or "red falcon," the full-grown male was termed "tiercel," or "tassel gentle," while the adult female alone arrived at the dignity of the term "falcon."—J. G. W.

LETTER LVII.

CHINESE DOGS—THE SPANIEL.

My near neighbour, a young gentleman in the service of the East India Company, has brought home a dog and a bitch of the Chinese breed from Canton; such as are fattened in that country for the purpose of being eaten. They are about the size of a moderate spaniel; of a pale yellow colour, with coarse bristling hair on their backs; sharp upright ears, and peaked heads, which give them a very fox-like appearance. Their hind legs are unusually straight, without any bend at the hock, or ham; to such a degree as to give them an awkward gait when they trot. When they are in motion, their tails are curved high over their backs like those of some hounds, and have a bare place each on the outside from the tip midway, that does not seem to be matter of accident, but somewhat singular. Their eyes are jet-black, small, and piercing: the insides of their lips and mouths of the same colour, and their tongues blue. The bitch has a dew-claw on each hind leg: the dog has none. When taken out into a field, the bitch showed some disposition for hunting, and dwelt on the scent of a covey of partridges till she sprung them, giving her tongue all the time. The dogs in South America are dumb: but these bark much in a short thick manner, like foxes, and have a surly, savage demeanour, like their ancestors, which are not domesticated, but bred up in sties, where they are fed for the table with rice-meal and other farinaceous food. These dogs, having been taken on board as soon as weaned, could not learn much from their dam; yet they did not relish flesh when they came to England. In the islands of the Pacific Ocean, the dogs are bred up on vegetables, and would not eat flesh when offered them by our circumnavigators.

We believe that all dogs, in a state of nature, have sharp, upright, fox-like ears; and that hanging ears, which are esteemed so graceful, are the effect of choice breeding and cultivation. Thus, in the Travels of Ysbrandt Ides from Muscovy to China, the dogs which draw the Tartars on snow-sledges near the river Oby, are engraved with prick-ears, like those from Canton. The Kamschatdales also train the same sort of sharp-eared, peak-nosed dogs to draw their sledges; as may be seen in an elegant print engraved for Captain Cook's last voyage round the world.

Now we are upon the subject of dogs, it may not be impertinent to add, that spaniels, as all sportsmen know, though

WATER-SPANIEL.

they hunt partridges and pheasants as it were by instinct,[1] and with much delight and alacrity, yet will hardly touch their bones when offered as food; nor will a mongrel dog of my own, though he is remarkable for finding that sort of game. But when we came to offer the bones of partridges to the two Chinese dogs, they devoured them with much greediness, and licked the platter clean.

No sporting dogs will flush woodcocks till inured to the scent, and trained to the sport, which they then pursue with vehemence and transport; but then they will not touch their bones, but turn from them with abhorrence, even when they are hungry.

(1) It is well known to all sportsmen, that the young pointer will often mark a bird the first time he is taken into the field, retaining not only the point, but even some of the peculiarities of his parents.—J. G. W.

Now, that dogs should not be fond of the bones of such birds as they are not disposed to hunt, is no wonder; but why they reject and do not care to eat their natural game, is not so easily accounted for, since the end of hunting seems to be, that the chase pursued should be eaten. Dogs, again, will not devour the more rancid water-fowls; nor indeed the bones of any wild-fowls; nor will they touch the fetid bodies of birds that feed on offal and garbage; and indeed there may be somewhat of providential instinct in this circumstance of dislike; for vultures,[1] and kites, and ravens, and crows, &c., were intended to be messmates with dogs over their carrion; and seem to be appointed by Nature as fellow-scavengers, to remove all cadaverous substances from the face of the earth.

LETTER LVIII.

FOSSIL WOOD—STONE-CURLEWS—ROOKS.

THE fossil wood buried in the bogs of Wolmer Forest, is not yet all exhausted; for the peat-cutters now and then stumble upon a log. I have just seen a piece which was sent by a labourer of Oakhanger to a carpenter of this village. This was the butt-end of a small oak, about five feet long, and about five inches in diameter. It had apparently been severed from the ground by an axe, was very ponderous, and as black as ebony. Upon asking the carpenter for what purpose he had procured it, he told me that it was to be sent to his brother, a joiner at Farnham, who was to make use of it in cabinet work, by inlaying it along with whiter woods.

Those that are much abroad on evenings after it is dark, in spring and summer, frequently hear a nocturnal bird passing by on the wing, and repeating often a short quick note. This bird I have remarked myself, but never could make it out till lately. I am assured now that it is the stone-curlew (*Charadrius œdicnemus*.) Some of them pass over or near my house almost every evening after it is dark, from the uplands

---

(1) Hasselquist, in his *Travels to the Levant*, observes that the dogs and vultures at Grand Cairo maintain such a friendly intercourse, as to bring up their young together in the same place.

of the hill and Northfield, away down towards Dorton; where, among the streams and meadows, they find a greater plenty of food. Birds that fly by night are obliged to be noisy; their notes, often repeated, become signals or watch-words to keep them together, that they may not stray or lose each other in the dark.

The evening proceedings and manœuvres of the rooks are curious and amusing in the autumn. Just before dusk, they return in long strings[1] from the foraging of the day, and ren-

ROOK. [2]

dezvous by thousands over Selborne-down, where they wheel round in the air, and sport and dive in a playful manner, all the while exerting their voices, and making a loud cawing, which, being blended and softened by the distance that we

(1) Every naturalist is a poet, and every poet is a naturalist. The reader will not fail to notice how Burns has observed the same habit of the rooks:—

"The weary beasts returning from the plough,
The blackening train of craws to their repose."—J. G. W.

(2) *Corvus frugilegus.* Although it displays such a horror of guns, the Rook is a very courageous bird, and will defend itself with great pertinacity. I once winged a rook, and on going up to take it, the wounded bird, which had previously made every effort to escape, suddenly turned round, and commenced a most determined attack on my ankles, forcing me to use the ramrod on his head before I could secure him.—J. G. W

at the village are below them, become a confused noise or chiding, or rather a pleasing murmur, very engaging to the imagination, and not unlike the cry of a pack of hounds in hollow echoing woods, or the rushing of the wind in tall trees, or the tumbling of the tide upon a pebbly shore. When this ceremony is over, with the last gleam of day, they retire for the night to the deep beechen woods of Tisted and Ropley. We remember a little girl, who, as she was going to bed, used to remark on such an occurrence, in the true spirit of physico-theology, that the rooks were saying their prayers; and yet this child was much too young to be aware that the Scriptures have said of the Deity, that "he feedeth the ravens who call upon him."

LETTER LIX.

RAIN—ECHOES—BAROMETER.

In reading Dr. Huxham's *Observationes de Aëre*, written at Plymouth, I find, by those curious and accurate remarks, which contain an account of the weather from the year 1727 to the year 1748, inclusive, that though there is frequent rain in that district of Devonshire, yet the quantity falling is not great; and that some years it has been very small: for in 1731, the rain measured only 17·266 inches, and in 1741, 20·354; and again, in 1743, only 20·908. Places near the sea have frequent scuds, that keep the atmosphere moist, yet do not reach far up into the country: making thus the maritime situations appear wet, when the rain is not considerable. In the wettest years at Plymouth, the doctor measured only once 36; and again once, viz. in 1734, 37·114; a quantity or rain that has twice been exceeded at Selborne in the short period of my observations. Dr. Huxham remarks, that frequent small rains keep the air moist; while heavy ones render it more dry, by beating down the vapours. He is also or opinion, that the dingy smoky appearance in the sky, in very dry seasons, arises from the want of moisture sufficient to let the light through, and render the atmosphere transparent; because he had observed several bodies more diaphanous when wet than dry; and did never recollect that the air had that look in rainy seasons.

My friend, who lives just beyond the top of the down, brought his three swivel guns to try them in my outlet, with their muzzles towards the Hanger, supposing that the report would have had a great effect; but the experiment did not answer his expectation. He then removed them to the alcove on the Hanger, when the sound, rushing along the Lythe and Comb-wood, was very grand; but it was at the Hermitage that the echoes and repercussions delighted the hearers; not only filling the Lythe with the roar, as if all the beeches were tearing up by the roots, but, turning to the left, they pervaded the vale above Comb-wood ponds; and, after a pause, seemed to take up the crash again, and to extend round Harteley Hangers, and to die away at last among the coppices and coverts of Ward-le-ham. It has been remarked before, that this district is an *Anathoth*, a place of responses, or echoes, and, therefore, proper for such experiments. We may further add, that the pauses in echoes, when they cease, and yet are taken up again, like the pauses in music, surprise the hearers, and have a fine effect on the imagination.

The gentleman above mentioned has just fixed a barometer in his parlour at Newton Valence. The tube was first filled here (at Selborne), twice with care, when the mercury agreed, and stood exactly with my own; but being filled again twice at Newton, the mercury stood, on account of the great elevation of that house, three-tenths of an inch lower than the barometers at this village, and so continues to do, be the weight of the atmosphere what it may. The plate of the barometer at Newton is figured as low as 27; because, in stormy weather, the mercury there will sometimes descend below 28. We have supposed Newton House to stand two hundred feet higher than this house; but if the rule holds good, which says that mercury in a barometer sinks one-tenth of an inch for every hundred feet elevation, then the Newton barometer, by standing three-tenths lower than that of Selborne, proves that Newton House must be three hundred feet higher than that in which I am writing, instead of two hundred.

It may not be impertinent to add, that the barometers at

Selborne stand three-tenths of an inch lower than the barometers at South Lambeth; whence we may conclude, that the former place is about three hundred feet higher than the latter; and with good reason, because the streams that rise with us run into the Thames at Weybridge, and so to London. Of course, therefore, there must be lower ground all the way from Selborne to South Lambeth; the distance between which, all the windings and indentings of the streams considered, cannot be less than a hundred miles.[1]

LETTER LX.

WEATHER—SHRUBS—FROSTS—BEES—THAWS.

SINCE the weather of a district is undoubtedly part of its natural history, I shall make no farther apology for the four following letters, which will contain many particulars concerning some of the great frosts, and a few respecting some very hot summers, that have distinguished themselves from the rest during the course of my observations.

As the frost in January, 1768, was, for the small time it lasted, the most severe that we had then known for many years, and was remarkably injurious to evergreens, some account of its rigour, and reason of its ravages, may be useful, and not unacceptable to persons that delight in planting and ornamenting: and may particularly become a work that professes never to lose sight of utility.

For the last two or three days of the former year, there were considerable falls of snow, which lay deep and uniform on the ground without any drifting, wrapping up the more humble vegetation in perfect security. From the first day to the fifth of the new year, more snow succeeded; but, from that day, the air became entirely clear, and the heat of the sun about noon had a considerable influence in sheltered situations.

---

(1) The aneroid barometer, an instrument made without mercury, and acting by the pressure of the atmosphere on a thin plate of metal, is so accurate and rapid in its working, that a gentleman who was travelling on one of the Lancashire Railways, was enabled to read off all the variations in level within a foot or so.—J. G. W.

It was in such an aspect, that the snow on the author's evergreens was melted every day, and frozen intensely every night; so that the laurustines, bays, laurels, and arbutuses, looked, in three or four days, as if they had been burnt in the fire; while a neighbour's plantation of the same kind, in a high, cold situation, where the snow was never melted at all, remained uninjured.

From hence I would infer, that it is the repeated melting and freezing of the snow that is so fatal to vegetation, rather than the severity of the cold. Therefore, it highly behoves every planter, who wishes to escape the cruel mortification of losing, in a few days, the labour and hopes of years, to bestir himself on such emergencies; and, if his plantations are small, to avail himself of mats, cloths, peas-haulm, straw, reeds, or any such covering, for a short time; or, if his shrubberies are extensive, to see that his people go about with prongs and forks, and carefully dislodge the snow from the boughs; since the naked foliage will shift much better for itself, than where the snow is partly melted and frozen again.

It may perhaps appear, at first, like a paradox; but doubtless the more tender trees and shrubs should never be planted in hot aspects: not only for the reason assigned above, but also, because, thus circumstanced, they are disposed to shoot earlier in the spring, and to grow on later in the autumn than they would otherwise do, and so are sufferers by lagging or early frosts. For this reason, also, plants from Siberia will hardly endure our climate; because, on the very first advances of spring, they shoot away, and so are cut off by the severe nights of March or April.

Dr. Fothergill and others have experienced the same inconvenience with respect to the more tender shrubs from North America; which they, therefore, plant under north walls. There should also, perhaps, be a wall to the east, to defend them from the piercing blasts from that quarter.

This observation might, without any impropriety, be carried into animal life; for discerning bee-masters now find that their hives should not, in the winter, be exposed to the hot sun, because such unseasonable warmth awakens the inhabitants too early from their slumbers; and, by putting their

juices into motion too soon, subjects them afterwards to
inconveniences when rigorous weather returns.

The coincidents attending this short but intense frost
were, that the horses fell sick, with an epidemic distemper,
which injured the winds of many, and killed some ; that
colds and coughs were general among the human species ;
that it froze under people's beds for several nights ; that
meat was so hard frozen that it could not be spitted, and
could not be secured but in cellars ; that several redwings
and thrushes were killed by the frost ; and that the large
titmouse continued to pull straws lengthwise from the eaves
of thatched houses and barns in a most adroit manner, for a
purpose that has been explained already.[1]

On the third of January, Benjamin Martin's thermometer,
within doors, in a close parlour where there was no fire, fell
in the night to 20, and on the fourth to 18, and on the seventeenth to $17\frac{1}{2}$, a degree of cold which the owner never since
saw in the same situation ; and he regrets much that he was
not able, at that juncture, to attend his instrument abroad.
All this time the wind continued north and north-east; and
yet on the eighth, roost-cocks, which had been silent, began
to sound their clarions, and crows to clamour, as prognostic
of milder weather ; and moreover, moles began to heave and

MOLE.[2]

(1) See Letter XLI. to Thomas Pennant, Esq.

(2) *Talpa Europæa.* The country people inform me that the mole works and
rests at regular intervals of three hours each.

No one would imagine from the appearance of the mole, that it was an animal of
so energetic, and even fierce a character. The appetite of the mole is perfectly
ravenous, and the animal dies if it has been forced to fast more than a few hours.
It also requires a constant supply of water, and if its residence does not happen to
be near a river or pond, it digs a number of little wells, which retain the rain and

work, and a manifest thaw took place. From the latter circumstance, we may conclude, that thaws often originate under ground from warm vapours which arise, else how should subterraneous animals receive such early intimations

MOLE-HILL.

dew. Its principal food is the earth-worm, but it has been known to catch small birds by surprise, and tear them to pieces with its powerful claws. It is also an exceedingly pugnacious animal, and engages in the most determined battles, which are usually fought on the surface of the ground.

The hunting grounds of the mole are carefully marked out, and no intruders are suffered on the domains, under penalty of a furious combat. There are certain high roads, by means of which it can reach any part of its domains without the trouble of making a fresh tunnel; and into this high road all the other passages open. The fortress of the mole, where it passes the intermediate time between the autumn and spring, is of a very complex and ingenious form, as will be seen by examination of the accompanying cut. A large mound is made in some secure place, and two galleries surround it, one near the bottom, and the other near the top. These two galleries are connected by five short passages. In the centre of the mound, at the level of the earth, it digs a circular hole, which is connected with the upper gallery by three passages. A communication is then made between the high road, and the circular cell, by a passage which dips under the lower gallery, and joins the high road a little outside it. Lastly, a number of passages are run from the lower gallery in all directions, so that if the mole be disturbed in his fortress he can escape in any direction he chooses.

The long nose of this creature is very useful in forcing its way through the ground. I have often seen a mole which had been captured above ground, making its way through the earth with its snout, by moving its head from side to side, and then enlarging the opening with its fore feet. It was astonishing to see in how short a space of time the creature would bury itself. It lived in a tub half full of earth, and was supplied with worms, which it soon caught. Its owners were accustomed to amuse themselves by taking it into the garden and giving chase to it, as it not only ran with considerable speed, but buried itself with such rapidity, that no small exertions were required to prevent it from escaping. In one of these excursions it did succeed in escaping, for it discovered a rather soft piece of ground, and was gone before any one had time to reach it. The gardener was greatly enraged at the escape of the mole into his garden, and prophesied all manner of evils. As, however, we never heard of the mole again, we supposed that it must have undermined the wall, and returned to its old haunts.

The fur of the mole has no particular grain, but is set on the skin quite perpendicularly. Its colour is usually a dark greyish brown, but varieties occur of an orange colour, and I have in my possession a mole-skin of a cream colour.

The mole is a good swimmer, and can pass rivers without difficulty.—J. G. W.

of their approach? Moreover, we have often observed
that cold seems to descend from above: for when a thermometer hangs abroad in a frosty night, the intervention of a
cloud shall immediately raise the mercury ten degrees; and a
clear sky shall again compel it to descend to its former gauge.[1]

And here it may be proper to observe, on what has been
said above, that though frosts advance to their utmost
severity by somewhat of a regular gradation, yet thaws do
not usually come on by as regular a declension of cold,
but often take place immediately from intense freezing;
as men in sickness often mend at once from paroxysm.

To the great credit of Portugal laurels and American
junipers, be it remembered, that they remained untouched
amidst the general havoc : hence men should learn to ornament chiefly with such trees as are able to withstand acci-
dental severities, and not subject themselves to the vexation
of a loss which may befal them once, perhaps, in ten years,
yet may hardly be recovered through the whole course of
their lives.

As it appeared afterwards, the ilexes were much injured,
the cypresses were half destroyed, the arbutuses lingered on
but never recovered; and the bays, laurustines, and laurels,
were killed to the ground! and the very wild hollies, in hot
aspects, were so much affected, that they cast all their
leaves.

By the fourteenth of January, the snow was entirely gone;
the turnips emerged, not damaged at all, save in sunny places:
the wheat looked delicately; and the garden plants were well
preserved; for snow is the most kindly mantle that infant
vegetation can be wrapped in: were it not for that friendly
meteor, no vegetable life could exist at all in northerly regions.
Yet in Sweden, the earth in April is not divested of snow for
more than a fortnight, before the face of the country is
covered with flowers.

(1) The reason of this phenomenon is evident. During the day, the earth absorbs
a large amount of heat from the sun's rays, and at night, after the sun has set, the
absorbed heat is given out again. If, then, the night be clear, the heat passes away
into the air, but if a cloud should come over, it radiates the heat back again to the
earth, and of course raises the temperature.—J. G. W.

LETTER LXI.

WINTER OF 1776.

THERE were some circumstances attending the remarkable frost of January 1776, so singular and striking, that a short detail of them may not be unacceptable.

The most certain way to be exact, will be to copy the passages from my journal, which were taken from time to time, as things occurred. But it may be proper previously to remark, that the first week in January was uncommonly wet, and drowned with vast rains from every quarter: from whence may be inferred, as there is great reason to believe is the case, that intense frosts seldom take place till the earth is completely glutted and chilled with water;[1] and hence dry autumns are seldom followed by rigorous winters.

January 7th.—Snow driving all the day, which was followed by frost, sleet, and some snow, till the twelfth, when a prodigious mass overwhelmed all the works of men, drifting over the tops of the gates, and filling the hollow lanes.

On the fourteenth, the writer was obliged to be much abroad: and thinks he never before, or since, has encountered such rugged Siberian weather. Many of the narrow roads are now filled above the tops of the hedges; through which the snow was driven in most romantic and grotesque shapes, so striking to the imagination, as not to be seen without wonder and pleasure. The poultry dared not to stir out of their roosting-places; for cocks and hens are so dazzled and confounded by the glare of the snow, that they would soon perish without assistance. The hares also lay sullenly in their seats, and would not move till compelled by hunger; being conscious, poor animals, that the drifts and heaps treacherously betray their footsteps, and prove fatal to numbers of them.

From the fourteenth the snow continued to increase, and

(1) The autumn preceding January, 1768, was very wet, and particularly the month of September, during which there fell at Lyndon, in the county of Rutland, six inches and a half of rain. And the terrible long frost in 1739-40, set in after a rainy season, and when the springs were very high.

began to stop the road-waggons and coaches, which could no longer keep on their regular stages ; and especially on the western roads, where the fall appears to have been greater than in the south. The company at Bath, that wanted to attend the Queen's birthday, were strangely incommoded ; many carriages of persons who got, in their way to town from Bath, as far as Marlborough, after strange embarrassments, here met with a *ne plus ultra*. The ladies fretted, and offered large rewards to labourers if they would shovel them a track to London ; but the relentless heaps of snow were too bulky to be removed ; and so the eighteenth passed over, leaving the company in very uncomfortable circumstances at the Castle and other inns.

On the twentieth, the sun shone out for the first time since the frost began ; a circumstance that has been remarked before, much in favour of vegetation. All this time the cold was not very intense, for the thermometer stood at 29, 28, 25, and thereabout : but on the twenty-first it descended to 20. The birds now began to be in a very pitiable and starving condition. Tamed by the season, sky-larks settled in the streets of towns, because they saw the ground was bare ; rooks frequented dunghills close to houses ; and crows watched horses as they passed, and greedily devoured what dropped from them ; hares now came into men's gardens, and scraping away the snow, devoured such plants as they could find.

On the twenty-second, the author had occasion to go to London : through a sort of Laplandian scene very wild and grotesque indeed. But the metropolis itself exhibited a still more singular appearance than the country ; for, being bedded deep in snow, the pavement could not be touched by the wheels or the horses' feet, so that the carriages ran about without the least noise.[1] Such an exemption from din and clatter was strange, but not pleasant ; it seemed to convey an uncomfortable idea of desolation :—

"Ipsa silentia terrent."

(1) This was the case at Oxford a few years since, when a stranger might think that he had entered a Russian city instead of an English one, for the streets were full of sledges drawn by gaily caparisoned horses, and profusely hung with bells. Skaters were also seen traversing the streets in all directions.—J. G. W.

On the twenty-seventh, much snow fell all day, and in the evening the frost became very intense. At South Lambeth, for the four following nights, the thermometer fell to 11, 7, 6, 6; and at Selborne to 7, 6, 10; and on the 31st of January, just before sunrise, with rime on the trees, and on the tube of the glass, the quicksilver sunk exactly to zero, being 32 degrees below the freezing point; but by eleven in the morning, though in the shade, it sprung up to $16\frac{1}{2}$[1]—a most unusual degree of cold this for the south of England! During these four nights, the cold was so penetrating, that it occasioned ice in warm chambers, and under beds: and in the day the wind was so keen, that persons of robust constitutions could scarcely endure to face it. The Thames was at once so frozen over, both above and below the bridge, that crowds ran about on the ice. The streets were now strangely encumbered with snow, which crumbled and trode dusty; and, turning gray, resembled bay-salt; what had fallen on the roofs was so perfectly dry, that from first to last it lay twenty-six days on the houses in the city; a longer time than had been remembered by the oldest house-keepers living. According to all appearances, we might now have expected the continuance of this rigorous weather for weeks to come, since every night increased in severity; but behold, without any apparent cause, on the first of February, a thaw took place, and some rain followed before night; making good the observation above, that frosts often go off, as it were at once, without any gradual declension of cold. On the second of February, the thaw persisted; and on the third, swarms of little insects were frisking and sporting in a court-yard at South Lambeth, as if they had felt no frost. Why the juices in the small bodies and smaller limbs of such minute beings are not frozen, is a matter of curious inquiry.[2]

---

(1) At Selborne, the cold was greater than at any other place that the author could hear of with certainty; though some reported at the time, that, at a village in Kent, the thermometer fell two degrees below zero, viz. thirty-four degrees below the freezing point.

The thermometer used at Selborne was graduated by Benjamin Martin.

(2) It is a well known fact that many animals may be actually frozen hard, so as to snap in pieces, and yet recover their vitality on being thawed. This is more particularly the case with the lower animals. There are some creatures to whom a low temperature is absolutely necessary, and who actually live in ice, while there are others which exist in water too hot for the hand to bear.—J. G. W.

Severe frosts seem to be partial, or to run in currents; for
at the same juncture, as the author was informed by accurate
correspondents at Lyndon, in the county of Rutland, the
thermometer stood at 19; at Blackburn, in Lancashire, at
19; and at Manchester at 21, 20, and 18. Thus does some
unknown circumstance strangely overbalance latitude, and
render the cold sometimes much greater in the southern than
the northern parts of this kingdom.

The consequences of this severity were, that in Hampshire,
at the melting of the snow, the wheat looked well, and the
turnips came forth little injured. The laurels and laurustines were somewhat damaged, but only in hot aspects. No
evergreens were quite destroyed; and not half the damage
sustained that befell in January, 1768. Those laurels that
were a little scorched on the south sides, were perfectly untouched on their north sides. The care taken to shake the
snow, day by day, from the branches, seemed greatly to avail
the author's evergreens. A neighbour's laurel hedge, in a
high situation, and facing to the north, was perfectly green
and vigorous; and the Portugal laurels remained unhurt.

As to the birds, the thrushes and blackbirds were mostly
destroyed: and the partridges, by the weather and poachers,
were so thinned, that few remained to breed the following year.

LETTER LXII.

WINTER OF 1784.

As the frost in December, 1784, was very extraordinary,
you, I trust, will not be displeased to hear the particulars;
and especially when I promise to say no more about the
severities of winter after I have finished this letter.

The first week in December was very wet, with the barometer very low. On the 7th, with the barometer at 28·5,
came on a vast snow, which continued all that day and the
next, and most part of the following night, so that, by the
morning of the 9th, the works of men were quite overwhelmed,
the lanes filled so as to be impassable, and the ground covered
twelve or fifteen inches without any drifting. In the evening
of the 9th, the air began to be so very sharp that we thought

it would be curious to attend to the motions of a thermometer; we therefore hung out two, one made by Martin and
one by Dollond, which soon began to show us what we were
to expect; for, by ten o'clock, they fell to 21, and, at eleven,
to 4, when we went to bed. On the 10th, in the morning,
the quicksilver of Dollond's glass was down to half a degree
below zero, and that of Martin's, which was absurdly graduated only to four degrees below zero, sunk quite into the
brass guard of the ball, so that, when the weather became
most interesting, this was useless. On the 10th, at eleven
at night, though the air was perfectly still, Dollond's glass
went down to one degree below zero! This strange severity
of the weather made me very desirous to know what degree
of cold there might be in such an exalted and near situation
as Newton. We had, therefore, on the morning of the 10th,
written to Mr. ———, and entreated him to hang out his thermometer, made by Adams, and to pay some attention to it
morning and evening, expecting wonderful phenomena in so
elevated a region, at two hundred feet, or more, above my
house; but, behold! on the 10th, at eleven at night, it was
down only to 17, and the next morning at 22, when mine
was at 10! We were so disturbed at this unexpected reverse
of comparative local cold, that we sent one of my glasses up,
thinking that of Mr.——— must, somehow, be wrongly constructed. But when the instruments came to be confronted,
they went exactly together, so that, for one night at least, the
cold at Newton was eighteen degrees less than at Selborne,
and, through the whole frost, ten or twelve degrees; and,
indeed, when we came to observe consequences, we could
readily credit this, for all my laurustines, bays, ilexes, arbutuses, cypresses, and even my Portugal laurels,[1] and, which
occasions more regret, my fine sloping laurel-hedge, were
scorched up, while, at Newton, the same trees have not lost
a leaf!

We had steady frost on the 25th, when the thermometer,
in the morning, was down to 10 with us, and at Newton

'(1) Mr. Miller, in his *Gardener's Dictionary*, says positively, that the Portugal
laurels remained untouched in the remarkable frost of 1739-40. So that either that
accurate observer was much mistaken, or else the frost of December, 1784, was
much more severe and destructive than that in the year above mentioned.

only to 21. Strong frost continued till the 31st, when some tendency to thaw was observed, and by January 3d, 1785, the thaw was confirmed, and some rain fell.

A circumstance that I must not omit, because it was new to us, is, that on Friday, December the 10th, being bright sunshine, the air was full of icy *spiculæ*, floating in all directions, like atoms in a sunbeam, let into a dark room. We thought them, at first, particles of the rime falling from my tall hedges, but were soon convinced to the contrary, by making our observations in open places, where no rime could reach us. Were they watery particles of the air frozen as they floated, or were they evaporations from the snow frozen as they mounted?

We were much obliged to the thermometers for the early information they gave us, and hurried our apples, pears, onions, potatoes, &c., into the cellar and warm closets; while those who had not, or neglected such warnings, lost all their stores of roots and fruits, and had their very bread and cheese frozen.

I must not omit to tell you, that during those two Siberian days my parlour cat was so electric, that had a person stroked her, and been properly insulated, the shock might have been given to a whole circle of people.

I forgot to mention before, that during the two severe days, two men, who were tracing hares in the snow, had their feet frozen; and two men, who were much better employed, had their fingers so affected by the frost, while they were thrashing in a barn, that mortification followed, from which they did not recover for many weeks.

The frost killed all the furze and most of the ivy, and in many places stripped the hollies of all their leaves. It came at a very early time of the year, before old November ended, and may yet be allowed, from its effects, to have exceeded any since 1739-40.

LETTER LXIII.

SUMMERS OF 1781 AND 1783.

As the effects of heat are seldom very remarkable in the northerly climate of England, where the summers are often so defective in warmth and sunshine, as not to ripen the fruits of the earth so well as might be wished, I shall be more concise in my account of the severity of a summer season, and so make a little amends for the prolix account of the degrees of cold and the inconveniences that we suffered from some late rigorous winters.

The summers of 1781 and 1783 were unusually hot and dry; to them, therefore, I shall turn back in my journals, without recurring to any more distant period. In the former of these years, my peach and nectarine-trees suffered so much from the heat, that the rind on the bodies was scalded and came off; since which, the trees have been in a decaying state. This may prove a hint to assiduous gardeners to fence and shelter their wall-trees with mats or boards, as they may easily do, because such annoyance is seldom of long continuance. During that summer, also, I observed that my apples were coddled, as it were, on the trees; so that they had no quickness of flavour, and would not keep in the winter. This circumstance put me in mind of what I have heard travellers assert, that they never ate a good apple or apricot in the south of Europe, where the heats were so great as to render the juices vapid and insipid.

The great pest of a garden are wasps, which destroy all the finer fruits just as they are coming into perfection. In 1781, we had none; in 1783, there were myriads, which would have devoured all the produce of my garden, had we not set the boys to take the nests, and caught thousands with hazel-twigs tipped with bird-lime; we have since employed the boys to take and destroy the large breeding wasps in spring. Such expedients have a great effect on these marauders, and will keep them under. Though wasps do not abound but in hot summers, yet they do not prevail in every hot summer, as I have instanced in the two years above mentioned.

In the sultry season of 1783, honey-dews were so frequent

WASPS.[1]

as to deface and destroy the beauties of my garden. My
honeysuckles, which were one week the most sweet and
lovely objects that eye could behold, became the next the
most loathsome, being enveloped in a viscous substance, and
loaded with black *aphides*, or smother-flies. The occasion of
this clammy appearance seems to be this, that in hot weather
the effluvia of flowers in fields, and meadows, and gardens,
are drawn up in the day by a brisk evaporation, and then in
the night fall down again with the dews in which they are
entangled; that the air is strongly scented, and therefore
impregnated with the particles of flowers in summer weather,
our senses will inform us; and that this clammy sweet substance is of the vegetable kind we may learn from bees, to
whom it is very grateful; and we may be assured that it falls in
the night, because it is always first seen in warm, still mornings.

On chalky and sandy soils, and in the hot villages about
London, the thermometer has been often observed to mount
as high as 83 or 84; but with us, in this hilly and woody
district, I have hardly ever seen it exceed 80, nor does it
often arrive at that pitch. The reason, I conclude, is, that
our dense clayey soil, so much shaded by trees, is not so
easily heated through as those above mentioned; and, besides,
our mountains cause currents of air and breezes; and the vast
effluvia from our woodlands temper and moderate our heats.

(1) *Vespa vulgaris.* The larger insect is the hornet, (*Vespa crabro.*)

LETTER LXIV.

SUMMER OF 1783.

THE summer of the year 1783 was an amazing and porten-
tous one, and full of horrible phenomena; for, besides the
alarming meteors and tremendous thunder-storms that
affrighted and distressed the different counties of this kingdom, the peculiar haze, or smoky fog, that prevailed for
many weeks in this island, and in every part of Europe, and
even beyond its limits, was a most extraordinary appearance,
unlike anything known within the memory of man. By my
journal, I find that I had noticed this strange occurrence
from June 23 to July 20 inclusive, during which period, the
wind varied to every quarter, without making any alteration
in the air. The sun, at noon, looked as black as a clouded
moon, and shed a rust-coloured ferruginous light on the
ground and floors of rooms, but was particularly lurid and
blood-coloured at rising and setting. All the time, the heat
was so intense that butchers' meat could hardly be eaten the
day after it was killed; and the flies swarmed so in the lanes
and hedges, that they rendered the horses half frantic, and
riding irksome. The country people began to look with a
superstitious awe at the red lowering aspect of the sun; and,
indeed, there was reason for the most enlightened person to
be apprehensive, for all the while Calabria, and part of the
isle of Sicily, were torn and convulsed with earthquakes;
and about that juncture, a volcano sprang out of the sea on
the coast of Norway. On this occasion, Milton's noble simile
of the sun, in his first book of Paradise Lost, frequently
occurred to my mind; and it is indeed particularly appli-
cable, because, towards the end, it alludes to a superstitious
kind of dread, with which the minds of men are always
impressed by such strange and unusual phenomena:—

> —— " As when the sun, new risen,
> Looks through the horizontal, misty air
> Shorn of his beams; or, from behind the moon,
> In dim eclipse, disastrous twilight sheds
> On half the nations, and with fear of change
> Perplexes monarchs." ——

LETTER LXV.

THUNDER-STORMS.

We are very seldom annoyed with thunder-storms; and it
is no less remarkable than true, that those which arise in the
south have hardly been known to reach this village; for,
before they get over us, they take a direction to the east or to
the west, or sometimes divide into two, and go in part to one
of those quarters, and in part to the other; as was truly the
case in the summer of 1783, when, though the country round
was continually harassed with tempests, and often from the
south, yet we escaped them all, as appears by my journal of
that summer. The only way that I can at all account for
this fact—for such it is—is, that on that quarter, between us
and the sea, there are continual mountains, hill behind hill,
such as Nore-hill, the Barnet, Burter-hill, and Portsdown,
which somehow divert the storms, and give them a different
direction. High promontories, and elevated grounds, have
always been observed to attract clouds, and disarm them of
their mischievous contents, which are discharged into the
trees and summits, as soon as they come into contact with
these turbulent meteors; while the humble vales escape,
because they are so far beneath them.

But when I say I do not remember a thunder-storm from
the south, I do not mean that we never have suffered from
thunder-storms at all; for on June 5th, 1784, the thermometer in the morning being at .64, and at noon at 70, the
barometer at $29\cdot6\frac{1}{2}$, and the wind north, I observed a blue
mist, smelling strongly of sulphur, hang along our sloping
woods, and seeming to indicate that thunder was at hand. I
was called in about two in the afternoon, and so missed seeing
the gathering of the clouds in the north, which they who
were abroad assured me had something uncommon in its
appearance. At about a quarter after two the storm began
in the parish of Hartley, moving slowly from north to south;
and from thence it came over Norton-farm, and so to Grange-
farm, both in this parish. It began with vast drops of rain,

which were soon succeeded by round hail, and then by convex pieces of ice, which measured three inches in girth.[1] Had it been as extensive as it was violent, and of any continuance (for it was very short), it must have ravaged all the neighbourhood. In the parish of Hartley, it did some damage to one farm; but Norton, which lay in the centre of the storm, was greatly injured; as was Grange, which lay next to it. It did but just reach to the middle of the village, where the hail broke my north windows, and all my garden lights, and hand-glasses, and many of my neighbours' windows. The extent of the storm was about two miles in length, and one in breadth. We were just sitting down to dinner; but were soon diverted from our repast by the clattering of tiles and the jingling of glass. There fell at the same time prodigious torrents of rain on the farms above mentioned, which occasioned a flood as violent as it was sudden; doing great damage to the meadows and fallows, by deluging the one, and washing away the soil of the other. The hollow lane towards Alton was so torn and disordered as not to be passable till mended, rocks being removed that weighed two hundred weight. Those that saw the effect which the great hail had on the ponds and pools, say that the dashing of the water made an extraordinary appearance, the froth and spray standing up in the air three feet above the surface. The rushing and roaring of the hail, as it approached, was truly tremendous.

Though the clouds at South Lambeth, near London, were at that juncture thin and light, and no storm was in sight,

---

(1) On the 5th of July, 1852, a similar event occurred at Oxford. At the time that the storm came on, I was with a water party at Nuneham. The whole sky was of a dark leaden colour, and on looking up, there appeared what I thought were the white bellies of martins flying about in the air, until one came down close by me, whizzing through the leaves of a tree, and cutting off a small branch. On picking it up, it turned out to be a hailstone two inches in length, and about one in breadth. They soon became more plentiful, but still fell so sparingly that they could be noticed when at a very great height. There was no wind, and they all fell perpendicularly, or the damage would have been very much greater than it was. The hailstones fell in such numbers, that the members of the water party, who were driven to stand under shelter, collected them and used them for an extempore sherry-cobbler. The branches of the trees were considerably damaged, and the leaves flying about in all directions. In the summer of the same year, there was a furious hailstorm at Cuddesdon, which so stripped the trees of their leaves, that the air was as thickly filled with them as it is with snow flakes during a storm, and the doors on the windward side of the house were completely blocked up with hail and leaves.—J. G. W.

nor within hearing, yet the air was strongly electric; for the bells of an electric machine at that place rang repeatedly, and fierce sparks were discharged.

When I first took the present work in hand, I proposed to have added an *Annus-Historico-Naturalis,* or the Natural History of the Twelve Months of the Year; which would have comprised many incidents and occurrences that have not fallen into my way to be mentioned in my series of letters; —but as Mr. Aiken of Warrington has lately published somewhat of this sort, and as the length of my correspondence has sufficiently put your patience to the test, I shall here take a respectful leave of you and Natural History together. And am, with all due deference and regard,

Your most obliged, and most humble servant,

GIL. WHITE.

SELBORNE, *June* 25, 1787.

OBSERVATIONS

ON

VARIOUS PARTS OF NATURE,

FROM MR. WHITE'S MSS.

WITH REMARKS BY MR. MARKWICK.

OBSERVATIONS ON BIRDS.

BIRDS IN GENERAL.

In severe weather, fieldfares, redwings, sky-larks, and tit-larks, resort to watered meadows for food; the latter wades up to its belly in pursuit of the pupæ of insects, and runs along upon the floating grass and weeds. Many gnats are on the snow near the water; these support the birds in part.

SKY-LARK [1]

Birds are much influenced in their choice of food by colour; for though white currants are much sweeter fruit than red, yet they seldom touch the former till they have devoured every branch of the latter.

(1) *Alauda arvensis.*—J. G. W.

Redstarts, fly-catchers, and black-caps, arrive early in April. If these little delicate beings are birds of passage, (as we have reason to suppose they are, because they are never seen in winter,) how could they, feeble as they seem, bear up against such storms of snow and rain, and make their way through such meteorous turbulence, as one should suppose would embarrass and retard the most hardy and resolute of the winged nation? Yet they keep their appointed times and seasons; and, in spite of frosts and winds, return to their stations periodically, as if they had met with nothing to obstruct them. The withdrawing and appearance of the short-winged summer birds is a very puzzling circumstance in natural history!

When the boys bring me wasps' nests, my bantam fowls fare deliciously, and, when the combs are pulled to pieces, devour the young wasps in their maggot state with the highest glee and delight. Any insect-eating bird would do the same; and therefore I have often wondered that the accurate Mr. Ray should call one species of buzzard *Buteo apivorus sive vespivorus*, or the *honey-buzzard*, because some combs of wasps happened to be found in one of their nests. The combs were conveyed thither doubtless for the sake of the maggots or nymphs, and not for their honey, since none is to be found in the combs of wasps. Birds of prey occasionally feed on insects; thus have I seen a tame kite picking up the female ants full of eggs, with much satisfaction.—WHITE.

That redstarts, fly-catchers, black-caps, and other slender-billed insectivorous small birds, particularly the swallow tribe, make their first appearance very early in the spring, is a well-known fact; though the fly-catcher is the latest of them all in its visit (as this accurate naturalist observes in another place), for it is never seen before the month of May. If these delicate creatures come to us from a distant country, they will probably be exposed in their passages, as Mr. White justly remarks, to much greater difficulties from storms and tempests than their feeble powers appear to be able to surmount: on the other hand, if we suppose them to pass the winter in a dormant state, in this country, concealed in caverns, or other hiding-places, sufficiently guarded from the extreme cold of our winter to preserve their life, and that, at the approach of spring, they revive from their torpid state, and re-assume their usual powers of action, it will entirely remove the first difficulty, arising from the storms and tempests they are liable to meet with in their passage: but how are we to get over the still greater difficulty of their revivification from their torpid state? What degree of warmth in the temperature of the air is necessary to produce that effect, and how it operates on the functions of animal life, are questions not easily answered.

How could Mr. White suppose that Ray named this species the honey-buzzard because it fed on honey, when he not only named it in Latin *Buteo apivorus sive vespivorus*, but expressly says, that "it feeds on insects, and brings up its young with the maggots, or nymphs, of wasps?"

That birds of prey, when in want of their proper food, flesh, sometimes feed on insects, I have little doubt, and think I have observed the common buzzard (*Falco buteo*) to settle on the ground and pick up insects of some kind or other.—Markwick.

Rooks.—Rooks are continually fighting, and pulling each other's nests to pieces: these proceedings are inconsistent with living in such close community. And yet, if a pair offer to build in a single tree, the nest is plundered and demolished at once. Some rooks roost on their nest trees. The twigs which the rooks drop in building, supply the poor with brushwood to light their fires. Some unhappy pairs are not permitted to finish any nest till the rest have completed their building. As soon as they get a few sticks together, a party comes and demolishes the whole. As soon as rooks have finished their nests, and before they lay, the cocks begin to feed the hens, who receive their bounty with a fondling, tremulous voice, and fluttering wings, and all the little blandishments that are expressed by the young, while in a helpless state. This gallant deportment of the male is continued through the whole season of incubation.—White.

After the first brood of rooks are sufficiently fledged, they all leave their nest-trees in the day-time, and resort to some distant place in search of food, but return regularly every evening, in vast flights, to their nest-trees, where, after flying round several times, with much noise and clamour, till they are all assembled together, they take up their abode for the night.—MARKWICK.

THRUSHES.—Thrushes, during long droughts, are of great service in hunting out shell-snails, which they pull in pieces for their young, and are thereby very serviceable in gardens. Missel-thrushes do not destroy the fruit in gardens like the other species of *turdi*, but feed on the berries of misseltoe, and in the spring on ivy berries, which then begin to ripen. In the summer, when their young become fledged, they leave neighbourhoods, and retire to sheep-walks and wild commons.

The magpies, when they have young, destroy the broods of missel-thrushes, though the dams are fierce birds, and fight boldly in defence of their nests. It is probably to avoid such insults, that this species of thrush, though wild at other times, delights to build near houses, and in frequented walks and gardens.—WHITE.

Of the truth of this I have been an eye-witness, having seen the common thrush feeding on the shell-snail.

In the very early part of this spring (1797), a bird of this species used to sit every morning on the top of some high elms close to my windows, and delight me with its charming song,[1] attracted thither, probably, by some ripe ivy berries that grew near the place.

I have remarked something like the latter fact; for I remember, many years ago, seeing a pair of these birds fly up repeatedly, and attack some larger bird, which I suppose disturbed their nest in my orchard, uttering, at the same time, violent shrieks.—Since writing the above, I have seen, more than once, a pair of these birds attack some magpies that had disturbed their nest, with great violence, and loud shrieks.— MARKWICK.

POULTRY.[2]—Many creatures are endowed with a ready

(1)   " . . .   dew-drops thick as early blossoms hung,
                And trembled as the minstrel sweetly sung."—BLOOMFIELD.
(2) Poultry of the ordinary kind agree together very well, but the great Cochin-China fowls lead a miserable life among the smaller but more pugnacious fowls.

discernment to see what will turn to their own advantage and emolument; and often discover more sagacity than could be expected. Thus, my neighbour's poultry watch for waggons

POULTRY.

loaded with wheat, and, running after them, pick up a number of grains which are shaken from the sheaves by the agitation of the carriages. Thus, when my brother used to take down his gun to shoot sparrows, his cats would run out before him, to be ready to catch up the birds as they fell.

The earnest and early propensity of the *gallinæ* to roost on high is very observable; and discovers a strong dread impressed on their spirits respecting vermin that may annoy them on the ground during the hours of darkness. Hence poultry, if left to themselves and not housed, will perch the winter through on yew-trees and fir-trees; and turkeys and

Curiously enough, the smallest bantam generally takes upon himself the duty of fighting the ungainly Cochin-China fowl, and always comes off conqueror; indeed, the Cochin-China seldom makes any defence at all, but runs off as fast as its legs can carry it.—J. G. W.

guinea fowls, heavy as they are, get up into apple-trees;
pheasants also, in woods, sleep on trees to avoid foxes;
while pea-fowls climb to the tops of the highest trees round
their owner's house for security, let the weather be ever so
cold or blowing. Partridges, it is true, roost on the ground,
not having the faculty of perching: but then the same fear
prevails in their minds; for, through apprehensions from
polecats and stoats, they never trust themselves to coverts,
but nestle together in the midst of large fields, far removed
from hedges and coppices, which they love to haunt in the
day, and where, at that season, they can skulk more secure
from the ravages of rapacious birds.

As to the ducks and geese, their awkward, splay, web-feet
forbid them to settle on trees; they therefore, in the hours
of darkness and danger, betake themselves to their own
element, the water, where, amidst large lakes and pools, like
ships riding at anchor, they float the whole night long in
peace and security.—WHITE.

GUINEA-FOWL.[1]

Guinea fowls not only roost on high, but in hard weather
resort, even in the day-time, to the very tops of highest
trees.

Last winter, when the ground was covered with snow, I discovered all my guinea fowls, in the middle of the day, sitting

(1) *Numida meleagris.*—J. G. W.

on the highest boughs of some very tall elms, chattering and making a great clamour : I ordered them to be driven down, lest they should be frozen to death in so elevated a situation; but this was not effected without much difficulty, they being very unwilling to quit their lofty abode, notwithstanding one of them had its feet so much frozen, that we were obliged to kill it. I know not how to account for this, unless it was occasioned by their aversion to the snow on the ground, they being birds that came originally from a hot climate.

Notwithstanding the awkward, splay, web-feet, as Mr. White calls them, of the duck genus, some of the foreign species have the power of settling on the boughs of trees, apparently with great ease ; an instance of which I have seen in the Earl of Ashburnham's menagerie, where the summer duck (*Anas sponsa*) flew up and settled on the branch of an oak-tree in my presence ; but whether any of them roost on trees in the night, we are not informed by any author that I am acquainted with. I suppose not ; but that, like the rest of the genus, they sleep on the water, where the birds of this genus are not always perfectly secure, as will appear from the following circumstances, which happened in this neighbourhood a few years since, as I was credibly informed. A female fox was found in the morning drowned in the same pond in which were several geese ; and it was supposed, that in the night, the fox swam into the pond to devour the geese, but was attacked by the gander, which being the most powerful in its own element, buffeted the fox with its wings about the head till it was drowned.[1]—MARKWICK.

(1) Duck. When ducks and barn-door fowls associate together, there is often a struggle as to the amount of food to be appropriated to each party. In a farm-yard with which I was once intimately acquainted, there were several ducks who were shut up at night in a very spacious coop, but who were not at all satisfied with the provender given to them, but yearned for some of that given to the fowls. So impatient were they of their imprisonment, that directly they saw any of their acquaintances in the farm-yard, they used to set up a most clamorous quacking, in hopes of being released. There were several grand battles between the ducks and the master cock of the yard, which invariably terminated in the victory of the ducks. The mode of combat was as follows. The poultry would be pecking up the grain thrown to them, when in would rush a duck, scooping up with its broad beak more at one sweep, than the fowls could take in a dozen pecks. This behaviour naturally incenses the cock, who accordingly flies at the duck and pecks it. The duck crouches down and makes no resistance, but contrives to get behind the cock and to give him a very hard peck, at the same time turning round and looking innocent. Round jumps the

HEN PARTRIDGE.—A hen partridge came out of a ditch, and ran along shivering with her wings, and crying out as if wounded and unable to get from us. While the dam acted

PARTRIDGE. [1]

this distress, the boy who attended me saw her brood, that was small and unable to fly, run for shelter into an old fox-earth under the bank. So wonderful a power is instinct.— WHITE.

cock, intent upon vengeance, but seeing nothing to account for the blow that he has just received, he puts it down to the charge of a stray stone, or other such misfortune, and returns to his meal. No sooner has his attention been fixed upon his food, when he receives another hard peck, jumps round and sees the duck looking innocent as before. This time, however, he suspects something, and while he pecks at the barley, keeps a look-out from the corner of his eye. Soon comes another peck; but this time the duck is seen, and aggrieved Chanticleer dashes at him with all the anger of the three assaults combined. Down flops the duck on the ground, tucking his head under his wing; the cock runs over him in triumph, walking once or twice over his prostrate enemy, and returns to his meal in high spirits. Presently the duck draws out his head, opens first one eye and then the other, gets up cautiously, saunters behind the cock and salutes him with another peck. The insulted bird again attacks his foe, again meets with no opposition, again returns to his food, and is again attacked in a similar manner, until he is completely wearied out, and fairly takes to flight, pursued by the triumphant duck, who has won, like Fabius, by delay.—J. G. W.

(1) *Perdix cinerea.*—J. G. W.

It is not uncommon to see an old partridge feign itself wounded, and run along on the ground fluttering and crying, before either dog or man, to draw them away from its help-less unfledged young ones. I have seen it often ; and once, in particular, I saw a remarkable instance of the old bird's solicitude to save its brood. As I was hunting with a young pointer, the dog ran on a brood of very small partridges; the old bird cried, fluttered, and ran tumbling along, just before the dog's nose, till she had drawn him to a considerable dis-tance, when she took wing and flew still farther off, but not out of the field : on this the dog returned to me, near which place the young ones lay concealed in the grass, which the old bird no sooner perceived, than she flew back again to us, settled just before the dog's nose again, and, by rolling and tumbling about, drew off his attention from her young, and thus preserved her brood a second time. I have also seen, when a kite has been hovering over a covey of young par-tridges, the old birds fly up at the bird of prey, screaming and fighting with all their might, to preserve their brood.— MARKWICK.

A HYBRID PHEASANT.—Lord Stawell sent me, from the great lodge in the Holt, a curious bird for my inspection. It was found by the spaniels of one of his keepers in a coppice, and shot on the wing. The shape, air, and habit of the bird, and the scarlet ring round the eyes, agreed well with the appearance of a cock pheasant ; but then the head and neck, and breast and belly, were of a glossy black ; and though it weighed three pounds three ounces and a half,[1] the weight of a large full-grown cock pheasant, yet there was no sign of any spurs on the legs, as is usual with all grown cock pheasants, who have long ones. The legs and feet were naked of feathers, and therefore it could be nothing of the grouse kind. In the tail were no long, bending feathers, such as cock pheasants usually have, and are characteristic of the sex. The tail was much shorter than the tail of a hen pheasant, and blunt and square at the end. The back, wing-feathers, and tail, were all of a pale russet, curiously streaked,

(1) Hen pheasants usually weigh only two pounds ten ounces.

somewhat like the upper parts of a hen partridge. I returned it with my verdict, that it was probably a spurious, or hybrid hen-bird, bred between a cock pheasant and some domestic fowl. When I came to talk with the keeper who brought it, he told me that some pea-hens had been known last summer to haunt the coppices and coverts where this mule was found.

PHEASANT [1]

Mr. Elmer, of Farnham, the famous game-painter, was employed to take an exact copy of this curious bird.

N.B. It ought to be mentioned, that some good judges have imagined this bird to have been a stray grouse or black cock ; it is, however, to be observed, that Mr. W. remarks that its legs and feet were naked, whereas those of the grouse are feathered to the toes.—WHITE.

Mr. Latham observes, that "pea-hens, after they have done laying, sometimes assume the plumage of the male bird," and has given a figure of the male-feathered pea-hen now to

(1) *Phasianus Colchicus.*—J. G. W.

be seen in the Leverian Museum; and M. Salerne remarks
that "the hen pheasant, when she has done laying and sitting,
will get the plumage of the male." May not this hybrid
pheasant, as Mr. White calls it, be a bird of this kind? that
is, an old hen pheasant which has just begun to assume the
plumage of the cock.[1]—MARKWICK.

LAND-RAIL.[2]—A man brought me a land-rail, or daker-hen,
a bird so rare in this district that we seldom see more than
one or two in a season, and these only in autumn. This is
deemed a bird of passage by all the writers; yet, from its
formation seems to be poorly qualified for migration; for its
wings are short, and placed so forward, and out of the centre
of gravity, that it flies in a very heavy and embarrassed
manner, with its legs hanging down; and can hardly be
sprung a second time, as it runs very fast, and seems to
depend more on the swiftness of its feet than on its flying.
When we came to draw it, we found the entrails so soft

(1) When surprised by a dog, the two sexes of this bird act very differently.
The hen bird squats closely to the ground, hoping to escape observation by the
similarity of its colour to that of the herbage; while the male bird runs away as
fast as he can, keeping close under a hedge, or running at the bottom of a dry
ditch. I once saw a boy, a very swift runner, chase a pheasant for nearly a hundred
yards along a hedge, until the bird, finding its pursuer gaining on it, made a full
stop and shot through the hedge. The boy threw himself on his face after the bird,
making a grasp at it at the same time, but only succeeded in seizing the tail
feathers, which came out in his hand, while the pheasant rose at the other side of
the hedge, screaming dolefully at the loss of his tail, which was carried off as a
trophy by his enemy.

In some well-kept preserves the pheasants are very tame. At Ashdown, I have
walked close by them without their moving away. Some had attached themselves
to the keeper, and lived in a kind of semi-domesticated state. Among them was a
remarkably fine specimen of the white variety.

Hybrids between pheasants and other gallinaceous birds are not at all uncommon.
I have seen a brood of chickens, whose male parent was a pheasant; and they had
the distinguishing neck-feathers of the pheasant, besides partaking in some degree
of the general form of that bird.—J. G. W.

(2) The corn-crake or land-rail is very common about Oxford, and is found in
most parts of England. It is very difficult to get a sight of the bird, as it
seldom flies, but trusts to its remarkably swift legs for escape. The young of
this bird, when taken, feign death admirably, and will suffer themselves to be
taken into the hand or rolled about with the foot without giving the least sign of
life. The sound of the land-rail's cry may be exactly imitated by drawing a very
stiff quill across a comb, and the bird may even be decoyed by employing this
stratagem.

The nest is made of hay worked into a depression on the ground, and contains from
eight to twelve eggs.—J. G. W.

and tender, that in appearance they might have been dressed
like the ropes of a woodcock.   The craw, or crop, was small
and lank, containing a mucus ; the gizzard thick and strong,
and filled with small shell-snails, some whole, and many
ground to pieces, through the attrition which is occasioned
by the muscular force and motion of that intestine.   We
saw no gravels among the food ; perhaps the shell-snails

LAND-RAIL.[2]

might perform the functions of gravels or pebbles, and might
grind one another.   Land-rails used to abound formerly, I
remember, in the low, wet bean-fields of Christian Malford,
in North Wilts, and in the meadows near Paradise Gardens,
at Oxford, where I have often heard them cry, Crex, crex.
The bird mentioned above weighed $7\frac{1}{2}$ oz., was fat and tender,
and in flavour like the flesh of a woodcock.   The liver was
very large and delicate. — WHITE.

Land-rails are more plentiful with us than in the neighbourhood of Selborne.   I have found four brace in an afternoon, and a friend of mine lately shot nine in two adjoining
fields ; but I never saw them in any other season than the
autumn.

That it is a bird of passage there can be little doubt,
though Mr. White thinks it poorly qualified for migration,

(1) *Ortygometra crex.*—J. G. W.

on account of the wings being short, and not placed in the exact centre of gravity: how that may be, I cannot say, but I know that its heavy sluggish flight is not owing to its inability of flying faster, for I have seen it fly very swiftly; although in general its actions are sluggish. Its unwillingness to rise proceeds, I imagine, from its sluggish disposition, and its great timidity; for it will sometimes squat so close to the ground as to suffer itself to be taken up by the hand, rather than rise; and yet it will at times run very fast.

What Mr. White remarks respecting the small shell-snails found in its gizzard, confirms my opinion, that it frequents corn-fields, seed clover, and brakes or fern, more for the sake of snails, slugs, and other insects which abound in such places, than for the grain or seeds; and that it is entirely an insectivorous bird.—MARKWICK.

FOOD FOR THE RING-DOVE.—One of my neighbours shot a ring-dove on an evening as it was returning from feed and going to roost. When his wife had picked and drawn it, she found its craw stuffed with the most nice and tender tops of turnips. These she washed and boiled, and so sat down to a choice and delicate plate of greens, culled and provided in this extraordinary manner.

Hence we may see that graminivorous birds, when grain fails, can subsist on the leaves of vegetables. There is reason to suppose that they would not long be healthy without; for turkeys, though corn-fed, delight in a variety of plants, such as cabbage, lettuce, endive, &c.; and poultry pick much grass; while geese live for months together on commons by grazing alone.

> " Nought is useless made :————
> ———— On the barren heath
> The shepherd tends his flock, that daily crop
> Their verdant dinner from the mossy turf
> Sufficient: after them, the cackling *goose*,
> Close grazer, finds wherewith to ease her want."—PHILIPS' *Cyder*.

WHITE.

That many graminivorous birds feed also on the herbage, or leaves of plants, there can be no doubt; partridges and larks frequently feed on the green leaves of turnips, which give a peculiar flavour to their flesh, that is to me very palatable ; the flavour also of wild ducks and geese greatly depends on the nature of their food ; and their flesh frequently contracts a rank unpleasant taste, from their having lately fed on strong marshy aquatic plants, as I suppose.

That the leaves of vegetables are wholesome, and conducive to the health of birds, seems probable, for many people fat their ducks and turkeys with the leaves of lettuce chopped small.—MARKWICK.

HEN-HARRIER.[1] — A neighbouring gentleman sprung a pheasant in a wheat stubble, and shot at it ; when, notwithstanding the report of the gun, it was immediately pursued by the blue hawk, known by the name of the hen-harrier, but escaped into some covert. He then sprung a second, and a third, in the same field, that got away in the same manner ; the hawk hovering round him all the while that he was beating the field, conscious, no doubt, of the game that lurked in the stubble. Hence we may conclude that this bird of prey was rendered very daring and bold by hunger, and that hawks cannot always seize their game when they please. We may farther observe, that they cannot pounce their quarry on the ground, where it might be able to make a stout resistance, since so large a fowl as a pheasant could not but be visible to the piercing eye of a hawk, when hovering over the field. Hence that propensity of cowering and squatting, till they are almost trod on, which, no doubt, was intended as a mode of security : though long rendered destructive to the whole race of *gallinæ* by the invention of nets and guns.—WHITE.

Of the great boldness and rapacity of birds of prey, when urged on by hunger, I have seen several instances ; particularly, when shooting in the winter, in company with two friends, a woodcock flew across us, closely pursued by a small

(1) *Circus cyaneus.* See p. 269. The hen-harrier is thinly distributed in all the British isles, and is found in most parts of the world. A variety of this bird inhabits America. The two birds called hen-harrier and the ring-tail hawk have been discovered to be the adult male and female.—J. G. W.

hawk ; we all three fired at the woodcock instead of the hawk, which, notwithstanding the report of three guns close by it, continued its pursuit of the woodcock, struck it down, and carried it off, as we afterwards discovered.

At another time, when partridge-shooting with a friend, we saw a ring-tail hawk rise out of a pit with some large bird in its claws ; though at a great distance, we both fired, and obliged it to drop its prey, which proved to be one of the partridges which we were in pursuit of : and lastly, in an evening, I shot at and plainly saw that I had wounded a partridge ; but, it being late, was obliged to go home without finding it again. The next morning, I walked round my land without any gun ; but a favourite old spaniel followed my heels. When I came near the field where I wounded the bird the evening before, I heard the partridges call, and they seemed to be much disturbed. On my approaching the bar-way, they all rose, some on my right and some on my left hand ; and just before and over my head, I perceived (though indistinctly, from the extreme velocity of their motion) two birds fly directly against each other, when instantly, to my great astonishment, down dropped a partridge at my feet ; the dog immediately seized it, and, on examination, I found the blood flow very fast from a fresh wound in the head, but there was some dry clotted blood on its wings and side ; whence I concluded, that a hawk had singled out my wounded bird as the object of his prey, and had struck it down the instant that my approach had obliged the birds to rise on the wing ; but the space between the hedges was so small, and the motion of the birds so instantaneous and quick, that I could not distinctly observe the operation.—MARKWICK.

GREAT SPECKLED DIVER, OR LOON.—As one of my neighbours was traversing Wolmer Forest, from Bramshot across the moors, he found a large uncommon bird fluttering in the heath, but not wounded, which he brought home alive. On examination it proved to be *Colymbus glacialis*, Linn., the great speckled diver. or loon, which is most excellently described in Willughby's *Ornithology*.

Every part and proportion of this bird is so incomparably

adapted to its mode of life, that in no instance do we see the wisdom of God in the creation to more advantage. The head is sharp, and smaller than the part of the neck adjoining, in order that it may pierce the water; the wings are

GREAT NORTHERN DIVER.[1]

placed forward, and out of the centre of gravity, for a purpose which shall be noticed hereafter; the thighs quite at the podex, in order to facilitate diving; and the legs are flat, and as sharp backwards almost as the edge of a knife, that, in striking, they may easily cut the water; while the feet are palmated and broad for swimming, yet so folded up, when advanced to take a fresh stroke, as to be full as narrow as the shank. The two exterior toes of the feet are longest; the nails flat and broad, resembling the human, which give strength, and increase the power of swimming. The foot, when expanded, is not at right angles with the leg

(1) *Columbus glacialis.*—J. G. W.

or body of the bird; but the exterior part, inclining towards the head, forms an acute angle with the body; the intention being, not to give motion in the line of the legs themselves, but, by the combined impulse of both in an intermediate line, the line of the body.

Most people know, that have observed at all, that the swimming of birds is nothing more than a walking in the water, where one foot succeeds the other as on the land; yet no one, as far as I am aware, has remarked that diving fowls, while under water, impel and row themselves forward by a motion of their wings, as well as by the impulse of their feet: but such is really the case, as any person may easily be convinced, who will observe ducks when hunted by dogs in a clear pond. Nor do I know that any one has given a reason why the wings of diving fowls are placed so forward: doubtless, not for the purpose of promoting their speed in flying, since that position certainly impedes it; but probably for the increase of their motion under water, by the use of four oars instead of two; yet were the wings and feet nearer together, as in land birds, they would, when in action, rather hinder than assist one another.[1]

This *colymbus* was of considerable bulk, weighing only three drachms short of three pounds avoirdupois. It measured in length, from the bill to the tail (which was very short) two feet, and to the extremities of the toes four inches more: and the breadth of the wings expanded was 42 inches. A person attempted to eat the body, but found it very strong and rancid, as is the flesh of all birds living on fish. Divers, or loons, though bred in the most northerly parts of Europe, yet are seen with us in very severe winters; and on the Thames are called sprat-loons, because they prey much on that sort of fish.

The legs of the *colymbi* and *mergi* are placed so very backward, and so out of all centre of gravity, that these birds cannot walk at all. They are called by Linnæus *compedes*, because they move on the ground as if shackled or fettered. —WHITE.

(1) The wings of many of the sea birds act as forelegs when they are on land, and so rapidly do these birds run among the grass, that they might be taken for quadrupeds.—J. G. W.

These accurate and ingenious observations, tending to set forth in a proper light the wonderful works of God in the creation, and to point out his wisdom in adapting the singular form and position of the limbs of this bird to the particular mode in which it is destined to pass the greatest part of its life, in an element much denser than the air, do Mr. White credit, not only as a naturalist, but as a man and as a philosopher, in the truest sense of the word, in my opinion; for, were we enabled to trace the works of Nature minutely and accurately, we should find, not only that every bird, but every creature, is equally well adapted to the purpose for which it was intended; though this fitness and propriety of form is more striking in such animals as are destined to any uncommon mode of life.

I have had in my possession two birds, which, though of a different genus, bear a great resemblance to Mr. White's *colymbus* in their manner of life, which is spent chiefly in the water, where they swim and dive with astonishing rapidity; for which purpose their fin-toed feet, placed far behind, and very short wings, are particularly well adapted, and show the wisdom of God in the creation as conspicuously as the bird before mentioned. These birds were the greater and lesser crested grebe (*Podiceps cristatus et auritus*). What surprised me most was, that the first of these birds was found alive on dry ground, about seven miles from the sea, to which place there was no communication by water. How did it get so far from the sea, its wings and legs being so ill adapted either to flying or walking? The lesser crested grebe was also found in a fresh-water pond, which had no communication with other water, at some miles distance from the sea.—MARKWICK.

STONE-CURLEW.—On the 27th of February, 1788, stone-curlews were heard to pipe; and on March 1st, after it was dark, some were passing over the village, as might be perceived by their quick short note, which they use in their nocturnal excursions by way of watch-word, that they may not stray and lose their companions.

Thus we see that, retire whithersoever they may in the winter, they return again early in the spring, and are, as it now appears, the first summer birds that come back. Perhaps the mildness of the season may have quickened the emigration of the curlews this year.

They spend the day in high elevated fields and sheep-walks; but seem to descend, in the night, to streams and meadows, perhaps for water, which their upland haunts do not afford them.—WHITE.

On the 31st of January, 1792, I received a bird of this species, which had been recently killed by a neighbouring farmer, who said that he had frequently seen it in his fields during the former part of the winter: this perhaps was an occasional straggler, which, by some accident, was prevented from accompanying its companions in their migration.

MARKWICK.

THE SMALLEST UNCRESTED WILLOW-WREN.—The smallest uncrested willow-wren, or chiff-chaff, is the next early summer

CHIFF-CHAFF [1]

bird which we have remarked: it utters two sharp piercing notes, so loud in hollow woods as to occasion an echo, and is usually first heard about the 20th of March.—WHITE.

(1) *Sylvia hippolais.* It derives its English name from its curious little song which resembles the two words chiff-chaff. It is tamed without difficulty, and has been taught to perch on the hand. It is found throughout all the southern counties of England, but is not common north of Northumberland. It is said to feed on the leaf roller caterpillars that do such infinity of mischief, especially to our noblest of forest trees the oak. The nest is entered by a hole at the side, like that of the wren, and is placed either on the ground, in a hedge-bank, or on the lower branches of a bush.—J. G. W.

This bird, which Mr. White calls the smallest willow-wren, or chiff-chaff, makes its appearance very early in the spring, and is very common with us; but I cannot make out the three different species of willow-wrens, which he assures us he has discovered. Ever since the publication of his *History of Selborne*, I have used my utmost endeavours to discover his three birds, but hitherto without success. I have frequently shot the bird which "haunts only the tops of trees, and makes a sibilous noise," even in the very act of uttering that sibilous note; but it always proved to be the common willow-wren, or his chiff-chaff. In short, I never could discover more than one species, unless my greater pettichaps (*Sylvia hortensis* of Latham) is his greatest willow-wren.—MARKWICK.

FERN-OWL, OR GOAT-SUCKER.—The country people have a notion that the fern-owl, or churn-owl, or eve-jarr, which they also call a puckeridge, is very injurious to weanling calves by inflicting, as it strikes at them, the fatal distemper known to cow-leeches by the name of puckeridge. Thus does this harmless, ill-fated bird fall under a double imputation, which it by no means deserves,—in Italy, of sucking the teats of goats, whence it is called *caprimulgus;* and with us, of communicating a deadly disorder to cattle. But the truth of the matter is, the malady above mentioned is occasioned by the *Œstrus bovis,* a dipterous insect, which lays its eggs along the chines of kine, where the maggots, when hatched, eat their way through the hide of the beast, into the flesh, and grow to a very large size. I have just talked with a man, who says he has more than once stripped calves who have died of the puckeridge; that the ail or complaint lay along the chine, where the flesh was much swelled, and filled with purulent matter. Once I myself saw a large rough maggot of this sort squeezed out of the back of a cow. These maggots in Essex are called wornils.

The least observation and attention would convince men that these birds neither injure the goatherd nor the grazier, but are perfectly harmless. and subsist alone, being night birds, on night-insects, such as *scarabœi* and *phalœnœ;* and through the month of July, mostly on the *Scarabœus solstitialis,* which in many districts abounds at that season. Those that we have opened have always had their craws stuffed with large night-moths and their eggs, and pieces of chaffers; nor does it anywise appear how they can, weak and unarmed as they seem, inflict any harm upon kine, unless they possess the powers of animal magnetism, and can affect them by fluttering over them.

A fern-owl this evening (August 27) showed off in a very unusual and entertaining manner, by hawking round and round the circumference of my great spreading oak, for twenty times following, keeping mostly close to the grass, but occasionally glancing up amidst the boughs of the tree. This amusing bird was then in pursuit of a brood of some particular *phalænæ* belonging to the oak, of which there are several sorts; and exhibited on the occasion a command of wing superior, I think, to that of the swallow itself.

When a person approaches the haunt of fern-owls in an evening, they continue flying round the head of the obtruder; and, by striking their wings together above their backs, in the manner that the pigeons called smiters are known to do, make a smart snap; perhaps at that time they are jealous for their young; and their noise and gesture are intended by way of menace.

Fern-owls have attachment to oaks, no doubt on account of food; for the next evening we saw one again several times among the boughs of the same tree; but it did not skim round its stem over the grass, as on the evening before. In May, these birds find the *Scarabaus melolontha* on the oak: and the *Scarabæus solstitialis* at midsummer. These peculiar birds can only be watched and observed for two hours in the twenty-four: and then in a dubious twilight, an hour after sun-set, and an hour before sun-rise.

On this day (July 14, 1789), a woman brought me two eggs of a fern-owl, or eve-jarr, which she found on the verge of the Hanger, to the left of the Hermitage, under a beechen shrub. This person, who lives just at the foot of the Hanger, seems well acquainted with these nocturnal swallows, and says she has often found their eggs near that place, and that they lay only two at a time on the bare ground. The eggs were oblong, dusky, and streaked somewhat in the manner of the plumage of the parent bird, and were equal in size at each end. The dam was sitting on the eggs when found, which contained the rudiments of young, and would have been hatched, perhaps, in a week. From hence we may see the time of their breeding, which corresponds pretty well with the swift, as does also the period of their arrival. Each species is usually seen about the beginning of May; each breeds but once in a summer; each lays only two eggs.

July 4, 1790.—The woman who brought me two fern-owls' eggs last year, on July 14, on this day produced me two more, one of which had been laid this morning, as appears plainly, because there was only one in the nest the evening before. They were found, as last July, on the verge of the Down above the Hermitage, under a beechen shrub, on the naked ground. Last year, those eggs were full of young, and just ready to be hatched.

These circumstances point out the exact time when these curious nocturnal migratory birds lay their eggs and hatch their young. Fern-owls, like snipes, stone-curlews, and some other birds, make no nests. Birds that build on the ground do not make much of nests.—WHITE.

No author that I am acquainted with has given so accurate and pleasing an account of the manners and habits of the goat-sucker as Mr. White, taken entirely from his own observations. Its being a nocturnal bird, has prevented my having many opportunities of observing it. I suspect that it passes the day in concealment amidst the dark and shady gloom of deep-wooded dells, or, as they are called here, gills; having more than once seen it roused from such solitary places by my dogs, when shooting in the day-time. I have also sometimes seen it in an evening, but not long enough to take notice of its habits and manners. I have never seen it but in the summer, between the months of May and September.—MARKWICK.

SAND-MARTINS.—March 23, 1788.—A gentleman, who was this week on a visit at Waverley, took the opportunity of examining some of the holes in the sand-banks with which that district abounds. As these are undoubtedly bored by bank-martins, and are the places where they avowedly breed, he was in hopes they might have slept there also, and that he might have surprised them just as they were awaking from their winter slumbers. When he had dug for some time, he found the holes were horizontal and serpentine, as I had observed before; and that the nests were deposited at the inner end, and had been occupied by broods in former summers; but no torpid birds were to be found. He opened and examined about a dozen holes. Another gentleman made the same search many years ago, with as little success. These holes were in depth about two feet.

March 21, 1790.—A single bank or sand-martin was seen hovering and playing round the sand-pit at Short Heath, where in the summer they abound.

April 9, 1793.—A sober hind assures us, that this day, on Wish-Hanger Common, between Hedleigh and Frinsham, he saw several bank-martins playing in and out, and hanging before some nest holes in a sand hill, where these birds usually nestle.

This incident confirms my suspicions that this species of *hirundo* is to be seen first of any; and gives great reason to suppose that they do not leave their wild haunts at all, but are secreted amidst the clefts and caverns of those abrupt cliffs where they usually spend their summers.

The late severe weather considered, it is not very probable that these birds should have migrated so early from a tropical region, through all these cutting winds and pinching frosts: but it is easy to suppose that they may, like bats and flies, have been awakened by the influence of the sun amidst their secret *latebræ*, where they have spent the uncomfortable foodless months in a torpid state, and the profoundest of slumbers.

There is a large pond at Wish-Hanger, which induces these sand-martins to frequent that district. For I have ever remarked that they haunt near great waters, either rivers or lakes.—WHITE.

Here, and in many other passages of his writings, this very ingenious naturalist favours the opinion that part, at least, of the swallow tribe pass their winter in a torpid state, in the same manner as bats and flies, and revive again on the approach of spring.

I have frequently taken notice of all these circumstances, which induced Mr. White to suppose that some of the *hirundines* lie torpid during winter. I have seen, so late as November, on a finer day than usual at that season of the year, two or three swallows flying backwards and forwards under a warm hedge, or on the sunny side of some old building; nay, I once saw, on the 8th of December, two martins flying about very briskly, the weather being mild. I had not seen any considerable number, either of swallows or martins, for a good while before: from whence, then could these few birds come, if not from some hole or cavern where they had laid themselves up for the winter? Surely it will not be asserted that these birds migrate back again, from some distant tropical region, merely on the appearance of a fine day or two at this late season of the year. Again, very early in the spring, and sometimes immediately after very cold, severe weather, on its growing a little warmer, a few of these birds suddenly make their appearance, long before the generality of them are seen. These appearances certainly favour the opinion of their passing the winter in a torpid state, but do not absolutely prove the fact: for who ever saw them reviving of their own accord from their torpid state, without being first brought to the fire, and, as it were, forced into life again; soon after which revivification, they constantly die?—MARKWICK.

SWALLOWS, CONGREGATING AND DISAPPEARANCE OF.—During the severe winds that often prevail late in the spring, it is not easy to say how the *hirundines* subsist; for they withdraw themselves, and are hardly ever seen, nor do any insects appear for their support. That they can retire to rest, and sleep away these uncomfortable periods, as bats do, is a matter rather to be suspected than proved: or do they not rather spend their time in deep and sheltered vales near waters, where insects are more likely to be found? Certain it is, that hardly any individuals of this genus have, at such times, been seen for several days together.

September 13, 1791.—The congregating flocks of *hirundines* on the church and tower are very beautiful and amusing! When they fly off together from the roof, on any alarm, they quite swarm in the air. But they soon settle in heaps, and, preening their feathers, and lifting up their wings to admit the sun, seem highly to enjoy the warm situation. Thus they spend the heat of the day, preparing for their emigration, and, as it were, consulting when and where they are to go. The flight about the church seems to consist chiefly of house-martins, about four hundred in number: but there are other places of rendezvous about the village frequented at the same time.

It is remarkable, that though most of them sit on the battlements and roof, yet many hang or cling for some time by their claws against the surface of the walls, in a manner not practised by them at any other time of their remaining with us.

The swallows seem to delight more in holding their assemblies on trees.

November 3, 1789.—Two swallows were seen this morning at Newton Vicarage House, hovering and settling on the roofs and out-buildings. None have been observed at Selborne since October 11. It is very remarkable, that after the *hirundines* have disappeared for some weeks, a few are occasionally seen again; sometimes, in the first week in November, and that only for one day. Do they not withdraw and slumber in some hiding-place during the interval? for we cannot suppose they had migrated to warmer climes, and so returned again for one day. Is it not more probable that they are awakened from sleep, and, like the bats, are come forth to collect a little food? Bats appear at all seasons through the autumn and spring months, when the thermometer is at 50, because then *phalænæ* and moths are stirring. These swallows looked like young ones.—WHITE.

Of their migration, the proofs are such as will scarcely admit of a doubt. Sir Charles Wager and Captain Wright saw vast flocks of them at sea, when on their passage from one country to another. Our author, Mr. White, saw what he deemed the actual migration of these birds, and which he has described at p. 94 of his *History of Selborne;* and of their congregating together on the roofs of churches and other buildings, and on trees, previous to their departure, many instances occur; particularly, I once observed a large flock of house-martins on the roof of the church here at Catsfield, which acted exactly in the manner here described by Mr. White, sometimes preening their feathers, and spreading their wings to the sun, and then flying off all together, but soon returning to their former situation. The greatest part of these birds seemed to be young ones.—MARKWICK.

WAGTAILS.—While the cows are feeding in the moist low pasture, broods of wagtails, white and grey, run round them, close up to their noses, and under their very bellies, availing themselves of the flies that settle on their legs, and probably

PIED WAGTAIL [1]

finding worms and *larvæ* that are roused by the trampling of their feet. Nature is such an economist, that the most incongruous animals can avail themselves of each other! Interest makes strange friendships.—WHITE.

Birds continually avail themselves of particular and unusual circumstances to procure their food; thus wagtails keep playing about the noses and legs of cattle as they feed, in quest of flies and other insects which abound near those animals; and great numbers of them will follow close to the plough to devour the worms, &c., that are turned up by that instrument. The red-breast attends the gardener in digging his borders; and will, with great familiarity and tameness,

(1) *Motacilla Yarrelli.*—J. G. W.

YELLOW WAGTAIL.[1]

pick out the worms almost close to his spade, as I have frequently seen. Starlings and magpies very often sit on the backs of sheep and deer to pick out their ticks.—MARKWICK.

WRYNECK.[2]

WRYNECKS.[1]—These birds appear on the grass-plots and walks; they walk a little as well as hop, and thrust their bills

(1) *Motacilla flava.*—J. G. W.

(2) *Yunx torquilla.* The name of this bird is derived from its habit of twisting its neck about in various directions, from which peculiarity it is called the snake bird in some parts of England. The insect food of the wryneck is taken by means of

into the turf, in quest, I conclude, of ants, which are their food. While they hold their bills in the grass, they draw out their prey with their tongues, which are so long as to be coiled round their heads.—WHITE.

GROSBEAK.—Mr. B. shot a cock grosbeak, which he had observed to haunt his garden for more than a fortnight. I began to accuse this bird of making sad havoc among the buds of the cherries, gooseberries, and wall-fruit of all the neighbouring orchards. Upon opening its crop, or craw, no buds were to be seen; but a mass of kernels of the stones of fruits. Mr. B. observed, that this bird frequented the spot where plum-trees grow; and that he had seen it with somewhat hard in its mouth, which it broke with difficulty; these were the stones of damsons. The Latin ornithologists call this bird *coccothraustes*, i. e. berry-breaker, because with its large horny beak it cracks and breaks the shells of stone fruits for the sake of the seed or kernel. Birds of this sort are rarely seen in England, and only in winter.—WHITE.

I have never seen this rare bird but during the severest cold of the hardest winters: at which season of the year, I have had in my possession two or three that were killed in this neighbourhood in different years.—MARKWICK.

OWLS. — Mr. White has observed, p. 197, that the owl returns to its young with food once in five minutes. Mr. Montague has observed, that the wren returns once in two minutes, or, upon an average, thirty-six times in an hour; and this continued full sixteen hours in a day, which, if equally divided between eight young ones, each would receive seventy-two feeds in the day, the whole amounting to five hundred and seventy-six. See *Ornitholog. Dict.* p. 35. To this I will add, that the swallow never fails to return to its nest at the expiration of every second or third minute.—MITFORD.

the long and flexible tongue, which is armed at the tip with a hairy point covered with a mucus to which the insects adhere. The structure of the tongue of the woodpecker and of the wryneck is very similar.

The egg is rather small, and of a delicate glossy white colour.—J. G. W.

Cuckoos.—Since Mr. White's time, much has been added to our knowledge of the cuckoo, by the patient attention of Dr. Jenner. Concerning the singing of the cuckoo, mentioned by Mr. White, at p. 187, I will add the following curious memoranda from the 7th volume of the *Transactions of the Linnæan Society*. "The cuckoo begins early in the season with the interval of a minor third, the bird then proceeds to a major third, next to a fourth, then a fifth, after which his voice breaks without attaining a minor sixth." This curious circumstance was, however, observed very long ago; and it forms the subject of an epigram in that scarce black-letter volume, the *Epigrams* of John Heywood, 1587.— Mitford.

OBSERVATIONS ON QUADRUPEDS.

Sheep.—The sheep on the downs this winter (1769) are very ragged, and their coats much torn; the shepherds say, they tear their fleeces with their own mouths and horns, and they are always in that way in mild wet winters, being teased and tickled with a kind of lice.

After ewes and lambs are shorn, there is great confusion and bleating, neither the dams nor the young being able to distinguish one another as before. This embarrassment seems not so much to arise from the loss of the fleece, which may occasion an alteration in their appearance, as from the defect of that *notus odor*, discriminating each individual personally: which also is confounded by the strong scent of the pitch and tar wherewith they are newly marked; for the brute creation recognise each other more from the smell than the sight; and in matters of identity and diversity, appeal much more to their noses than their eyes. After sheep have been washed, there is the same confusion, from the reason given above.— White.

RABBIT. [1]

RABBITS.—Rabbits make incomparably the finest turf, for they not only bite closer than larger quadrupeds, but they allow no bents to rise; hence warrens produce much the most delicate turf for gardens. Sheep never touch the stalks of grasses.—WHITE.

CAT AND SQUIRRELS.—A boy has taken three little young squirrels in their nest, or drey, as it is called in these parts. These small creatures he put under the care of a cat who had lately lost her kittens, and finds that she nurses and suckles them with the same assiduity and affection as if they were her own offspring. This circumstance corroborates my suspicion, that the mention of exposed and deserted children being nurtured by female beasts of prey who had lost their young, may not be so improbable an incident as many have supposed; and therefore may be a justification of those authors who have gravely mentioned what some have deemed to be a wild and improbable story.

So many people went to see the little squirrels suckled by a cat, that the foster-mother became jealous of her charge, and in pain for their safety; and therefore hid them over the ceiling, where one died. This circumstance shows her affection for these foundlings, and that she supposed the squirrels to be her own young. Thus hens, when they have hatched ducklings, are equally attached to them as if they were their own chickens.—WHITE.

HORSE.—An old hunting mare, which ran on the common, being taken very ill, ran down into the village, as it were to implore the help of men, and died the night following in the street.—WHITE.

(1) *Lepus cuniculus.*—J. G. W.

HOUNDS.—The king's stag-hounds came down to Alton, attended by a huntsman and six yeomen prickers, with horns, to try for the stag that has haunted Harteley Wood for so long a time. Many hundreds of people, horse and foot, attended the dogs to see the deer unharboured; but though the huntsman drew Harteley Wood, and Long Coppice, and Shrubwood, and Temple Hangers, and, in their way back, Harteley and Ward-le-ham Hangers, yet no stag could be found.

The royal pack, accustomed to have the deer turned out before them, never drew the coverts with any address and spirit, as many people that were present observed; and this remark the event has proved to be a true one: for as a person was lately pursuing a pheasant that was wing-broken, in Harteley Wood, he stumbled upon the stag by accident, and ran in upon him as he lay concealed amidst a thick brake of brambles and bushes.—WHITE.

OBSERVATIONS ON INSECTS AND VERMES.

INSECTS IN GENERAL.

THE day and night insects occupy the annuals alternately; the *papilios, muscæ,* and *apes,* are succeeded at the close of day by *phalænæ,* earwigs, woodlice, &c. In the dusk of the

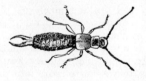

EARWIG. [1]

evening, when beetles begin to buzz, partridges begin to call: these two circumstances are exactly coincident.

(1) *Forficula forcipata.* This is rather a rare species. It is found in the New Forest, and also in Coombe Wood.—J. G. W.

Ivy is the last flower that supports the hymenopterous and dipterous insects. On sunny days, quite on to November, they swarm on trees covered with this plant; and when they disappear, probably retire under the shelter of its leaves, concealing themselves between its fibres and the trees which it entwines.—WHITE.

This I have often observed, having seen bees and other winged insects swarming about the flowers of the ivy very late in the autumn.—MARKWICK.

Spiders, woodlice, *lepismæ* in cupboards and among sugar, some *empedes*, gnats, flies of several species, some *phalænæ* in hedges, earth-worms, &c., are stirring at all times, when winters are mild; and are of great service to those soft-billed birds that never leave us.

HEAD AND FOOT OF EMPIS TESSELATA.

On every sunny day, the winter through, clouds of insects, usually called gnats (I suppose *tipulæ* and *empedes*),[1] appear sporting and dancing over the tops of the evergreen trees in the shrubbery, and frisking about as if the business of generation was still going on. Hence it appears that these *diptera* (which by their sizes appear to be of different species) are not subject to a torpid state in the winter, as most winged insects are. At night, and in frosty weather, and when it rains and blows, they seem to retire into those trees. They often are out in a fog.—WHITE.

(1) The *Empedes* are very curious little flies, not unlike gnats in appearance, but may readily be distinguished from them by the form of the head. One very common species, *Empis tesselata*, feeds on little moths, particularly the green oak moth, *Tortrix clorana*. I have seen hundreds of these *empedes* flying about, each with an unfortunate oak moth in its grasp.—J. G. W.

This I have also seen, and have frequently observed swarms of little winged insects playing up and down in the air in the middle of the winter, even when the ground has been covered with snow.—MARKWICK.

HUMMING IN THE AIR.—There is a natural occurrence to be met with upon the highest part of our down in hot summer days, which always amuses me much, without giving me any satisfaction with respect to the cause of it ; and that is, a loud audible humming of bees in the air, though not one insect is to be seen.   This sound is to be heard distinctly the whole common through, from the Money-dells, to Mr. White's avenue gate.   Any person would suppose that a large swarm of bees was in motion, and playing about over his head.   This noise was heard last week, on June 28th.[1]

> " Resounds the living surface of the ground,
>   Nor undelightful is the ceaseless *hum*
>   To him who muses ———— at noon.——
>   Thick in yon stream of light, a thousand ways,
>   Upward and downward, thwarting and convolved,
>   The quivering nations sport."—THOMSON's *Seasons*.

WHITE.

CHAFFERS.—Cockchaffers[2] seldom abound oftener than once in three or four years ; when they swarm, they deface the trees and hedges.   Whole woods of oaks are stripped bare by them.

Chaffers are eaten by the turkey, the rook, and the house-sparrow.

The *Scarabœus solstitialis* first appears about June 26th ;

(1) This is most probably caused by gnats.—J. G. W.
(2) The mischief done by the larva of this beetle is perfectly incredible ; at all events, to any one who has not dissected it.   The larva feeds upon roots of grasses and such like nutriment.   To enable it to crop these, it is furnished with sickle-shaped jaws, which cut like a pair of scissors.   Moreover, the whole interior of the creature is apparently composed of nothing but a huge stomach, surrounded with fat, and so obese do these grubs become, that when they have arrived at their full growth, they are forced to lie on their sides.

Scarcely less destructive are they when they have assumed the perfect form, seeing that whereas in their larval state they devoured the roots, in their perfect state they eat the leaves.   So we may well be grateful to any creature who frees us from these pests.   The number of this insect varies very considerably.   Some years the country is overwhelmed with them, and in others there is hardly a chaffer to be seen.   Scotland is fortunately almost exempt from this destructive beetle, and in the northern parts it is hardly ever seen.—J. G. W.

they are very punctual in their coming out every year. They are a small species, about half the size of a May-chaffer, and are known in some parts by the name of the fern-chaffer.— WHITE.

COCKCHAFFER.[1]

A singular circumstance relative to the cockchaffer, or, as it is called here, the May-bug (*Scarabæus melolontha*), happened this year (1800):—My gardener, in digging some ground, found, about six inches under the surface, two of these insects alive and perfectly formed, so early as the 24th of March. When he brought them to me, they appeared to be as perfect and as much alive as in the midst of summer, crawling about as briskly as ever : yet I saw no more of this insect till the 22d of May, when it began to make its appearance. How comes it, that though it was perfectly formed so early as the 24th of March, it did not show itself above ground till nearly two months afterwards?—MARKWICK.

PTINUS PECTINICORNIS.[2]—Those maggots that make worm-

(1) *Melolontha vulgaris.* Of the three beetles represented in the cut, the lower figure is the cockchaffer ; that on the left-hand the stag-beetle, (*Lucanus cervus,*) and that on the right the common dor-beetle, (*Geotrupes stercorarius.*) All these beetles are lamellicorn, that is, their antennæ are divided into lamelli, or little plates at their extremities.—J. G. W.

(2) *Ptilinus pectinicornis.*—J. G. W.

holes in tables, chairs, bed-posts, &c., and destroy wooden furniture, especially where there is any sap, are the *larvæ* of the *Plinus pectinicornis*. This insect, it is probable, deposits its eggs on the surface, and the worms eat their way in.

In their holes, they turn into their *pupæ* state, and so come forth winged in July : eating their way through the valances or curtains of a bed, or any other furniture that happens to obstruct their passage.

They seem to be most inclined to breed in beech ; hence beech will not make lasting utensils or furniture. If their eggs are deposited on the surface, frequent rubbing will preserve wooden furniture.—WHITE.

BLATTA ORIENTALIS (COCKROACH). — A neighbour complained to me that her house was overrun with a kind of black beetle, or, as she expressed herself, with a kind of black-bob, which swarmed in her kitchen when they got up in the morning before daybreak.

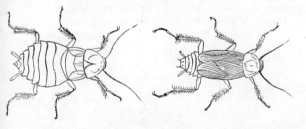

COCKROACH.[1]

Soon after this account, I observed an unusual insect in one of my dark chimney closets, and find since, that in the night they swarm also in my kitchen. On examination, I soon ascertained the species to be the *Blatta orientalis* of Linnæus, and the *Blatta molendinaria* of Mouffet. The male

(1) The eggs of this very unpleasant insect are compound, that is, are enclosed in a case. The case is ridged and toothed, and the young escape by the ridge. The two sexes of the cockroach differ in the number of abdominal segments, those of the male being eight in number, while those of the female are never more than seven. The right-hand figure is the male, that on the left-hand is the female.— J. G. W.

is winged; the female is not, but shows somewhat like the rudiments of wings, as if in the *pupa* state.

These insects belonged originally to the warmer parts of America, and were conveyed from thence by shipping to the East Indies ; and, by means of commerce, begin to prevail in the more northern parts of Europe, as Russia, Sweden, &c. How long they have abounded in England I cannot say ; but have never observed them in my house till lately.

They love warmth, and haunt chimney closets and the backs of ovens. Poda says that these and house-crickets will not associate together ; but he is mistaken in that assertion, as Linnæus suspected he was. They are altogether night-insects, *lucifugæ*, never coming forth till the rooms are dark and still, and escaping away nimbly at the approach of a candle. Their antennæ are remarkably long, slender, and flexile.

October, 1790.—After the servants are gone to bed, the kitchen hearth swarms with young crickets, and young *Blattæ molendinariæ* of all sizes, from the most minute growth to their full proportions. They seem to live in a friendly manner together, and not to prey the one on the other.

August, 1792.—After the destruction of many thousands of *Blattæ molendinariæ*, we find that at intervals a fresh detachment of old ones arrives, and particularly during this hot season ; for, the windows being left open in the evenings, the males come flying in at the casements from the neighbouring houses, which swarm with them. How the females, that seem to have no perfect wings that they can use, can contrive to get from house to house, does not so readily appear. These, like many insects, when they find their present abodes overstocked, have powers of migrating to fresh quarters. Since the *blattæ* have been so much kept under, the crickets have greatly increased in number.—WHITE.

GRYLLUS DOMESTICUS (HOUSE CRICKET).—November.— After the servants are gone to bed, the kitchen hearth swarms with minute crickets, not so large as fleas, which must have been lately hatched. So that these domestic insects, cherished by the influence of a constant large fire, regard not the season of the year, but produce their young at times when their congeners are either dead or laid up for the winter, to pass away the uncomfortable months in the profoundest slumbers, and a state of torpidity.

When house-crickets are out and running about in a room in the night, if surprised by a candle, they give two or three shrill notes, as it were for a signal to their fellows, that they may escape to their crannies and lurking holes, to avoid danger.—WHITE.

CIMEX LINEARIS.—August 12, 1775.—From the multitude of minute young, of all gradations of sizes, these insects seem without doubt to be viviparous.[1]—WHITE.

PHALÆNA QUERCUS. — Most of our oaks are naked of leaves, and even the Holt in general, having been ravaged by the caterpillars of a small *phalæna*, which is of a pale yellow colour. These insects, though a feeble race, yet, from their infinite numbers, are of wonderful effect, being able to destroy the foliage of whole forests and districts. At this season they leave their *aurelia*, and issue forth in their fly state, swarming and covering the trees and hedges.

(1) *Gerris lacustris.* The Water-flea.—This insect belongs to the order Hemiptera, or half-winged insects; so called because the upper wings are partly horny, like those of the beetles, and partly transparent, like those of the bees. There are several species of these insects, resembling each other very closely, and which, on account of their resemblance, are often thought to be the same insect. They all are found living on the banks of rivers and ponds, and may be seen shooting over the surface of the water in defiance of the hungry fish below them. The fish never attempt to seize a water-flea when it is sporting about on the surface of the water.

Being rather struck by this fact, I one day threw a gerris, that had been accidentally maimed, into the water. The creature began to kick and plunge, but was unable to execute those swift movements, by which these insects generally skim over the water. A bleak, seeing the struggles of the wounded insect, took it, as I suppose, for another insect, and rising to the surface, caught it after several unavailing attempts, and carried it down to that mid-water in which bleaks love to swim. Hardly had the fish sunk two feet, before I saw the gerris slip out of its mouth and rise to the surface. The fish soon followed, and again carried it down. Several times the insect escaped in this manner, nor was it until the fish had carried it off five or six times, that the insect was finally swallowed.

The gerris being an aquatic insect, born and bred on the water, is of course to be found near water. I once, however, found a male gerris under a stone, on the summit of Shotover Hill, some distance from any water. Its hiding-place was sandy, and quite dry.

Mr. White is mistaken in imagining that this insect is viviparous. It is oviparous, and deposits its eggs in the substance of the water plants. As to the difference in size, White was deceived by the similarity which these creatures bear, in their imperfect state, to the perfect insect, the only evident distinction being the undeveloped wings.—J. G. W.

In a field near Greatham I saw a flight of swifts busied in catching their prey near the ground ; and found they were hawking after these *phalænæ*. The *aurelia* of this moth is shining, and as black as jet ; and lies wrapped up in a leaf of the tree, which is rolled round it, and secured at the ends by a web, to prevent the maggot from falling out.—WHITE.

I suspect that the insect here meant is not the *Phalæna quercus*, but the *Phalæna viridata*, concerning which I find the following note in my *Naturalist's Calendar* for the year 1785 :—

About this time, and for a few days last past, I observed the leaves of almost all the oak-trees in Denn copse to be eaten and destroyed, and on examining more narrowly, saw an infinite number of small beautiful pale green moths [1] flying about the trees ; the leaves of which, that were not quite destroyed, were curled up, and withinside were the *exuviæ*, or remains, of the *chrysalis*, from whence I suppose the moths had issued, and whose caterpillar had eaten the leaves.— MARKWICK.

EPHEMERA CAUDA BISETA (MAY FLY).—June 10, 1771. Myriads of May flies appeared for the first time on the Alresford stream. The air was crowded with them, and the surface of the water covered. Large trouts sucked them in as they lay struggling on the surface of the stream, unable to rise till their wings were dried.

This appearance reconciled me in some measure to the wonderful account that Scopoli gives of the quantities emerging from the rivers of Carniola. Their motions are very peculiar, up and down for many yards almost in a perpendicular line.—WHITE.

I once saw a swarm of these insects playing up and down over the surface of a pond in Denn park, exactly in the manner described by this accurate naturalist. It was late in the evening of a warm summer day when I observed them.— MARKWICK.

(1) *Tortrix clorana.* These little creatures proceed from those leaf-rolling caterpillars that do so much damage to the oaks.—J. G. W.

SPHYNX OCELLATA.[1]—A vast insect appears after it is dusk, flying with a humming noise, and inserting its tongue into the bloom of the honeysuckle ; it scarcely settles upon the plants, but feeds on the wing in the manner of humming-birds.—WHITE.

HUMMING-BIRD MOTH.[2]

I have frequently seen the large bee-moth (*Sphynx stellatarum*) inserting its long tongue, or proboscis, into the centre of flowers, and feeding on their nectar without settling on them, but keeping constantly on the wing.—MARKWICK.

WILD BEE.—There is a sort of wild bee frequenting the garden-campion for the sake of its tomentum, which probably it turns to some purpose in the business of nidification. It is very pleasant to see with what address it strips off the *pubes*, running from the top to the bottom of a branch, and shaving it bare with all the dexterity of a hoop shaver. When it has got a vast bundle, almost as large as itself, it flies away, holding it secure between its chin and its fore legs.

(1) There are so many of the *sphingidæ* that feed in this manner, that it is difficult to say which of them this may be. At all events the eyed-hawk moth (*Smerinthus ocellatus*) cannot be considered as a vast insect.—J. G. W.

(2) *Macroglossa stellatarum.* This active and beautiful moth may be taught to know those who do not hurt it. When it comes to sip its nectareous food, it hovers on the wing, and if but a finger is raised it darts off in the most inconceivably rapid manner. If, however, it finds that no harm is meant, it comes back again, and will in a short time permit a spectator to examine it quite closely while it is hovering over the flowers.—J. G. W.

There is a remarkable hill on the downs near Lewes, in Sussex, known by the name of Mount Carburn, which overlooks that town, and affords a most engaging prospect of all the country round, besides several views of the sea. On the very summit of this exalted promontory, and amidst the trenches of its Danish camp, there haunts a species of wild bee, making its nest in the chalky soil. When people approach the place, these insects begin to be alarmed, and, with a sharp and hostile sound, dash and strike round the heads and faces of intruders. I have often been interrupted myself, while contemplating the grandeur of the scenery around me, and have thought myself in danger of being stung.[1]—WHITE.

WASPS.—Wasps abound in woody wild districts, far from neighbourhoods. They feed on flowers, and catch flies and caterpillars to carry to their young. Wasps make their nests with the raspings of sound timber; hornets with what they gnaw from decayed. These particles of wood are kneaded up with a mixture of saliva from their bodies and moulded into combs.

When there is no fruit in the gardens, wasps eat flies, and suck the honey from flowers, from ivy-blossoms, and umbellated plants. They carry off also flesh from the butchers' shambles.—WHITE.

In the year 1775, wasps abounded so prodigiously in this neighbourhood, that, in the month of August, no less than seven or eight of their nests were ploughed up in one field; of which there were several instances, as I was informed.

In the spring, about the beginning of April, a single wasp is sometimes seen, which is of a larger size than usual. This, I imagine, is the queen, or female wasp, the mother of the future swarm.[2]—MARKWICK.

(1) Probably *Bombus lapidarius*, in which case Mr. White may have thought himself fortunate that he was not stung, for this insect is very ferocious, and will often attack and drive away any intruder upon their domains.—J. G. W.

(2) Mr. Smee gives a singular account of the power of insects to alter their usual plans of workmanship when pressed by circumstances. It is well known that the cells of a wasp's nest are built in horizontal tiers, each tier being joined to that on either side of it by little pillars made of the same materials as the cells. In one instance, after the wasps had made several tiers of cells, some accident happened to

ŒSTRUS CURVICAUDA.—This insect lays its nits, or eggs, on horses' legs, flanks, &c. each on a single hair. The maggots, when hatched, do not enter the horses' skins, but fall to the ground. It seems to abound most in moist, moorish places, though sometimes seen in the uplands.—WHITE.

NOSE FLY.—About the beginning of July, a species of fly (*musca*) obtains, which proves very tormenting to horses, trying still to enter their nostrils and ears, and actually laying their eggs in the latter of those organs, or perhaps in both. When these abound, horses in woodland districts become very impatient at their work, continually tossing their heads and rubbing their noses on each other, regardless of the driver; so that accidents often ensue. In the heat of the day, men are often obliged to desist from ploughing. Saddle-horses are also very troublesome at such seasons. Country people call this insect the nose fly.—WHITE.

Is not this insect the *Œstrus nasalis* of Linnæus, so well described by Mr. Clark, in the third volume of the *Linnæan Transactions*, under the name of *Œstrus veterinus?*[1]—MARKWICK.

ICHNEUMON FLY.—I saw lately a small ichneumon fly attack a spider much larger than itself, on a grass walk. When the spider made any resistance, the ichneumon applied her tail to him, and stung him with great vehemence, so that he soon became dead and motionless. The ichneumon then, running backwards, drew her prey very nimbly over the walk into the standing grass. This spider would be deposited in some hole where the ichneumon would lay

the nest which made it fall on one side, thereby bringing the mouths of the cells into a horizontal instead of a vertical position. The wasps, finding that their nest was not in its right direction, discontinued the formation of cells in the same line as those already made, and commenced a fresh series of tiers. So that the nest consisted of four vertical, and three horizontal tiers, standing nearly at right angles to each other.

Mr. White is correct in his conjecture respecting the large wasp found in the spring. It is the queen of the future kingdom, but has not only to find a country, but to bring forth subjects and to provide homes for them. Those who wish to procure female wasps for dissection, will find that the spring is the best time, as in the other seasons the various organs are not nearly so highly developed.—J. G. W.

(1) *Gasterophilus nasalis.*—J. G. W.

some eggs ; and as soon as the eggs were hatched, the carcase would afford ready food for the maggots.

Perhaps some eggs might be injected into the body of the spider, in the act of stinging. Some ichneumons deposit their eggs in the *aurelia* of moths and butterflies.— WHITE.

In my *Naturalist's Calendar* for 1795, July 21st, I find the following note :—

It is not uncommon for some of the species of ichneumon flies to deposit their eggs in the chrysalis of a butterfly.[1] Some time ago, I put two of the chrysales of a butterfly into a box, and covered it with gauze, to discover what species of butterfly they would produce; but instead of a butterfly, one of them produced a number of small ichneumon flies.

There are many instances of the great service these little insects are to mankind in reducing the number of noxious insects, by depositing their eggs in the soft bodies of their *larvæ;* but none more remarkable than that of the ichneumon *tipula*, which pierces the tender body, and deposits its eggs in the *larva* of the *Tipula tritici*, an insect which, when it abounds greatly, is very prejudicial to the grains of wheat. This operation I have frequently seen it perform with wonder and delight.—MARKWICK.

BOMBYLIUS MEDIUS. — The *Bombylius medius* is much about in March and the beginning of April, and soon seems to retire. It is a hairy insect, like a humble-bee, but with only two wings, and a long straight beak, with which it sucks the early flowers. The female seems to lay its eggs as it poises on its wings, by striking its tail on the ground, and against the grass that stands in its way, in a quick manner, for several times together.—WHITE.

(1) The egg is deposited in the creature during its larval state, and the time of depositing the eggs is so calculated, that the ichneumons attain the pupal state just as the caterpillar itself does. Sometimes they make their way through the caterpillar's skin, just as the change takes place, but sometimes, especially in the larger chrysalides, they remain within the skin until they attain the perfect state. The eggs are laid in the *caterpillar*, not in the chrysalis, as Markwick supposes, and the young derive their nourishment from its fat. Moreover, the skin of the chrysalis would be too hard to be pierced by the ovipositor of the ichneumon.—J. G. W.

I have often seen this insect fly with great velocity, stop on a sudden, hang in the air in a stationary position for

BOMBYLIUS MEDIUS.[1]

some time, and then fly off again ; but do not recollect having ever seen it strike its tail against the ground, or any other substance.—MARKWICK.

MUSCÆ (FLIES).—In the decline of the year, when the mornings and evenings become chilly, many species of flies (*muscæ*) retire into houses, and swarm in the windows.

At first they are very brisk and alert ; but, as they grow more torpid, one cannot help observing that they move with difficulty, and are scarce able to lift their legs, which seem as if glued to the glass ; and by degrees, many do actually stick on till they die in the place.

It has been observed that divers flies, besides their sharp hooked nails, have also skinny palms or flaps to their feet, whereby they are enabled to stick on glass and other smooth bodies, and to walk on ceilings with their backs downward, by means of the pressure of the atmosphere on those flaps ; the weight of which they easily overcome in warm weather, when they are brisk and alert.[2] But, in the decline of the year, this resistance becomes too mighty for their diminished

(1) The flight of this insect is very curious. In the spring, while one is admiring the beauty of the early spring flowers, suddenly, one knows not how it came, there is seen suspended over the flower a bombylius hovering with outstretched proboscis. If the slightest movement is made, it shoots off as mysteriously as it came. Its whole flight very closely resembles that of the humming-bird moth.—J. G. W.

(2) These flaps or suckers are composed of a great number of smaller suckers, each attached to a footstalk, and forming a most efficient organ of prehension.—J. G. W.

strength ; and we see flies labouring along, and lugging their feet in windows, as if they stuck fast to the glass, and it is with the utmost difficulty they can draw one foot after another, and disengage their hollow caps from the slippery surface.

Upon the same principle that flies stick and support themselves, do boys, by way of play, carry heavy weights by only a piece of wet leather, at the end of a string, clapped close on the surface of a stone.—WHITE.

TIPULÆ, OR EMPEDES.—May.—Millions of *empedes,* or *tipulæ,* come forth at the close of day, and swarm to such a degree as to fill the air. At this juncture they sport, and, as it grows more dark, they retire. All day they hide in the hedges. As they rise in a cloud, they appear like smoke.

I do not ever remember to have seen such swarms, except in the fens of the Isle of Ely. They appear most over grass grounds.—WHITE.

APHIDES.—On the first of August, about half an hour after three in the afternoon, the people of Selborne were surprised by a shower of *aphides,* which fell in these parts. They who were walking in the streets at that time, found themselves covered with these insects, which settled also on the trees and gardens, and blackened all the vegetables where they alighted. These armies, no doubt, were then in a state of emigration, and shifting their quarters ; and might, perhaps, come from the great hop plantations of Kent or Sussex, the wind being that day at north. They were observed at the same time at Farnham, and all along the Vale at Alton.—WHITE.

ANTS.—August 23.—Every ant-hill, about this time, is in a strange hurry and confusion ; and all the winged ants, agitated by some violent impulse, are leaving their homes, and, bent on emigration, swarm by myriads in the air, to the great emolument of the *hirundines,* which fare luxuriously. Those that escape the swallows, return no more to their nests, but, looking out for fresh settlements, lay a foundation for future colonies. All the females at this time are pregnant; the males that escape being eaten, wander away and die.[1]

October 2.—Flying ants, male and female, usually swarm and migrate on hot sunny days in August and September; but this day a vast emigration took place in my garden, and myriads came forth, in appearance, from the drain which goes under the fruit wall; filling the air and the adjoining trees and shrubs with their numbers. The females were full of eggs. This late swarming is probably owing to the backward wet season. The day following, not one flying ant was to be seen.

Horse ants travel home to their nests laden with flies, which they have caught, and the *aureliæ* of smaller ants, which they seize by violence.—WHITE.

In my *Naturalist's Calendar* for the year 1777, on September 6th, I find the following note to the article, Flying Ants :—

I saw a prodigious swarm of these ants flying about the top of some tall elm-trees close by my house: some were continually dropping to the ground as if from the trees, and others rising up from the ground: and I imagine their life is but short; for as soon as produced from the egg by the heat of the sun, they propagate their species, and soon after perish. They were black, somewhat like the small black ant, and had four wings. I saw, also, at another place, a large sort, which were yellowish. On the 8th of September, 1785, I again observed the same circumstance of a vast number of these insects flying near the tops of the elms, and dropping to the ground.

On the 2d of March, 1777, I saw great numbers of ants come out of the ground.—MARKWICK.

(1) It may as well be mentioned here, that the wheat-like whitish bodies about which the ants are so anxious, are not the eggs, but the pupæ, enclosed in their silken cells.

These pupæ were long mistaken for wheat, and led to the prevailing idea that ants laid up food for the winter. The ant, it must be remembered, is a carnivorous insect, devouring with great eagerness dead insects or smaller animals, and is very useful in cleaning the bones of a mouse or bird, if a skeleton is to be made, but by no means would eat the dry hard grains of wheat or barley, which it has been so often said to lay up in its store-houses.—J. G. W.

GLOW-WORMS.—By observing two glow-worms, which were
brought from the field to the bank in the garden, it appeared
to us that these little creatures put out their lamps between
eleven and twelve, and shine no more for the rest of the
night

GLOW-WORM. 1

Male glow-worms, attracted by the light of the candles,
come into the parlour.—WHITE.

EARTH-WORMS.—Earth-worms make their casts most in
mild weather, about March and April; they do not lie torpid
in winter, but come forth when there is no frost; they travel
about in rainy nights, as appears from their sinuous tracks on
the soft muddy soil, perhaps in search of food.

When earth-worms lie out a-nights on the turf, though
they extend their bodies a great way, they do not quite leave
their holes, but keep the ends of their tails fixed therein, so
that, on the least alarm, they can retire with precipitation
under the earth. Whatever food falls within their reach when
thus extended, they seem to be content with,—such as blades
of grass, straws, fallen leaves, the ends of which they often
draw into their holes.—WHITE.

SNAILS AND SLUGS.—The shell-less snails called slugs are
in motion all the winter, in mild weather, and commit great

(1) *Lampyris noctiluca.* Both sexes of this curious beetle give out the peculiar
greenish light, but that of the male is very faint indeed. As is not unfrequently the
case with insects, the male only is furnished with wings, the female compounding for
the plainness of her form by the brilliancy of her light. The left-hand figure is the
male, the right-hand figure represents the female insect.—J. G. W.

depredations on garden plants, and much injure the green wheat, the loss of which is imputed to earth-worms; while the shelled snail, the φερέοικος, does not come forth at all till about April 10th, and not only lays itself up pretty early in autumn, in places secure from frost, but also throws out round the mouth of its shell a thick *operculum* formed from its own saliva; so that it is perfectly secured, and corked up, as it were, from all inclemencies. The cause why the slugs are able to endure the cold so much better than shell-snails is, that their bodies are covered with slime, as whales are with blubber.

Soon after midsummer, snails deposit their eggs in the mould, by running their heads and bodies under ground. Hence, the way to be rid of them is, to kill as many as possible before they begin to breed.

Large, gray, shell-less cellar snails lay themselves up about the same time with those that live abroad; hence, it is plain that a defect of warmth is not the only cause that influences their retreat.—WHITE.

SNAKE'S SLOUGH.—

——— "There the snake throws her enamell'd skin."
SHAKSPEARE, *Mids. Night's Dream.*

About the middle of this month (September) we found, in a field near a hedge, the slough of a large snake, which seemed to have been newly cast. From circumstances, it appeared as if turned wrong side outward, and as drawn off backward, like a stocking, or woman's glove.[1] Not only the whole skin, but scales from the very eyes, are peeled off, and appear in the head of the slough like a pair of spectacles. The reptile, at the time of changing his coat, had entangled himself intricately in the grass and weeds, so that the friction of the stalks and blades might promote this curious shifting of his *exuviæ*—

———————"Lubrica serpens
Exuit in spinis vestem."—LUCRET.

"Smooth serpents that in thickets leave their skin."

_____

(1) So when the crustacea cast their skins, they throw off not only the outer case of their bodies, but even the lining membranes of the stomach, together with its teeth, the broad tendons of the claws, and the coverings of the eyes.—J. G. W.

It would be a most entertaining sight, could a person be an eye-witness to such a feat, and see the snake in the act of changing his garment. As the convexity of the scales of the eyes in the slough is now inward, that circumstance alone is a proof that the skin has been turned: not to mention that now the present inside is much darker than the outer. If you look through the scales of the snake's eyes from the concave side, viz. as the reptile used them, they lessen objects much. Thus it appears, from what has been said, that snakes crawl out of the mouth of their own sloughs, and quit the tail part last, just as eels are skinned by a cook-maid. While the scales of the eyes are growing loose, and a new skin is forming, the creature, in appearance, must be blind, and feel itself in an awkward, uneasy situation.—WHITE.

I have seen many sloughs, or skins of snakes, entire, after they have cast them off; and once, in particular, I remember to have found one of these sloughs so intricately interwoven amongst some brakes, that it was with difficulty removed without being broken: this undoubtedly was done by the creature to assist in getting rid of its encumbrance.

I have great reason to suppose that the eft, or common lizard, also casts its skin, or slough, but not entire like the snake; for, on the 30th of March, 1777, I saw one with something ragged hanging to it, which appeared to be part of its old skin.[1]—MARKWICK.

(1) The eft and the lizard are two different animals. The eft does cast its skin; see note on p. 74. I had an Indian lizard in a bottle of spirits, which cast several skins while in the spirit, each skin peeling off in flakes — J. G. W.

OBSERVATIONS ON VEGETABLES.

TREES, ORDER OF LOSING THEIR LEAVES.

ONE of the first trees that become naked is the walnut; the mulberry, the ash, especially if it bears many keys, and the horse-chestnut come next. All lopped trees, while their heads are young, carry their leaves a long while. Apple-trees and peaches remain green very late, often till the end of November: young beeches never cast their leaves till spring, till the new leaves sprout and push them off: in the autumn, the beechen leaves turn of a deep chestnut colour. Tall beeches cast their leaves about the end of October.—WHITE.

SIZE AND GROWTH.—Mr. Marsham, of Stratton, near Norwich, informs me by letter thus: "I became a planter early; so that an oak, which I planted in 1720, is become now, at one foot from the earth, 12 feet 6 inches in circumference, and at 14 feet (the half of the timber length), is 8 feet 2 inches So, if the bark were to be measured as timber, the tree gives 116½ feet, buyer's measure. Perhaps you never heard of a larger oak, while the planter was living. I flatter myself that I increased the growth by washing the stem, and digging a circle as far as I supposed the roots to extend, and by spreading sawdust, &c., as related in the *Phil. Trans.* I wish I had begun with beeches (my favourite trees, as well as yours); I might then have seen very large trees of my own raising. But I did not begin with beech till 1741, and then by seed; so that my largest is now, 5 feet from the ground, 6 feet 3 inches in girth, and, with its head, spreads a circle of 20 yards diameter. This tree was also dug round, washed, &c. *Stratton, 24th July,* 1790."

FIR. [1]

The circumference of trees planted by myself, at one foot from the ground (1790):—

		Feet.	Inches.
Oak in	1730 . . .	4	5
Ash.	1730 . . .	4	6½
Great fir	1751 . . .	5	0
Greatest beech	1751 . . .	4	0
Elm	1750 . . .	5	3
Lime	1756 . . .	5	5

(1) *Pinus sylvestris.*—J. G. W.

The great oak in the Holt, which is deemed by Mr. Marsham to be the biggest in this island, at 7 feet from the ground, measures, in circumference, 34 feet. It has, in old times, lost several of its boughs, and is tending to decay. Mr. Marsham computes that, at 14 feet length, this oak contains 1000 feet of timber.

It has been the received opinion that trees grow in height only by their annual upper shoot. But my neighbour over the way, whose occupation confines him to one spot, assures me that trees are expanded and raised in the lower parts also. The reason that he gives is this: the point of one of my firs began, for the first time, to peer over an opposite roof at the beginning of summer; but, before the growing season was over, the whole shoot of the year, and three or four joints of the body beside, became visible to him as he sits on his form in his shop. According to this supposition, a tree may advance in height considerably, though the summer shoot should be destroyed every year.—WHITE.

FLOWING OF SAP.—If the bough of a vine is cut late in the spring, just before the shoots push out, it will bleed considerably; but, after the leaf is out, any part may be taken off without the least inconvenience. So oaks may be barked while the leaf is budding; but, as soon as they are expanded, the bark will no longer part from the wood, because the sap that lubricates the bark, and makes it part, is evaporated off through the leaves.—WHITE.

RENOVATION OF LEAVES.—When oaks are quite stripped of their leaves by chaffers, they are clothed again soon after midsummer with a beautiful foliage; but beeches, horse-chestnuts, and maples, once defaced by those insects, never recover their beauty again for the whole season. —WHITE.

ASH-TREES.—Many ash-trees bear loads of keys every year; others never seem to bear any at all. The prolific ones are naked of leaves, and unsightly; those that are sterile abound in foliage, and carry their verdure a long while, and are pleasing objects.—WHITE.

ASH. [1]

BEECH.—Beeches love to grow in crowded situations, and will insinuate themselves through the thickest covert, so as to surmount it all : they are therefore proper to mend thin places in tall hedges.—WHITE.

SYCAMORE.—May 12.—The sycamore, or great maple, is in bloom, and at this season makes a beautiful appearance,

(1) *Fraxinus excelsior.*—J. G. W.

and affords much pabulum for bees, smelling strongly like honey. The foliage of this tree is very fine, and very ornamental to outlets. All the maples have saccharine juices.—WHITE.

SYCAMORE.[1]

GALLS OF LOMBARDY POPLAR.—The stalks and ribs of the leaves of the Lombardy poplar are embossed with large tumours of an oblong shape, which, by incurious observers, have been taken for the fruit of the tree. These galls are full of small insects, some of which are winged, and some

(1) *Acer pseudoplatanus.*—J. G. W.

not. The parent insect is of the genus *cynips*. Some poplars in the garden are quite loaded with these excrescences. —WHITE.

POPLAR [1]

CHESTNUT TIMBER.—John Carpenter brings home some old chestnut-trees, which are very long; in several places the woodpeckers had begun to bore them. The timber and bark of these trees are so very like oak, as might easily deceive an indifferent observer; but the wood is very shakey, and, towards the heart, *cup-shakey* (that is to say, apt to separate in round pieces like cups), so that the inward parts

(1) *Populus nigra.*—J. G. W.

are of no use.   They are bought for the purpose of cooperage, but must make but ordinary barrels, buckets, &c.   Chestnut sells for half the price of oak ; but has sometimes been sent into the king's dock, and passed off instead of oak.—WHITE.

CHESTNUT [1]

LIME BLOSSOMS.—Dr. Chandler tells, that in the south of France an infusion of the blossoms of the lime-tree (*tilia*) is in much esteem as a remedy for coughs, hoarsenesses, fevers, &c.: and that at Nismes, he saw an avenue of limes that was quite ravaged and torn in pieces by people greedily

(1) *Fagus castanea.*—J. G. W.

gathering the bloom, which they dried and kept for these purposes.

Upon the strength of the information, we made some tea

LIME [1]

of lime blossoms; and found it a very soft, well-flavoured, pleasant, saccharine julep, in taste much resembling the juice of liquorice.—WHITE.

BLACKTHORN.—This tree usually blossoms while cold N.E. winds blow; so that the harsh rugged weather obtaining at this season is called by the country people blackthorn winter.—WHITE.

(1) *Tilia Europæa.*—J. G. W.

Ivy Berries.—Ivy berries afford a noble and providential supply for birds in winter and spring; for the first severe frost freezes and spoils all the haws, sometimes by the middle of November. Ivy berries do not seem to freeze.— White.

Hops.—The culture of Virgil's vines corresponded very exactly with the modern management of hops. I might instance in the perpetual diggings and hoeings, in the tying to the stakes and poles, in pruning the superfluous shoots, &c.; but lately I have observed a new circumstance, which was, a neighbouring farmer's harrowing between the rows of hops with a small triangular harrow, drawn by one horse, and guided by two handles. This occurrence brought to my mind the following passage :—

——————— " ipsa
Flectere luctantes inter vineta juvencos."
Georgic II.

The struggling steers between the vine-rows bend.

Hops are diœcious plants : hence, perhaps, it might be proper, though not practised, to leave purposely some male plants in every garden, that their farina might impregnate the blossoms. The female plants, without their male attendants, are not in their natural state : hence we may suppose the frequent failure of crop so incident to hop-grounds. No other growth, cultivated by man, has such frequent and general failures as hops.

Two hop-gardens much injured by a hail-storm, June 5, show now (September 2) a prodigious crop, and larger and fairer hops than any in the parish. The owners seem now to be convinced that the hail by beating off the tops of the binds has increased the side-shoots, and improved the crop. Query. Therefore, should not the tops of hops be pinched off when the binds are very gross and strong ?—White.

Seed lying dormant.—The naked part of the Hanger is now covered with thistles of various kinds. The seeds of these thistles may have lain probably under the thick shade of the beeches for many years, but could not vegetate till the sun and air were admitted. When old beech-trees are cleared away, the naked ground, in a year or two, becomes covered with strawberry plants, the seeds of which must have lain in the ground for an age at least. One of the *slidders*, or trenches, down the middle of the Hanger, close covered over with lofty beeches near a century old, is still called *Strawberry-slidder*, though no strawberries have grown there in the memory of man. That sort of fruit did once, no doubt, abound there, and will again, when the obstruction is removed.—WHITE.

BEANS SOWN BY BIRDS.—Many horse-beans sprang up in my field-walks in the autumn, and are now grown to a considerable height. As the Ewel was in beans last summer, it is most likely that these seeds came from thence; but then the distance is too considerable for them to have been conveyed by mice. It is most probable, therefore, that they were brought by birds, and, in particular, by jays and pies, who seem to have hid them among the grass and moss, and then to have forgotten where they had stowed them. Some peas are growing also in the same situation, and probably under the same circumstances.—WHITE.

CUCUMBERS SET BY BEES.—If bees, who are much the best setters of cucumbers, do not happen to take kindly to the frames, the best way is to tempt them by a little honey put on the male and female bloom. When they are once induced to haunt the frames, they set all the fruit, and will hover with impatience round the lights in a morning, till the glasses are opened. Proved by experience.[1]—WHITE.

WHEAT.—A notion has always obtained, that, in England, hot summers are productive of fine crops of wheat; yet in the years 1780 and 1781, though the heat was intense, the wheat was much mildewed, and the crop light. Does not severe heat, while the straw is milky, occasion its juices to exude, which being extravasated, occasion spots, discolour the stems and blades, and injure the health of the plants?—WHITE.

(1) Cucumbers can be set by hand without troubling the bees.—J. G. W.

TRUFFLES.—August.—A truffle-hunter called on us, having in his pocket several large truffles found in this neighbourhood. He says, these roots are not to be found in deep woods, but in narrow hedge-rows and the skirts of coppices. Some truffles, he informed us, lie two feet within the earth, and some quite on the surface; the latter, he added, have little or no smell, and are not so easily discovered by the dogs as those that lie deeper. Half-a-crown a pound was the price which he asked for this commodity.

Truffles never abound in wet winters and springs. They are in season, in different situations, at least nine months in the year.—WHITE.

TREMELLA NOSTOC.—Though the weather may have been ever so dry and burning, yet after two or three wet days, this jelly-like substance abounds on the walks.—WHITE.

FAIRY RINGS.—The cause, occasion, call it what you will, of *fairy rings*, subsists in the turf, and is conveyable with it; for the turf of my garden-walks, brought from the down above, abounds with those appearances, which vary their shape, and shift situation continually, discovering themselves now in circles, now in segments, and sometimes in irregular patches and spots. Wherever they obtain, puff-balls abound; the seed of which were doubtless brought in the turf.[1]—WHITE.

(1) It is now considered settled that the Fairy Rings may be attributable to fungi. Some fields are entirely covered with them.—J. G. W.

METEOROLOGICAL OBSERVATIONS.

BAROMETER.—November 22, 1768.—A remarkable fall of the barometer all over the kingdom. At Selborne, we had no wind, and not much rain; only vast, swagging, rock-like clouds appeared at a distance.—WHITE.

PARTIAL FROST.—The country people, who are abroad in winter mornings long before sun-rise, talk much of hard frost in some spots, and none in others. The reason of these partial frosts is obvious, for there are at such times partial fogs about: where the fog obtains, little or no frost appears; but where the air is clear, there it freezes hard. So the frost takes place either on hill or in dale, wherever the air happens to be clearest and freest from vapour.—WHITE.

THAW.—Thaws are sometimes surprisingly quick, considering the small quantity of rain. Does not the warmth at such times come from below? The cold in still, severe seasons, seems to come down from above: for the coming over of a cloud in severe nights raises the thermometer abroad at once full ten degrees. The first notices of thaws often seem to appear in vaults, cellars, &c.

If a frost happens, even when the ground is considerably dry, as soon as a thaw takes place the paths and fields are all in a batter. Country people say that the frost draws moisture. But the true philosophy is, that the steam and vapours continually ascending from the earth, are bound in by the frost, and not suffered to escape, till released by the thaw. No wonder, then, that the surface is all in a float; since the quantity of moisture by evaporation that arises daily from every acre of ground is astonishing.—WHITE.

FROZEN SLEET.—January 20.—Mr. H.'s man says, that he caught this day, in a lane near Hackwood-park, many rooks, which, attempting to fly, fell from the trees with their wings frozen together by the sleet that froze as it fell. There were, he affirms, many dozen so disabled.—WHITE.

MIST, CALLED LONDON SMOKE.—This is a blue mist, which has somewhat the smell of coal-smoke, and as it always comes to us with a north-east wind, is supposed to come from London. It has a strong smell, and is supposed to occasion blights. When such mists appear, they are usually followed by dry weather.—WHITE.

REFLECTION ON FOG.[1]—When people walk in deep white fog by night with a lantern, if they will turn their backs to the light, they will see their shades impressed on the fog in rude gigantic proportions. This phenomenon seems not to have been attended to, but implies the great density of the meteor at that juncture.—WHITE.

HONEY DEW.[2]—June 4, 1783.--Vast honey dews this week. The reason of these seems to be, that in hot days the effluvia of flowers are drawn up by brisk evaporation, and then in the night fall down with the dews, with which they are entangled.

This clammy substance is very grateful to bees, who gather it with great assiduity; but it is injurious to the trees on which it happens to fall, by stopping the pores of the leaves. The greatest quantity falls in still, close weather; because winds disperse it, and copious dews dilute it, and prevent its ill effects. It falls mostly in hazy, warm weather.—WHITE.

(1) Some years since, I was standing on the summit of a lofty hill in Wiltshire, at an early hour in the morning, there being a dense fog. As the sun rose, my shadow was thrown forward on the fog, apparently at about fifty yards distance, the head of the reflection being surrounded with an umbra or glory of the prismatic colours. The figure was of gigantic proportions. A companion was with me on the hill, and although each saw his own image, neither of us could see the reflection of the other. The appearance lasted nearly half an hour, during which time the fog was so dense below us, that while we were standing in the full blaze of a morning sun, others who were just below us, but enveloped in the fog, could not see us at all. A similar occurrence is related as taking place in the Arctic regions.—J. G. W.

(2) Honey dew is now ascertained to be the rejectamenta of the aphides which throng the leaves, and from whose bodies it has been seen exuding.—J. G. W.

MORNING CLOUDS.—After a bright night and vast dews, the sky usually becomes cloudy by eleven or twelve o'clock in the forenoon, and clear again towards the decline of the day. The reason seems to be, that the dew drawn up by evaporation occasions the clouds; which, towards evening, being no longer rendered buoyant by the warmth of the sun, melt away, and fall down again in dews. If clouds are watched in a still, warm evening, they will be seen to melt away and disappear.—WHITE.

DRIPPING WEATHER AFTER DROUGHT.—No one that has not attended to such matters, and taken down remarks, can be aware how much ten days dripping weather will influence the growth of grass or corn after a severe dry season. This present summer, 1776, yielded a remarkable instance; for, till the 30th of May, the fields were burnt up and naked, and the barley not half out of the ground; but now, June 10, there is an agreeable prospect of plenty.—WHITE.

AURORA BOREALIS.—November 1, 1787.—The north aurora made a particular appearance, forming itself into a broad, red, fiery belt, which extended from east to west across the welkin: but the moon rising at about ten o'clock, in unclouded majesty, in the east, put an end to this grand, but awful, meteorous phenomenon.—WHITE.

BLACK SPRING, 1771.—Dr. Johnson says, that "in 1771 the season was so severe in the Island of Skye, that it is remembered by the name of the *black spring*. The snow, which seldom lies at all, covered the ground for eight weeks; many cattle died, and those that survived were extremely emaciated." The case was just the same with us here in the south; never were so many barren cows known as in the spring following that dreadful period. Whole dairies missed being in calf together.

At the end of March, the face of the earth was naked to a surprising degree: wheat hardly to be seen, and no signs of any grass; turnips all gone, and sheep in a starving way; all provisions rising in price. Farmers cannot sow for want of rain.—WHITE.

SUMMARY OF THE WEATHER.

MEASURE OF RAIN IN INCHES AND HUNDREDS.

Year.	Jan.	Feb.	Mar.	April.	May.	June.	July.	Aug.	Sept.	Oct.	Nov.	Dec.	Total.
1782	4.64	1.98	6.54	4.57	6.34	1.75	7.09	8.28	3.72	1.93	2.51	0.91	50.26
1783	4.43	5.54	2.16	0.88	2.84	2.82	1.45	2.24	5.52	1.71	3.01	1.10	33.71
1784	3.18	0.77	3.82	3.92	1.52	3.65	2.40	3.88	2.51	0.39	4.70	3.06	33.80
1785	2.84	1.80	0.30	0.17	0.60	1.39	3.80	3.21	5.94	5.21	2.27	4.20	31.55
1786	6.91	1.42	1.62	1.81	2.40	1.20	1.99	4.34	4.79	5.40	4.38	—	—
1787	0.88	3.67	4.28	0.74	2.60	1.50	6.53	0.83	1.56	5.40	4.9	5.60	36.24
1788	1.60	3.37	1.31	0.61	0.76	1.27	3.58	3.22	5.71	—	0.86	0.23	22.50
1789	4.48	4.11	2.47	1.81	4.50	4.24	3.69	0.99	2.82	5.40	3.67	4.62	42.00
1790	1.99	0.49	0.45	3.64	4.38	0.13	3.24	2.30	0.66	2.10	6.95	5.94	32.27
1791	6.73	4.64	1.59	1.13	1.33	0.91	5.66	1.73	1.73	6.49	8.16	4.93	44.93
1792	6.70	1.68	6.70	4.08	3.00	2.78	5.16	4.25	5.53	5.55	1.65	2.11	48.56
1793	3.71	2.32	3.33	3.19	1.21								

SUMMARY OF THE WEATHER.

—•—

1768. BEGINS with a fortnight's frost and snow; rainy during February. Cold and wet spring; wet season from the beginning of June to the end of harvest. Latter end of September foggy, without rain. All October and the first part of November rainy; and thence to the end of the year alternate rains and frosts.

1769. January and February, frosty and rainy, with gleams of fine weather in the intervals. To the middle of March, wind and rain. To the end of March dry and windy. To the middle of April stormy, with rain. To the end of June, fine weather, with rain. To the beginning of August, warm, dry weather. To the end of September, rainy, with short intervals of fine weather. To the latter end of October, frosty mornings, with fine days. The next fortnight rainy; thence to the end of November, dry and frosty. December, windy, with rain and intervals of frost, and the first fortnight very foggy.

1770. Frost for the first fortnight; during the 14th and 15th, all the snow melted. To the end of February, mild, hazy weather. The whole of March, frosty, with bright weather. April cloudy, with rain and snow. May began with summer showers, and ended with dark cold rains. June rainy, chequered with gleams of sunshine. The first fortnight in July, dark and sultry; the latter part of the month, heavy rain. August, September, and the first fortnight in October, in general fine weather, though with frequent interruptions of rain; from the middle of October to the end of the year, almost incessant rains.

1771. Severe frosts till the last week in January. To the first week in February, rain and snow; to the end of February, spring weather. To the end of the third week in April, frosty weather. To the end of the first fortnight in

May, spring weather with copious showers To the end of June, dry, warm weather. The first fortnight in July, warm, rainy weather. To the end of September, warm weather, but in general cloudy, with showers. October rainy. November frost, with intervals of fog and rain. December, in general, bright, mild weather, with hoar frosts.

1772. To the end of the first week in February, frost and snow. To the end of the first fortnight in March, frost, sleet, rain, and snow. To the middle of April, cold rains. To the middle of May, dry weather, with cold piercing winds. To the end of the first week in June, cool showers. To the middle of August, hot, dry, summer weather. To the end of September, rain, with storms and thunder. To December 22, rain, with mild weather. December 23, the first ice. To the end of the month, cold, foggy weather.

1773. The first week in January, frost; thence to the end of the month, dark, rainy weather. The first fortnight in February, hard frost. To the end of the first week in March, misty showery weather. Bright spring days to the close of the month. Frequent showers to the latter end of April. To the end of June, warm showers, with intervals of sunshine. To the end of August, dry weather, with a few days of rain. To the end of the first fortnight in November, rainy. The next four weeks, frost; and thence to the end of the year, rainy.

1774. Frost and rain to the end of the first fortnight in March ; thence to the end of the month, dry weather. To the 15th of April, showers ; thence to the end of April, fine spring days. Du ng May, showers and sunshine in about an equal proportion. Dark, rainy weather to the end of the third week in July ; thence to the 24th of August, sultry, with thunder and occasional showers. To the end of the third week in November, rain, with frequent intervals of sunny weather. To the end of December, dark, dripping fogs.

1775. To the end of the first fortnight in March, rain almost every day. To the first week in April, cold winds, with showers of rain and snow. To the end of June, warm, bright weather, with frequent showers. The first fortnight in July almost incessant rains. To the 26th of August, sultry weather, with frequent showers. To the end of the third week in September, rain, with a few intervals of fine weather. To the end of the year, rain, with intervals of hoar-frost and sunshine.

1776. To January 24, dark, frosty weather, with much snow. March 24, to the end of the month, foggy, with hoar-frost. To the 30th of May, dark, dry, harsh weather, with cold winds. To the end of the first fortnight in July, warm, with much rain. To the end of the first week in August, hot and dry, with intervals of thunder showers. To the end of October, in general fine seasonable weather, with a considerable proportion of rain. To the end of the year, dry, frosty weather, with some days of hard rain.

1777. To the 10th of January, hard frost. To the 20th of January, foggy, with frequent showers. To the 18th of February, hard, dry frost, with snow. To the end of May, heavy showers, with intervals of warm dry spring days. To the 8th of July, dark, with heavy rain. To the 18th of July, dry warm weather. To the end of July, very heavy rains. To the 12th of October, remarkably fine, warm weather. To the end of the year, gray mild weather, with but little rain, and still less frost.

1778. To the 13th of January, frost, with a little snow; to the 24th of January, rain; to the 30th, hard frost. To the 23d February, dark, harsh, foggy weather, with rain. To the end of the month, hard frost with snow. To the end of the first fortnight in March, dark, harsh weather. From the 1st to the end of the first fortnight in April, spring weather. To the end of the month, snow and ice. To the 11th of June, cool, with heavy showers. To the 19th of July, hot, sultry, parching weather. To the end of the month, heavy showers. To the end of September, dry, warm weather. To the end of the year, wet, with considerable intervals of sunshine.

1779. Frost and showers to the end of January. To the 21st of April, warm, dry weather. To 8th of May, rainy. To the 7th June, dry and warm. To the 6th July, hot weather, with frequent rain. To the 18th July, dry, hot weather. To August 8th, hot weather, with frequent rains. To the end of August, fine dry harvest weather. To the end of November, fine autumnal weather, with intervals of rain. To the end of the year, rain with frost and snow.

1780. To the end of January, frost. To the end of February, dark, harsh weather, with frequent intervals of frost. To the end of March, warm, showery, spring weather. To the end of April, dark, harsh weather, with rain and frost. To the end of the first fortnight in May, mild, with rain. To the end of August, rain and fair weather in pretty equal proportions. To the end of October, fine autumnal weather, with intervals of rain. To the 24th of November, frost. To December 16, mild, dry, foggy weather. To the end of the year, frost and snow.

1781. To January 25, frost and snow. To the end of February, harsh and windy, with rain and snow. To April 5, cold, drying winds. To the end of May, mild spring weather, with a few light showers. June began with heavy rain, but thence to the end of October, dry weather, with a few flying showers. To the end of the year, open weather, with frequent rains.

1782. To February 4, open, mild weather. To February 22, hard frost. To the end of March, cold, blowing weather, with frost, and snow, and rain. To May 7, cold, dark rains. To the end of May, mild, with incessant rains. To the end of June, warm and dry. To the end of August, warm, with almost perpetual rains. The first fortnight in September, mild and dry: thence to the end of the month, rain. To the end of October, mild, with frequent showers. November began with hard frost, and continued throughout, with alternate frost and thaw. The first part of December frosty; the latter part mild.

1783. To January 16, rainy, with heavy winds. To the 24th hard frost. To the end of the first fortnight in February, blowing, with much rain. To the end of February, stormy, dripping weather. To the 9th of May, cold, harsh winds (thick ice on 5th of May). To the end of August, hot weather, with frequent showers. To the 23d September, mild, with heavy driving rains. To November 12th, dry, mild weather. To the 18th December, gray, soft weather, with a few showers. To the end of the year, hard frost.

1784. To February 19, hard frost, with two thaws; one

the 14th January, the other 5th February. To February 28, mild, wet fogs. To the 3d March, frost, with ice. To March 10, sleet and snow. To April 2, snow and hard frost. To April 27, mild weather, with much rain. To May 12, cold, drying winds. To May 20, hot, cloudless weather. To June 27, warm, with frequent showers. To July 18, hot, and dry. To the end of August, warm, with heavy rains. To November 6, clear, mild, autumnal weather, except a few days of rain at the latter end of September. To the end of the year, fog, rain, and hard frost (on December 10, the therm. 1 degree below 0).

1785. A thaw began on the 2d of January, and rainy weather, with wind, continued to January 28. To 15th March, very hard frost. To 21st March, mild, with sprinkling showers. To April 7, hard frost. To May 17, mild, windy weather, without a drop of rain. To the end of May, cold, with a few showers. To June 9, mild weather, with frequent soft showers. To July 13, hot, dry weather, with a few showery intervals. To July 22, heavy rain. To the end of September, warm, with frequent showers. To the end of October, frequent rain. To 18th of November, dry, mild weather. (Hay-making finished November 9, and the wheat harvest November 14.) To December 23, rain. To the end of the year, hard frost.

1786. To the 7th January, frost and snow. To January 13, mild, with much rain. To the 21st January, deep snow. To February 11, mild, with frequent rains. 21st February, dry, with high winds. To 10th March, hard frost. To 13th April, wet, with intervals of frost. To the end of April, dry, mild weather. On the 1st and 2d May, thick ice. To 10th May, heavy rain. To June 14, fine, warm, dry weather. From the 8th to the 11th July, heavy showers. To October 13, warm, with frequent showers. To October 19, ice. To October 24, mild, pleasant weather. To November 3, frost. To December 16, rain, with a few detached days of frost. To the end of the year, frost and snow.

1787. To January 24, dark, moist, mild weather. To January 28, frost and snow. To February 16, mild, showery weather. To February 28, dry, cool weather. To March 10, stormy, with driving rain. To March 24, bright frosty weather. To the end of April, mild, with frequent rain. To May 22, fine, bright weather. To the end of June, mostly warm, with frequent showers (on June 7, ice as thick as a crown piece). To the end of July, hot and sultry, with copious rain. To the end of September, hot, dry weather, with occasional showers. To November 23, mild, with light frosts and rain. To the end of November, hard frost. To December 21, still and mild, with rain. To the end of the year, frost.

1788. To January 13, mild and wet. To January 18, frost. To the end of the month, dry, windy weather. To the end of February, frosty, with frequent showers. To March 14, hard frost. To the end of March, dark, harsh weather, with frequent showers. To April 4, windy, with showers. To the end of May, bright, dry, warm weather, with a few occasional showers. From June 28 to July 17, heavy rains. To August 12, hot dry weather. To the end of September, alternate showers and sunshine. To November 22, dry, cool weather. To the end of the year, hard frost.

1789. To January 13, hard frost. To the end of the month, mild, with showers. To the end of February, frequent rain, with snow showers, and heavy gales of wind. To 13th March, hard frost, with snow. To April 18, heavy rain, with frost, and snow, and sleet. To the end of April, dark, cold weather, with frequent rains. To June 9, warm spring weather, with brisk winds, and frequent showers. From June 4, to the end of July, warm, with much rain. To August 29, hot, dry, sultry weather. To September 11, mild, with frequent showers. To the end of September, fine autumnal weather, with occasional showers. To November 17, heavy rain, with violent gales of wind. To December 18, mild, dry weather, with a few showers. To the end of the year, rain and wind.

1790. To January 16, mild, foggy weather, with occasional rains. To January 21, frost. To January 28, dark with driving rains. To February 14, mild, dry weather. To February 22, hard frost. To April 5, bright, cold weather, with a few showers. To April 15, dark and harsh, with a deep snow. To April 21, cold, cloudy weather, with ice. To June 6, mild, spring weather, with much rain. From

July 3 to July 14, cool, with heavy rain. To the end of July, warm dry weather. To August 6, cold, with wind and rain. To August 24, fine harvest weather. To September 5, strong gales, with driving showers. To November 26, mild autumnal weather, with frequent showers. To December 1, hard frost and snow. To the end of the year, rain and snow, and a few days of frost.

1791. To the end of January, mild, with heavy rains. To the end of February, windy, with much rain and snow. From March to the end of June, mostly dry, especially June. March and April, rather cold and frost. May and June, hot. July, rainy. Fine harvest weather, and pretty dry, to the end of September. Wet October, and cold towards the end. Very wet and stormy in November. Much frost in December.

1792. Some hard frost in January, but mostly wet and mild. February, some hard frost, and a little snow. March, wet and cold. April, great storms on the 13th, then some very warm weather. May and June, cold and dry. July, wet and cold; indifferent harvest, rather late and wet. September, windy and wet. October, showery and mild November, dry and fine. December, mild.

A

COMPARATIVE VIEW

OF

THE NATURALIST'S CALENDAR,

AS KEPT AT SELBORNE, IN HAMPSHIRE,

BY THE LATE REV. GILBERT WHITE, M.A.

AND AT CATSFIELD, NEAR BATTLE, IN SUSSEX,

BY WILLIAM MARKWICK, ESQ. F.L.S.

FROM THE YEAR 1768, TO THE YEAR 1793.

N.B.—The dates in the following Calendars, when more than one, express the *earliest* and *latest* times in which the circumstance noted was observed.

A COMPARATIVE VIEW

OF

WHITE'S AND MARKWICK'S CALENDARS.

Of the abbreviations used, fl. signifies *flowering*; l. *leafing*; and
ap. the first *appearance*.

	WHITE.	MARKWICK.
Redbreast (*sylvia rubecula*) sings . . .	Jan. 1—12.	Jan. 3—31, and again Oct. 6.
Larks (*alauda arvensis*) congregate . .	Jan. 1—18.	Oct. 16, Feb. 9.
Nuthatch (*sitta europæa*) heard . . .	Jan. 1—14.	March 3, April 10.
Winter Aconite (*helleborus hiemalis*) fl. .	Jan. 1, Feb. 18.	Feb. 28, April 17.
Shell-less Snail or Slug (*limax*) ap. . .	Jan. 2.	Jan. 16, May 31.
Grey and } Wagtail {(*motacilla boarula*) ap. White } {(*motacilla alba*) ap.	} Jan. 2—11.	{ Jan. 24, March 26. { Dec. 12, Feb. 23.
Missel Thrush (*turdus viscivorus*) sings .	Jan. 2—14.	Feb. 19, April 14.
Bearsfoot (*helleborus fœtidus*) fl. . . .	Jan. 2, Feb. 14.	March 1, May 5.
Polyanthus (*primula polyantha*) fl. . .	Jan. 2, April 12.	Jan. 1, April 9.
Double Daisy (*bellis perennis plena*) fl. .	Jan. 2, Feb. 1.	March 17, April 29.
Mezereon (*daphne mezereum*) fl. . . .	Jan. 3, Feb. 16.	Jan. 2, April 4.
Pansy (*viola tricolor*) fl.	Jan. 3.	Jan. 1, May 10.
Red Dead-nettle (*lamium purpureum*) fl. .	Jan. 3—21.	Jan. 1, April 5.
Groundsel (*senecio vulgaris*) fl. . . .	Jan. 3—15.	Jan. 1, April 9.
Hazel (*corylus avellana*) fl.	Jan. 3, Feb. 28.	Jan. 21, March 11.
Hepatica (*anemone hepatica*) fl. . . .	Jan. 4, Feb. 18.	Jan. 17, April 9.
Hedge-Sparrow (*sylvia modularis*) sings. .	Jan. 5—12.	Jan. 16, March 13.
Common Flies (*musca domestica*) seen in numbers	} Jan. 5, Feb. 3.	May 15.
Greater Titmouse (*parus major*) sings .	Jan. 6, Feb. 6.	Feb. 17, March 17.
Thrush (*turdus musicus*) sings	Jan. 6—22.	Jan. 15, April 4.
Insects swarm under sunny hedges . .	Jan. 6.	
Primrose (*primula vulgaris*) fl. . . .	Jan. 6, April 7.	Jan. 3, March 22.
Bees (*apis mellifica*) ap.	Jan. 6, March 19.	{ Jan. 31, April 11; { last seen Dec. 30.
Gnats play about	Jan. 6, Feb. 3.	
Chaffinches, male and female (*fringilla cœlebs*) seen in equal numbers . .	} Jan. 6—11.	Dec. 2, Feb. 3.
Furze or Gorse (*ulex europæus*) fl. . .	Jan. 8, Feb. 1.	Jan. 1, March 27.
Wallflower (*cheiranthus cheiri; seu fruticulosus* of Smith) fl.	} Jan. 8, April 1.	Feb. 21, May 9.
Stock (*cheiranthus incanus*) fl. . . .	Jan. 8—12.	Feb. 1, June 3.
Emberiza alba (bunting) in great flocks. .	Jan. 9.	
Linnets (*fringilla linota*) congregate .	Jan. 9.	Jan. 11.
Lambs begin to fall	Jan. 9—11.	Jan. 6, Feb. 21.
Rooks (*corvus frugilegus*) resort to their nest trees	} Jan. 10, Feb. 11.	Jan. 23.
Black Hellebore (*helleborus niger*) fl. . .	Jan. 10.	April 27.
Snowdrop (*galanthus nivalis*) fl. . . .	Jan. 10, Feb. 5.	Jan. 18, March 1.

	WHITE.	MARKWICK.
White Dead-nettle (*lamium album*) fl. . .	Jan. 13.	March 23, May 10.
Trumpet Honeysuckle, fl.	Jan. 13.	
Common Creeping Crowfoot (*ranunculus repens*) fl.	Jan. 13.	April 10, May 12.
House-Sparrow (*fringilla domestica*) chirps	Jan. 14.	Feb. 17, May 9.
Dandelion (*leontodon taraxacum*) fl. . . .	Jan. 16, March 11.	Feb. 1, April 17.
Bat (*vespertilio*) ap.	Jan. 16, March 24.	{Feb. 6, June 1; last seen Nov. 20.
Spiders shoot their webs	Jan. 16.	
Butterfly, ap.	Jan. 16.	{Feb. 21, May 8; last seen Dec. 22.
Brambling (*fringilla montifringilla*) ap. . .	Jan. 16.	Jan. 10—31.
Blackbird (*turdus merula*) whistles . . .	Jan. 17.	Feb. 15, May 13.
Wren (*sylvia troglodytes*) sings	Jan. 17.	Feb. 7, June 12.
Earth-Worms lie out	Jan. 18, Feb. 8.	
Crocus (*crocus vernus*) fl.	Jan. 13, March 18.	Jan. 20, March 19.
Skylark (*alauda arvensis*) sings . . .	Jan. 21.	{Jan. 12, Feb. 27; sings till Nov. 13.
Ivy casts its leaves	Jan. 22.	
Helleborus hiemalis fl.	Jan. 22—24.	Feb. 28, April 17.
Common Dor, or Clock (*scarabæus stercorarius*)	Jan. 23.	{Feb. 12, April 12; last seen Nov. 24.
Peziza acetabulum ap.	Jan. 23.	
Helleborus virid. fl.	Jan. 23, March 5.	
Hazel (*corylus avellana*) fl.	Jan. 23, Feb. 1.	Jan. 27, March 11.
Woodlark (*alauda arborea*) sings . . .	Jan. 24, Feb. 21.	Jan. 28, June 5.
Chaffinch (*fringilla cœlebs*) sings . . .	Jan. 24, Feb. 15.	Jan. 21, Feb. 26.
Jackdaws begin to come to churches . .	Jan. 25, March 4.	
Yellow Wagtail (*motacilla flava*) ap. . .	Jan. 25, April 14.	{April 13, July 3; last seen Sept. 8.
Honeysuckle (*lonicera periclymenum*) l. .	Jan. 25.	Jan. 1, April 9.
Field or Procumbent Speedwell (*veronica agrestis*) fl.	Jan. 27, March 15.	Feb. 12, March 29.
Nettle Butterfly (*papilio urticæ*) ap. . .	Jan. 27, April 2.	{March 5, April 24; last seen June 6.
White Wagtail (*motacilla alba*) chirps .	Jan. 28.	March 16.
Shell-Snail (*helix nemoralis*) ap. . . ,	Jan. 28, Feb. 24.	April 2, June 11.
Earth-Worms engender	Jan. 30.	
Barren Strawberry (*fragaria sterilis*) fl. .	Feb. 1, March 26.	Jan. 13, March 26.
Blue Titmouse (*parus cæruleus*) chirps .	Feb. 1.	April 27.
Brown Wood-Owls hoot	Feb. 2.	
Hen (*phasianus gallus*) sits	Feb. 3.	March 8, hatches.
Marsh Titmouse begins his two harsh sharp notes	Feb. 3.	
Gossamer floats	Feb. 4, April 1.	
Musca tenax ap.	Feb. 4, April 8.	
Laurustine (*viburnum tinus*) fl.	Feb. 5.	Jan. 1, April 5.
Butcher's Broom (*ruscus aculeatus*) fl. . .	Feb. 5.	Jan. 1, May 10.
Fox (*canis vulpes*) smells rank	Feb. 7.	{May 19, young brought forth.
Turkey-Cocks strut and gobble	Feb. 10.	
Yellow-Hammer (*emberiza citrinella*) sings	Feb. 12.	Feb. 18, April 28.
Brimstone Butterfly (*papilio ramni*) ap. .	Feb. 13, April 2.	{Feb. 13, March 8; last seen Dec. 24.
Green Woodpecker (*picus viridis*) makes a loud cry	Feb. 13, March 23.	Jan. 1, April 17.
Raven (*corvus corax*) builds	Feb. 14—17.	{April 1, has young ones June 1.

	WHITE.	MARKWICK.
Yew-tree (*taxus baccata*) fl.	Feb. 14, March 27.	Feb. 2, April 11.
Coltsfoot (*tussilago farfara*) fl.	Feb. 15, March 23.	Feb. 18, April 13.
Rooks (*corvus frugilegus*) build	Feb. 16, March 6.	Feb. 28, March 5.
Partridges (*perdix cinerea*) pair	Feb. 17.	Feb. 16, March 20.
Peas (*pisum sativum*) sown	Feb. 17, March 8.	Feb. 8, March 31.
House-Pigeon (*columba domestica*) has young ones	Feb. 18.	Feb. 8.
Field-Crickets open their holes	Feb. 20, March 30.	
Common Flea (*pulex irritans*) ap.	Feb. 21—26.	
Pilewort (*ficaria verna*) fl.	Feb. 21, April 13.	Jan. 25, March 26.
Goldfinch (*fringilla carduelis*) sings	Feb. 21, April 5.	Feb. 28, May 5.
Viper (*coluber berus*) ap.	Feb. 22, March 26.	Feb. 23, May 6; last seen Oct. 28.
Wood-louse (*oniscus asellus*) ap.	Feb. 23, April 1.	April 27, June 17.
Missel-Thrushes pair	Feb. 24.	
Daffodil (*narcissus pseudonarcissus*) fl.	Feb. 24, April 7.	Feb. 26, April 18.
Willow (*salix alba*) fl.	Feb. 24, April 2.	Feb. 27, April 11.
Frogs (*rana temporaria*) croak	Feb. 25.	March 9, April 20.
Sweet Violet (*viola odorata*) fl.	Feb. 26, March 31.	Feb. 7, April 5.
Phalæna tinea vestianella ap.	Feb. 26.	
Stone-Curlew (*otis œdicnemus*) clamorous	Feb. 27, April 24.	June 17.
Filbert (*corylus sativus*) fl.	Feb. 27.	Jan. 25, March 26.
Ring-Dove coos	Feb. 27, April 5.	March 2, August 10.
Apricot-Tree (*prunus armeniaca*) fl.	Feb.	Feb. 28, April 5.
Toad (*rana bufo*) ap.	Feb. 28, March 24.	March 15, July 1.
Frogs (*rana temporaria*) spawn	Feb. 28, March 22.	Feb. 9, April 10, tadpoles March 19.
Ivy-leaved Speedwell (*veronica hederifolia*) fl.	March 1, April 2.	Feb. 16, April 10.
Peach (*amygdalus persica*) fl.	March 2, April 17.	March 4, April 29.
Frog (*rana temporaria*) ap.	March 2, April 6.	March 9.
Shepherd's Purse (*thlaspi bursa pastoris*) fl	March 3.	Jan. 2, April 16.
Pheasant (*phasianus colchicus*) crows	March 3—29.	March 1, May 22.
Land-Tortoise comes forth	March 4, May 8.	
Lungwort (*pulmonaria officinalis*) fl.	March 4, April 16.	March 2, May 19.
Podura fimetaria ap.	March 4.	
Aranea scenica saliens ap.	March 4.	
Scolopendra forficata ap.	March 5—16.	
Wryneck (*junx torquilla*) ap.	March 5, April 25.	March 26, April 23; last seen Sept. 14.
Goose (*anas anser*) sits on its eggs	March 5.	March 21.
Duck (*anas boschas*) lays	March 5.	March 28.
Dog's Violet (*viola canina*) fl.	March 6, April 18.	Feb. 28, April 22.
Peacock Butterfly (*papilio io*) ap.	March 6.	Feb. 13, April 20; last seen Dec. 25.
Trouts begin to rise	March 7—14.	
Field Beans (*vicia faba*) planted	March 8.	April 29, emerge.
Blood-Worms appear in the water	March 8.	
Crow (*corvus corone*) builds	March 10.	July 1, has young ones.
Oats (*avena sativa*) sown	March 10—18.	March 16, April 13.
Golden-crowned Wren (*sylvia regulus*) sings	March 12, Apr. 30.	April 15, May 22; seen Dec.23, Jan. 26.
Asp (*populus tremula*) fl.	March 12.	Feb. 26, March 28.
Common Elder (*sambucus nigra*) l.	March 13—20.	Jan. 24, April 22.
Laurel (*prunus laurocerasus*) fl.	March 15, May 21.	April 2, May 27.
Chrysomela Gotting ap.	March 15.	
Black Ants (*formica nigra*) ap.	March 15, Apr. 22.	March 2, May 18.

	WHITE.	MARKWICK.
Ephemeræ bisetæ ap.	March 16.	
Gooseberry (*ribes grossularia*) l. . .	March 17, Apr. 11.	Feb. 26, April 9.
Common Stitchwort (*stellaria holostea*) fl. . .	March 17, May 19.	March 8, May 7.
Wood Anemone (*anemone nemorosa*) fl. . .	March 17, Apr. 22.	Feb. 27, April 10.
Blackbird (*turdus merula*) lays	March 17.	April 14, young ones May 19.
Raven (*corvus corax*) sits	March 17.	April 1, builds.
Wheatear (*sylvia œnanthe*) ap. . . .	March 18—30.	March 13, May 23; last seen Oct. 26.
Musk-wood Crowfoot (*adoxa moschatellina*) fl.	March 18, Apr. 13.	Feb. 23, April 28.
Willow-Wren (*sylvia trochilus*) ap. . .	March 19, Apr. 13.	March 30, May 16, sits May 27; last seen Oct. 23.
Fumaria bulbosa fl.	March 19.	
Elm (*ulmus campestris*) fl.	March 19, April 4.	Feb. 17, April 25.
Turkey (*meleagris gallopavo*) lays . . .	March 19, April 7.	March 18—25, sits April 4, young ones April 30.
House-Pigeons (*columba domestica*) sit .	March 20.	March 20, young hatched.
Marsh Marigold (*caltha palustris*) fl. . . .	March 20, Apr. 14.	March 22, May 8.
Buzz-Fly (*bombylius medius*) ap . . .	March 21, Apr. 28.	March 15, April 30.
Sand-Martin (*hirundo riparia*) ap. . . .	March 21, Apr. 12.	April 8, May 16; last seen Sept. 8.
Snake (*coluber natrix*) ap.	March 22—30.	March 3, April 29; last seen Oct. 2.
Horse-Ant (*formica herculeana*) ap. . .	March 22, Apr. 18.	Feb. 4, March 26; last seen Nov. 12.
Greenfinch (*loxia chloris*) sings . . .	March 22, Apr. 22.	March 6, April 26.
Ivy (*hedera helix*) berries ripe	March 23, Apr. 14.	Feb. 16, May 19.
Periwinkle (*vinca minor*) fl.	March 25.	Feb. 6, May 7.
Spurge Laurel (*daphne laureola*) fl. . . .	March 25, Apr. 1.	April 12—22.
Swallow (*hirundo rustica*) ap.	March 26, Apr. 20.	April 7—27; last seen Nov. 16.
Blackcap (*sylvia atricapilla*) heard . . .	March 26, May 4.	Apr. 14, May 18; seen April 14, May 20; last seen Sept. 19.
Young Ducks hatched	March 27.	April 6, May 16.
Golden Saxifrage (*chrysosplenium oppositifolium*) fl.	March 27, Apr. 9.	Feb. 7, March 27.
Martin (*hirundo urbica*) ap.	March 28, May 1.	April 14, May 8; last seen Dec. 8.
Double Hyacinth (*hyacinthus orientalis*) fl.	March 29, Apr. 22.	March 13, April 24.
Young Geese (*anas anser*)	March 29.	March 29, April 19.
Wood Sorrel (*oxalis acetosella*) fl. . . .	March 30, Apr. 22.	Feb. 26, April 26.
Ringousel (*turdus torquatus*) seen . .	March 30, Apr. 17.	October 11.
Barley (*hordeum sativum*) sown . . .	March 31, Apr. 30.	April 12, May 20.
Nightingale (*sylvia luscinia*) sings . . .	April 1, May 1.	April 5, July 4; last seen Aug. 29.
Ash (*fraxinus excelsior*) fl.	April 1, May 4.	March 16, May 8.
Spiders' Webs on the surface of the ground	April 1.	
Chequered daffodil (*fritillaria meleagris*) fl.	April 2—24.	April 15, May 1.
Julus terrestris ap.	April 2.	
Cowslip (*primula veris*) fl.	April 3—24.	March 3, May 17.
Ground-Ivy (*glecoma hederacea*) fl. . .	April 3—15.	March 2, April 16.
Snipe pipes	April 3.	
Box-Tree (*buxus sempervirens*) fl. . . .	April 3.	March 27, May 8.

	WHITE.	MARKWICK.
Elm (*ulmus campestris*) l.	April 3.	April 2, May 19.
Gooseberry (*ribes hortense*) fl.	April 3—14.	March 21, May 1.
Currant (*ribes hortense*) fl.	April 3—5.	March 24, April 28.
Pear-Tree (*pyrus communis*) fl.	April 3, May 29.	March 30, April 30.
Lacerta vulgaris (newt or eft) ap.	April 4.	{ Feb. 17, April 15 ; { last seen Oct. 9.
Dog's Mercury (*mercurialis perennis*) fl.	April 5—19.	Jan. 20, April 16.
Wych Elm (*ulmus glabra seu montana* of } Smith) fl. . . . }	April 5.	April 19, May 10.
Ladysmock (*cardamine pratensis*) fl.. . .	April 6—20.	Feb. 21, April 26.
Cuckoo (*cuculus canorus*) heard . . .	April 7—26.	{ April 15, May 3 ; last { heard June 28.
Black-Thorn (*prunus spinosa*) fl. . . .	April 7, May 10.	March 16, May 8.
Deathwatch (*termes pulsatorius*) beats .	April 7.	March 28, May 28.
Gudgeon spawns	April 7.	
Redstart (*sylvia phœnicurus*) ap. . . .	April 8—28.	{ Apr. 5, sings Apr. 25 ; { last seen Sept. 30.
Crown Imperial (*fritillaria imperialis*) fl. .	April 8—24.	April 1, May 13.
Tit-Lark (*alauda pratensis*) sings	April 9—19.	{ April 14—29, sits { June 16—17.
Beech (*fagus sylvatica*) l.	April 10, May 8.	April 24, May 25.
Shell-Snail (*helix memoralis*) comes out in } troops . . }	April 11, May 9.	May 17, June 11 ap.
Middle Yellow Wren ap.	April 11.	
Swift (*hirundo apus*) ap.	April 13, May 7.	April 28, May 19.
Stinging-Fly (*canops calcitrans*) ap. . .	April 14, May 17.	
Whitlow-Grass (*draba verna*) fl. . . .	April 14.	Jan. 15, March 24.
Larch-Tree (*pinus-larix rubra*) l. . . .	April 14.	April 1, May 9.
Whitethroat (*sylvia cinerea*) ap.	April 14, May 14.	{ Apr. 14, May 5 ; sings { May 3—10 ; last { seen Sept. 23.
Red Ant (*formica rubra*) ap.	April 14.	April 9, June 26.
Mole Cricket (*gryllus gryllotalpa*) churs	April 14.	
Second Willow or Laughing Wren ap. . .	April 14—19—23.	
Red Rattle (*pedicularis sylvatica*) fl. . .	April 15—19.	April 10, June 4.
Common Flesh-Fly (*musca carnaria*) ap. .	April 15.	
Lady-Cow (*coccinella bipunctata*) ap. . .	April 16.	
Grasshopper Lark (*alauda locustæ voce*) ap.	April 16—30.	
Willow Wren, its shivering note heard . .	April 17, May 7.	April 28, May 14.
Middle Willow Wren (*regulus non cristatus* } *medius*) ap. . . . }	April 17—27.	
Wild Cherry (*prunus cerasus*) fl. . . .	April 18, May 12.	March 30, May 10.
Garden Cherry (*prunus cerasus*) fl. . . .	April 18, May 11.	March 25, May 6.
Plum (*prunus domestica*) fl.	April 18, May 5.	March 24, May 6.
Harebell (*hyacinthus non-scriptus seu scilla* } *nutans* of Smith) fl. . . . }	April 19—25.	March 27, May 8.
Turtle (*columba turtur*) coos	April 20—27.	May 14, Aug. 10, seen.
Hawthorn (*cratægus seu mespilus oxy-* } *cantha* of Smith) fl. . . . }	April 20, June 11.	April 19, May 26.
Male Fool's Orchis (*orchis mascula*) fl. .	April 21.	March 29, May 13.
Blue Flesh-fly (*musca vomitoria*) ap. . .	April 21, May 23.	
Black Snail or Slug (*limax ater*) abounds .	April 22.	Feb. 1, Oct. 24 ap.
Apple-Tree (*pyrus malus sativus*) fl. . .	April 22, May 25.	April 11, May 26.
Large Bat ap.	April 22, June 11.	
Strawberry, Wild Wood (*fragaria vesca* } *sylv.*) fl. . . . . }	April 23—29.	April 8—9.
Sauce-alone (*erysimum alliaria*) fl. . .	April 23.	March 31, May 8.
Wild or Bird Cherry (*prunus avium*) fl. .	April¹ 24.	March 30, May 10.

	WHITE.	MARKWICK.
Apis hypnorum ap.	April 24.	
Musca meridiana ap.	April 24, May 28.	
Wolf-Fly (*asilus*) ap.	April 25.	
Cabbage Butterfly (*papilio brassicæ*) ap. .	April 28, May 20.	April 29, June 10.
Dragon-Fly (*libellula*) ap.	April 30, May 21.	{ April 18, May 13; last seen Nov. 10.
Sycamore (*acer pseudoplatanus*) fl. . .	April 30, June 6.	April 20, June 4.
Bombylius minor ap.	May 1.	
Glow-Worm (*lampyris noctiluca*) shines	May 1, June 11.	June 19, Sept. 28.
Fern-Owl or Goatsucker (*caprimulgus europæus*) ap.	May 1—26.	May 16, Sept. 14.
Common Bugle (*ajuga reptans*) fl. . .	May 1.	March 27, May 10.
Field Crickets (*gryllus campestris*) crink	May 2—24.	
Chaffer or May-Bug (*scarabæus melolontha*) ap.	May 2—26.	May 2, July 7.
Honeysuckle (*lonicera periclymenum*) fl. .	May 3—30.	April 24, June 21.
Toothwort (*lathræa squamaria*) fl. . .	May 4—12.	
Shell-Snails copulate	May 4, June 17.	
Sedge Warbler (*sylvia salicaria*) sings .	May 4.	June 2—30.
Mealy Tree (*viburnum lantana*) fl. . .	May 5—17.	April 25, May 22.
Flycatcher (*stoparolas. muscicapa grisola*) ap.	May 10—30.	April 29. May 21.
Apis longicornis ap.	May 10, June 9.	
Sedge Warbler (*sylvia salicaria*) ap. . .	May 11—13.	August 2.
Oak (*quercus robur*) fl.	May 13—15.	April 29, June 4.
Admiral Butterfly (*papilio atalanta*) ap.	May 13.	
Orange-tip (*papilio cardamines*) ap. . .	May 14.	March 30, May 19.
Beech (*fagus sylvatica*) fl.	May 15—26.	April 23, May 28.
Common Maple (*acer campestre*) fl. . .	May 16.	April 24, May 27.
Barberry-Tree (*berberis vulgaris*) fl. . .	May 17—26.	April 28, June 4.
Wood Argus Butterfly (*papilio ægeria*) ap.	May 17.	
Orange Lily (*lilium bulbiferum*) fl. . .	May 18, June 11.	June 14, July 22.
Burnet Moth (*sphinx filipendulæ*) ap. .	May 18, June 13.	May 24, June 26.
Walnut (*juglans regia*) l.	May 18.	April 10, June 1.
Laburnum (*cytisus laburnum*) fl. . . .	May 18, June 5.	May 1, June 23.
Forest Fly (*hippobosca equina*) ap. . .	May 18, June 9.	
Saintfoin (*hedysarum onobrychis*) fl. . .	May 19, June 8.	May 21, July 28.
Peony (*pæonia officinalis*) fl.	May 20, June 15.	April 18, May 26.
Horse Chestnut (*æsculus hippocastanum*) fl.	May 21, June 9.	April 19, June 7.
Lilac (*syringa vulgaris*) fl.	May 21.	April 15, May 30.
Columbine (*aquilegia vulgaris*) fl. . . .	May 21—27.	May 6, June 13.
Medlar (*mespilus germanica*) fl. . . .	May 21, June 20.	April 8, June 19.
Tormentil (*tormentilla erecta seu officinalis* of Smith) fl.	May 21.	April 17, June 11.
Lily of the Valley (*convallaria majalis*) fl.	May 22.	April 27, June 13.
Bees (*apis mellifica*) swarm	May 22, July 22.	May 12, June 23.
Woodroof (*asperula odorata*) fl. . . .	May 22—25.	April 14, June 4.
Wasp, female (*vespa vulgaris*) ap. . . .	May 23.	{ April 2, June 4; last seen Nov. 2.
Mountain Ash (*sorbus seu pyrus aucuparia* of Smith) fl.	May 23, June 8.	April 20, June 8.
Bird's-nest Orchis (*ophrys nidus avis*) fl.	May 24, June 11.	May 18, June 12.
White-beam Tree (*cratægus seu pyrus aria* of Smith) l.	May 24, June 4.	May 3.
Milkwort (*polygala vulgaris*) fl. . . .	May 24, June 7.	April 13, June 2.
Dwarf Cistus (*cistus helianthemum*) fl. .	May 25.	May 4, August 8.
Guelder Rose (*viburnum opulus*) fl. . .	May 26.	May 10, June 8.
Common Elder (*sambucus nigra*) fl. . .	May 26, June 25.	May 6, June 17.

	WHITE.	MARKWICK.
Cantharis noctiluca ap.	May 26.	
Apis longicornis bores holes in walks . .	May 27, June 9.	
Mulberry-Tree (*morus nigra*) l. . . .	May 27, June 13.	May 20, June 11.
Wild Service-Tree (*cratægus seu pyrus tor-* *minalis* of Smith) fl. . . . . . . }	May 27.	May 13, June 19.
Sanicle (*sanicula europæa*) fl.	May 27, June 13.	April 23, June 4.
Avens (*geum urbanum*) fl.	May 28.	May 9, June 11.
Female Fool's Orchis (*orchis morio*) fl. . .	May 28.	April 17, May 20.
Ragged Robin (*lychnis flos cuculi*) fl. . .	May 29, June 1.	May 12, June 8.
Burnet (*poterium sanguisorba*) fl. . . .	May 29.	April 30, August 7.
Foxglove (*digitalis purpurea*) fl. . . .	May 30, June 22.	May 23, June 15.
Cornflag (*gladiolus communis*) fl. . . .	May 30, June 20.	June 9, July 8.
Serapias longifol. fl.	May 30, June 13.	
Raspberry (*rubus idæus*) fl.	May 30, June 21.	May 10, June 16.
Herb Robert (*geranium Robertianum*) fl.	May 30.	March 7, May 16.
Figwort (*scrophularia nodosa*) fl. . . .	May 31.	May 12, June 20.
Gromwell (*lithospermum officinale*) fl. .	May 31.	May 10—24.
Wood Spurge (*euphorbia amygdaloides*) fl. .	June 1.	March 23, May 13.
Ramsons (*allium ursinum*) fl.	June 1.	April 21, June 4.
Mouse-ear Scorpion Grass (*myosotis scor-* *pioides*) fl. . . . . }	June 1.	April 11, June 1.
Grasshopper (*gryllus grossus*) ap. . . .	June 1—14.	{ March 25, July 6 ; { last seen Nov. 3.
Rose (*rosa hortensis*) fl.	June 1—21.	June 7, July 1.
Mouse-ear Hawkweed (*hieracium pilo-* *sella*) fl. . . . . }	June 1, July 16.	April 19, June 12.
Buckbean (*menyanthes trifoliata*) fl. . .	June 1.	April 20, June 8.
Rosechaffer (*scarabæus auratus*) ap. . .	June 2—8.	April 18, August 4.
Sheep (*ovis aries*) shorn	June 2—23.	May 23, June 17.
Waterflag (*iris pseudo-acorus*) fl. . . .	June 2.	May 8, June 9.
Cultivated Rye (*secale cereale*) fl. . . .	June 2.	May 27.
Hound's-tongue (*cynoglossum officinale*) fl.	June 2.	May 11, June 7.
Helleborine (*serapias latifolia*) fl. . . .	June 2, Aug. 6.	July 22, Sept. 6.
Green Goldfly (*musca cæsar*) ap. . . .	June 2.	
Argus Butterfly (*papilia moera*) ap. . .	June 2.	
Spearwort (*ranunculus flammula*) fl. . .	June 3.	April 25, June 13.
Birdsfoot Trefoil (*lotus corniculatus*) fl. .	June 3.	April 10, June 3.
Fraxinella, or White Dittany (*dictamnus* *albus*) fl. . . . . }	June 3—11.	June 9, July 24.
Phryganea nigra ap.	June 3.	
Angler's May-fly (*ephemera vulg.*) ap. . .	June 3—14.	
Ladies' Finger (*anthyllis vulneraria*) fl. . .	June 4.	June 1, August 16
Bee Orchis (*ophrys apifera*) fl.	June 4, July 4.	
Pink (*dianthus deltoides*) fl.	June 5—19.	May 26, July 6.
Mock Orange (*philadelphus coronarius*) fl.	June 5.	May 16, June 23.
Libellula virgo ap.	June 5—20.	
Vine (*vitis vinifera*) fl.	June 7, July 30.	June 18, July 29.
Portugal Laurel (*prunus lusitanicus*) fl. .	June 8, July 1.	June 3, July 16.
Purple-spotted Martagon (*lilium marta-* *gon*) fl. . . . . }	June 8—25.	June 18, July 19.
Meadow Crane's-bill (*geranium pratense*) fl.	June 8, Aug. 1.	
Black Bryony (*tamus communis*) fl. . .	June 8.	May 15, June 21.
Field Pea (*pisum sativum arvense*) fl. . .	June 9.	May 15, June 21.
Bladder Campion (*cucubalus behen seu si-* *lene inflata* of Smith) fl. . . . . }	June 9.	May 4, July 13.
Bryony (*brionia alba*) fl.	June 9.	May 13, August 17.
Hedge-Nettle (*stachys sylvatica*) fl. . .	June 10.	May 28, June 24.
Bittersweet (*solanum dulcamara*) fl. . .	June 11.	May 15, June 20.

	WHITE.	MARKWICK.
Walnut (*juglans regia*) fl.	June 12.	April 18, June 1.
Phallus impudicus ap.	June 12, July 23.	
Rosebay Willow-Herb (*epilobium angusti-folium*) fl.	June 12.	June 4, July 28.
Wheat (*triticum hyburnum*) fl.	June 13, July 22.	June 4—30.
Comfrey (*symphytum officinale*) fl.	June 13.	May 4, June 23.
Yellow Pimpernel (*lysimachia nemorum*) fl.	June 13—30.	April 10, June 12.
Tremella nostoc ap.	June 15, Aug. 24.	
Buckthorn (*rhamnus catharticus*) l.	June 16.	May 25.
Cuckoo-spit Insect (*cicada spumaria*) ap.	June 16.	June 2—21.
Dog-Rose (*rosa canina*) fl.	June 17, 18.	May 24, June 21.
Puff-Ball (*lycoperdon bovista*) ap.	June 17, Sept. 3.	May 6, August 19.
Mullein (*verbascum thapsus*) fl.	June 18.	June 10, July 22.
Viper's Bugloss (*echium anglicum seu vulgare* of Smith) fl.	June 19.	May 27, July 3.
Meadow Hay cut	June 19, July 20.	June 13, July 7.
Stag-Beetle (*lucanus cervus*) ap.	June 19.	June 14—21.
Borage (*borago officinalis*) fl.	June 20.	April 22, July 26.
Spindle-Tree (*euonymus europæus*) fl.	June 20.	May 11, June 25.
Musk Thistle (*carduus nutans*) fl.	June 20, July 4.	June 4, July 25.
Dogwood (*cornus sanguinea*) fl.	June 21.	May 28, June 27.
Field Scabious (*scabiosa arvensis*) fl.	June 21.	June 16, August 14.
Marsh Thistle (*carduus palustris*) fl.	June 21—27.	May 15, June 19.
Dropwort (*spiræa filipendula*) fl.	June 22, July 9.	May 8, Sept. 3.
Great Wild Valerian (*valeriana officinalis*) fl.	June 22, July 7.	May 22, July 21.
Quail (*perdix coturnix*) calls	June 22, July 4.	July 23; seen Sept. 1—18.
Mountain Willow-Herb (*epilobium montanum*) fl.	June 22.	June 5—21.
Thistle upon Thistle (*carduus crispus*) fl.	June 23—29.	May 22, July 22.
Cow-Parsnip (*heracleum sphondylium*) fl.	June 23.	May 27, July 12.
Earth-Nut (*bunium bulbocastanum seu flexuosum* of Smith) fl.	June 23.	May 4—31.
Young Frogs migrate	June 23, Aug. 2.	
Œstrus curvicauda ap.	June 24.	
Vervain (*verbena officinalis*) fl.	June 24.	June 10, July 17.
Corn Poppy (*papaver rhoeas*) fl.	June 24.	April 30, July 15.
Self-heal (*prunella vulgaris*) fl.	June 24.	June 7—23.
Agrimony (*agrimonia eupatoria*) fl.	June 24—29.	June 7, July 9.
Great Horse-fly (*tabanus bovinus*) ap.	June 24, Aug. 2.	
Greater Knapweed (*centaurea scabiosa*) fl.	June 25.	June 7, August 14.
Mushroom (*agaricus campestris*) ap.	June 26, Aug. 30.	April 16, August 16.
Common Mallow (*malva sylvestris*) fl.	June 26.	May 27, July 13.
Dwarf Mallow (*malva rotundifolia*) fl.	June 26.	May 12, July 20.
St. John's Wort (*hypericum perforatum*) fl.	June 26.	June 15, July 12.
Broom-Rape (*orobanche major*) fl.	June 27, July 4.	May 9, July 25.
Henbane (*hyoscyamus niger*) fl.	June 27.	May 13, June 19.
Goat's-beard (*tragopogon pratense*) fl.	June 27.	June 5—14.
Deadly Nightshade (*atropa belladonna*) fl.	June 27.	May 22, August 14.
Truffles begin to be found	June 28, July 29.	
Young Partridges fly	June 28, July 31.	July 8—28.
Lime-Tree (*tilia europæa*) fl.	June 28, July 31.	June 12, July 30.
Spearthistle (*carduus lanceolatus*) fl.	June 28, July 12.	June 27, July 18.
Meadow-sweet (*spiræa ulmaria*) fl.	June 28.	June 16, July 24.
Greenweed (*genista tinctoria*) fl.	June 28.	June 4, July 24.
Wild Thyme (*thymum serpyllum*) fl.	June 28.	June 6, July 19.
Stachys germanic. fl.	June 29, July 20.	
Day-Lily (*hemerocallis flava*) fl.	June 29, July 4.	May 29, June 9.

	WHITE.	MARKWICK.
Jasmine (*jasminum officinale*) fl. . . .	June 29, July 30.	June 27, July 21.
Hollyoak (*alcea rosea*) fl.	June 29, Aug. 4.	July 4, Sept. 7.
Monotropa hypopithys fl.	June 29, July 23.	
Ladies' Bedstraw (*galium verum*) fl. . .	June 29.	June 22, August 3.
Galium palustre fl.	June 29.	
Nipplewort (*lapsana communis*) fl. . .	June 29.	May 30, July 24.
Welted Thistle (*carduus acanthoides*) fl.	June 29.	
Sneezewort (*achillea ptarmica*) fl. . .	June 30.	June 22, August 3.
Musk Mallow (*malva moschata*) fl. . .	June 30.	June 9, July 14.
Pimpernel (*anagallis arvensis*) fl. . .	June 30.	May 4, June 22.
Hoary Beetle (*scarabæus solstit.*) ap. .	June 30, July 17.	
Corn Saw-wort (*serratula arvensis seu car-duus arvensis* of Smith) fl.	July 1.	June 15, July 15.
Pheasant's Eye (*adonis annua seu autumnalis* of Smith) fl.	July 1.	April 11, July 15.
Red Eyebright (*euphrasia seu bartsia odontites* of Smith) fl.	July 2.	June 20, August 10.
Thorough Wax (*bupleurum rotundifol.*) fl.	July 2.	
Cockle (*agrostemma githago*) fl. . . .	July 2.	May 14, July 25.
Ivy-leaved Wild Lettuce (*prenanthes muralis*) fl.	July 2.	June 2, July 25.
Feverfew (*matricaria seu pyrethrum parthenium* of Smith) fl.	July 2.	June 19, July 24.
Wall Pepper (*sedum acre*) fl.	July 3.	June 8, July 12.
Privet (*ligustrum vulgare*) fl.	July 3.	June 3, July 13.
Common Toadflax (*antirrhinum linaria*) fl.	July 3.	June 21, August 3.
Perennial Wild Flax (*linum perenne*) fl. .	July 4.	April 21, July 6.
Whortleberries ripe (*vaccinium ulig.*)	July 4—24.	
Yellow Base Rocket (*reseda lutea*) fl. .	July 5.	July 19.
Bluebottle (*centaurea cyanus*) fl. . . .	July 5.	May 15, Oct. 14.
Dwarf Carline Thistle (*carduus acaulis*) fl.	July 5—12.	June 30, August 4.
Bullrush, or Catstail (*typha latifolia*) fl.	July 6.	June 29, July 21.
Spiked Willow-herb (*lythrum salicaria*) fl.	July 6.	June 24, August 17.
Black Mullein (*verbascum niger*) fl. . .	July 6.	
Chrysanthemum coronarium fl.	July 6.	May 28, July 28.
Marigolds (*calendula officinalis*) fl. . .	July 6—9.	April 20, July 16.
Little Field Madder (*sherardia arvensis*) fl.	July 7.	Jan. 11, June 6.
Calamint (*melissa seu thymus calamintha* of Smith) fl.	July 7.	July 21.
Black Horehound (*ballota nigra*) fl. . .	July 7.	June 16, Sept. 12.
Wood Betony (*betonica officinalis*) fl. .	July 8—19.	June 10, July 15.
Round-leaved Bell-flower (*campanula rotundifolia*) fl.	July 8.	June 12, July 29.
All-good (*chenopodium bonus henricus*) fl.	July 8.	April 21, June 15.
Wild-Carrot (*daucus carota*) fl. . . .	July 8.	June 7, July 14.
Indian-Cress (*epopæolum majus*) fl. . .	July 8—20.	June 11, July 25.
Cat-Mint (*nepeta cataria*) fl.	July 9.	
Cow-Wheat (*melampyrum sylvaticum seu pratense* of Smith) fl	July 9.	May 2, June 22.
Crosswort (*valantia cruciata seu galium cruciatum* of Smith) fl.	July 9.	April 10, May 28.
Cranberries ripe	July 9—27.	
Tufted Vetch (*vicia cracca*) fl. . . .	July 10.	May 31, July 8.
Wood Vetch (*vicia sylvat.*) fl.	July 10.	
Little Throatwort (*campanula glomerata*) fl.	July 11.	July 28, August 18.
Sheep's Scabious (*jasione montana*) fl. .	July 11.	June 10, July 25.
Pastinaca sylv. fl.	July 12.	
White Lily (*lilium candidum*) fl. . . .	July 12.	June 21, July 22.

	WHITE.	MARKWICK.
Hemlock (*conium maculatum*) fl.	July 13.	June 4, July 20.
Caucalis anthriscus fl.	July 13.	
Flying Ants ap.	July 13—Aug. 11.	Aug. 20, Sept. 19.
Moneywort (*lysimachia nummularia*) fl.	July 13.	June 14, Aug. 16.
Scarlet Martagon (*lilium chalcedonicum*) fl.	July 14—Aug. 4.	June 21, Aug. 6.
Lesser Stitchwort (*stellaria graminea*) fl.	July 14.	May 8, June 23.
Fool's Parsley (*æthusa cynapium*) fl.	July 14.	June 9, Aug. 9.
Dwarf Elder (*sambucus ebulus*) fl.	July 14—29.	
Swallows and Martins congregate	July 14, Aug. 29.	Aug. 12, Sept. 8.
Potato (*solanum tuberosum*) fl.	July 14.	June 3, July 12.
Angelica sylv. fl.	July 15.	
Digitalis ferrugin. fl.	July 15—25.	
Ragwort (*senecio jacobæa*) fl.	July 15.	June 22, July 13.
Golden Rod (*solidago virgaurea*) fl.	July 15.	July 7, Aug. 29.
Star Thistle (*centaurea calcitrapa*) fl.	July 16.	July 16, Aug. 16.
Tree Primrose (*œnothera biennis*) fl.	July 16.	June 12, July 18.
Peas (*pisum sativum*) cut	July 17, Aug. 14.	July 13, Aug. 15.
Galega officin. fl.	July 17.	
Apricots (*prunus armeniaca*) ripe	July 17, Aug. 21.	July 5, Aug. 16.
Crown's Allheal (*stachys palustris*) fl.	July 17.	June 12, July 14.
Branching Willow-herb (*epilobium ramos.*) fl.	July 17.	
Rye Harvest begins	July 17, Aug. 7.	
Yellow Centaury (*chlora perfoliata*) fl.	July 18, Aug. 15.	June 15, Aug. 13.
Yellow Vetchling (*lathyrus aphaca*) fl.	July 18.	
Enchanter's Nightshade (*circæa lutetiana*) fl.	July 18.	June 20, July 27.
Water Hemp Agrimony (*eupatorium cannabinum*) fl.	July 18.	July 4, Aug. 6.
Giant Throatwort (*campanula trachelium*) fl.	July 19.	July 13, Aug. 14.
Eyebright (*euphrasia officinalis*) fl.	July 19.	May 28, July 19.
Hops (*humulus lupulus*) fl.	July 19, Aug. 10.	July 20, Aug. 17.
Poultry moult	July 19.	
Dodder (*cuscuta europæa seu epithymum* of Smith) fl.	July 20.	July 9, Aug. 7.
Lesser Centaury (*gentiana seu chironia centarium* of Smith) fl.	July 20.	June 3, July 19.
Creeping Water Parsnep (*sium nodiflorum*) fl.	July 20.	July 10, Sept. 11.
Common Spurrey (*spergula arvensis*) fl.	July 21.	April 10, July 16.
Wild Clover (*trifolium pratense*) fl.	July 21.	May 2, June 7.
Buckwheat (*polygonum fagopyrum*) fl.	July 21.	June 27, July 10.
Wheat Harvest begins	July 21, Aug. 23.	July 11, Aug. 26.
Great Burr-Reed (*sparganium erectum*) fl.	July 22.	June 10, July 23.
Marsh St. John's-Wort (*hypericum elodes*) fl.	July 22—31.	June 16, Aug. 10.
Sun-Dew (*drosera rotundifolia*) fl.	July 22.	Aug. 1.
March Cinquefoil (*comarum palustre*) fl.	July 22.	May 27, July 12.
Wild Cherries ripe	July 22.	
Lancashire Asphodel (*anthericum ossifragum*) fl.	July 22.	June 31, July 29.
Hooded Willow-Herb (*scutellaria galericulata*) fl.	July 23.	June 2, July 31.
Water Dropwort (*œnanthe fistulos.*) fl.	July 23.	
Horehound (*marrubium vulg.*) fl.	July 23.	
Seseli carnifol. fl.	July 24.	
Water Plantain (*alisma plantago*) fl.	July 24.	May 1, July 31.
Alopecurus myosuroides fl.	July 25.	
Virgin's Bower (*clematis vitalba*) fl.	July 25, Aug. 9.	July 13, Aug. 14.

	WHITE.	MARKWICK.
Bees kill the Drones	July 25.	
Teasel (*dipsacus sylvestris*) fl.	July 26.	July 16, Aug. 3.
Wild Marjoram (*origanum vulgare*) fl. .	July 26.	July 17, Aug. 29.
Swifts (*hirundo apus*) begin to depart .	July 27—29.	Aug. 5.
Small Wild Teasel (*dipsacus pilosus*) fl. .	July 28, 29.	
Wood Sage (*teucrium scorodonia*) fl. . .	July 28.	June 17, July 24.
Everlasting Pea (*lathyrus latifolius*) fl. .	July 28.	June 20, July 30.
Trailing St. John's-Wort (*hypericum humifusum*)	July 29.	May 20, June 22.
White Hellebore (*veratrum album*) fl. .	July 30.	July 18—22.
Camomile (*anthemis nobilis*) fl. . . .	July 30.	June 21, Aug. 20.
Lesser Field Scabious (*scabiosa columbaria*) fl.	July 30.	July 13, Aug. 9.
Sunflower (*helianthus multiflorus*) fl. . .	July 31, Aug. 6.	July 4, Aug. 22.
Yellow Loosestrife (*lysimachia vulgaris*) fl.	July 31.	July 2, Aug. 7.
Swift (*hirundo apus*) last seen . . .	July 31, Aug. 27.	Aug. 11.
Oats (*avena sativa*) cut	Aug. 1—16.	July 26, Aug. 19.
Barley (*hordeum sativum*) cut	Aug. 1—26.	July 27, Sept. 4.
Lesser Hooded Willow-Herb (*scutelaria minor*) fl.	Aug. 1.	Aug. 8, Sept. 7.
Middle Fleabane (*inula disinterica*) fl. .	Aug. 2.	July 7, Aug. 3.
Apis manicata ap.	Aug. 2.	
Swallow-tailed Butterfly (*papilio machaon*) ap.	Aug. 2.	{ April 20, June 7; last seen Aug. 28.
Whame or Burrel-Fly (*œstrus bovis*) lays eggs on horses .	Aug. 3—19.	
Sow Thistle (*sonchus arvensis*) fl. . . .	Aug. 3.	June 17, July 21.
Plantain Fritillary (*papilio cinxia*) ap. . .	Aug. 3.	
Yellow Succory (*picris hieracioides*) fl. . .	Aug. 4.	June 6—25.
Musca mystacea ap.		
Canterbury Bells (*campanula medium*) fl. .	Aug. 5.	June 5, Aug. 11.
Mentha longifol. fl.	Aug. 5.	
Carline Thistle (*carlina vulgaris*) fl. . .	Aug. 7.	July 21, Aug. 18.
Venetian Sumach (*rhus cotinus*) fl. . .	Aug. 7.	June 5, July 20.
Ptinus pectinicornis ap.	Aug. 7.	
Burdock (*arctium lappa*) fl.	Aug. 8.	June 17, Aug. 4.
Fell-Wort (*gentiana amarella*) fl. . . .	Aug. 8, Sept. 3.	
Wormwood (*artemisia absinthium*) fl. . .	Aug. 8.	July 22, Aug. 21.
Mugwort (*artemisia vulgaris*) fl. . . .	Aug. 8.	July 9, Aug. 10.
St. Barnaby's Thistle (*centauria solstit.*) fl. .	Aug. 10.	
Meadow Saffron (*colchicum autumnale*) fl. .	Aug. 10, Sept. 13.	Aug. 15, Sept. 29.
Michaelmas Daisy (*aster tradescantia*) fl. .	Aug. 12, Sept. 27.	Aug. 11, Oct. 8.
Meadow Rue (*thalictrum flavum*) fl. . .	Aug. 14.	
Sea Holly (*eryngium marit.*) fl. . . .	Aug. 14.	
China Aster (*aster chinensis*) fl. . . .	Aug. 14, Sept. 28.	Aug. 6, Oct. 2.
Boletus albus ap.	Aug. 14.	May 10.
Less Venus' Looking-glass (*campanula hybrida*) fl.	Aug. 15.	May 14.
Carthamus tinctor. fl.	Aug. 15.	
Goldfinch (*fringilla carduelis*) young broods ap.	Aug. 15.	June 15.
Lapwings (*tringa vanellus*) congregate . .	Aug. 15, Sept. 12.	Sept. 25, Feb. 4.
Black-eyed Marble Butterfly (*papilio semele*) ap.	Aug. 15.	
Birds reassume their spring notes. . . .	Aug. 16.	
Devil's Bit (*scabiosa succisa*) fl. . . .	Aug. 17.	June 22, Aug. 23.
Thistle-down floats	Aug. 17, Sept. 10.	

	WHITE.	MARKWICK.
Ploughman's Spikenard (*conyza squarrosa*) fl.	Aug. 18.	
Autumnal Dandelion (*leontodon autumnale*) fl.	Aug. 18.	July 25.
Flies about in Windows	Aug. 18.	
Linnets (*fringilla linota*) congregate	Aug. 18, Nov. 1.	Aug. 22, Nov. 8.
Bulls make their shrill autumnal noise	Aug. 20.	
Aster amellus fl.	Aug. 22.	
Balsam (*impatiens balsamina*) fl.	Aug. 23.	May 22, July 26.
Milk Thistle (*carduus marinus*) fl.	Aug. 24.	April 21, July 18.
Hop-picking begins	Aug. 24, Sept. 17.	Sept. 1—15.
Beech (*fagus sylvatica*) turns yellow	Aug. 24, Sept. 22.	Sept. 5—29.
Soapwort (*saponaria officinalis*) fl.	Aug. 25.	July 19, Aug. 23.
Ladies' Traces (*ophrys spiralis*) fl.	Aug. 27, Sept. 12.	Aug. 18, Sept. 18.
Small Golden Black-spotted Butterfly (*papilio phlæas*) ap.	Aug. 29.	
Swallow (*hirundo rustica*) sings	Aug. 29.	April 11, Aug. 20.
Althæa frutex (*hibiscus syriacus*) fl.	Aug. 30, Sept. 2.	July 20, Sept. 28.
Great Fritillary (*papilio paphia*) ap.	Aug. 30.	
Willow Red-under-wing Moth (*phalæna pacta*) ap.	Aug. 31.	
Stone Curlew (*otis œdicnemus*) clamours	Sept. 1, Nov. 7.	June 17.
Phalæna russula ap.	Sept. 1.	
Grapes ripen	Sept. 4, Oct. 24.	Aug. 31, Nov. 4.
Wood Owls hoot	Sept. 4, Nov. 9.	
Saffron Butterfly (*papilio hyale*) ap.	Sept. 4.	Aug. 5, Sept. 26.
Ring-ousel appears on his autumnal visit	Sept. 4—30.	
Flycatcher (*musicapa grisola*) last seen	Sept. 6—29.	Sept. 4—30.
Beans (*vicia faba*) cut	Sept. 11.	Aug. 9, Oct. 14.
Ivy (*hedera helix*) fl.	Sept. 12, Oct. 2.	Sept. 18, Oct. 28.
Stares congregate	Sept. 12, Nov. 1.	June 4, March 21.
Wild Honeysuckles fl. a second time	Sept. 25.	
Woodlark sings	Sept. 28, Oct. 24.	
Woodcock (*scolopax rusticola*) returns	Sept. 29, Nov. 11.	Oct. 1, Nov. 1; young ones April 28; last seen April.
Strawberry-Tree (*arbutus unedo*) fl.	Oct. 1.	May 21, Dec. 10.
Wheat sown	Oct. 3, Nov. 9.	Sept. 23, Oct. 19.
Swallows last seen. (N.B. The House Martin the latest)	Oct. 4, Nov. 5.	Nov. 16.
Redwing (*turdus iliacus*) comes	Oct. 10, Nov. 10.	Oct. 1, Dec. 18; sings Feb. 10, March 21; last seen April 13.
Fieldfare (*turdus pilaris*) returns	Oct. 12, Nov. 23.	Oct. 13, Nov. 18; last seen May 1.
Gossamer fills the air	Oct. 15—27.	
Chinese Hollyoak (*alcea rosea*) fl.	Oct. 19.	July 7, Aug. 21.
Hen Chaffinches congregate	Oct. 20, Dec. 31.	
Wood Pigeons come	Oct. 23, Dec. 27.	
Royston Crow (*corvus cornix*) returns	Oct. 23, Nov. 29.	Oct. 13, Nov. 17; last seen April 15.
Snipe (*scolopax gallinago*) returns	Oct. 25, Nov. 20.	Sept. 29, Nov. 11; last seen April 14.
Tortoise begins to bury himself	Oct. 27, Nov. 26.	
Rooks (*corvus frugilegus*) return to their nest trees	Oct. 31, Dec. 25.	June 29, Oct. 20.
Bucks grunt	Nov. 1.	
Primrose (*primula vulgaris*) fl.	Nov. 10.	Oct. 7, Dec. 30.

	WHITE.	MARKWICK.
Green Whistling Plover ap.	Nov. 13, 14.	
Helvella mitra ap.	Nov. 16.	
Greenfinches flock	Nov. 27.	Feb. 19.
Hepatica fl.	Nov. 30, Dec. 29	Dec. 16—31.
Furze (*ulex europæus*) fl.	Dec. 4—21.	Dec. 31.
Polyanthus (*primula polyanthus*) fl.	Dec. 7—16.	Dec. 12, Feb. 21.
Young Lambs dropped	Dec. 11—27.	
Moles work in throwing up hillocks	Dec. 12—23.	
Helleborus fœtidus fl.	Dec. 14—30.	Dec. 26—31.
Daisy (*bellis perennis*) fl.	Dec. 15.	Nov. 5.
Wallflower (*cheiranthus cheiri seu fruticulosus* of Smith) fl.	Dec. 15.	
Mezereon fl.	Dec. 15.	
Snowdrop fl.	Dec. 29.	

In sese vertitur annus.

I N D E X.[1]

---

[1] The numbers to which asterisks are prefixed, denote the page on which the object is illustrated.

THE END.